KARST HYDROLOGY

KARST HYDROLOGY

Concepts from the Mammoth Cave Area

Edited by

William B. White

Elizabeth L. White

The Pennsylvania State University

VNR SPRINGER SCIENCE+BUSINESS MEDIA, LLC

Originally published by Van Nostrand Reinhold in 1989
Softcover reprint of the hardcover 1st edition 1989

Library of Congress Catalog Card Number 88-20971

ISBN 978-1-4615-7319-7 ISBN 978-1-4615-7317-3 (eBook)
DOI 10.1007/978-1-4615-7317-3

16 15 14 13 12 11 10 9 8 7 6 5 4 3 2 1

Additional material to this book can be downloaded from http://extras.springer.com

Library of Congress Cataloging in Publication Data
Karst hydrology : concepts from the Mammoth Cave area / edited by
William B. White and Elizabeth L. White.
p. cm.
Bibliography: p.
Includes index.
ISBN 978-1-4615-7319-7
1. Hydrology, Karst—Kentucky—Mammoth Cave Region.
I. White, William B. (William Blaine), 1934– . II. White, Elizabeth L.
GB705.K4K37 1989
551.49—dc19 88-20971
CIP

CONTENTS

PREFACE

This volume has its roots in the distant past of more than 20 years ago, the International Hydrologic Decade (IHD), 1964–1974. One of the stated goals of the IHD was to promote research into groundwater situations for which the state of knowledge was hopelessly inadequate. One of these problem areas was the hydrology of carbonate terrains. Position papers published early in the IHD emphasized the special problems of karst; carbonate terrains were supposed to receive a substantial amount of attention during the IHD. There were indeed many new contributions from European colleagues but, unfortunately, in the United States the good intentions were not backed up by much in the way of federal funding. Some good and interesting work was published, particularly by the U.S. Geological Survey (USGS), but in the academic community the subject languished.

About this same time the Cave Research Foundation (CRF), organized in 1957 to promote the systematic exploration, survey, and scientific study of the great cave systems of Mammoth Cave National Park, was casting about for a broader scope for its research activities. Up until that time, CRF research had been largely restricted to detailed mineralogical and geological investigations within the caves, with the main part of the effort concentrated on exploration and survey. The decision to investigate the hydrology required a certain enlargement of vision because investigators then had to consider the entire karst drainage basin rather than isolated fragments of cave passage. Much of the CRF effort was directed toward understanding the fossil conduit system. In few other locations in the world was there such an extensive display of all the bits and pieces of the plumbing available for inspection.

As the IHD drew to a close in 1974, some of the contributors to this volume gave thought to pulling their work together for formal publication. At the same time, the National Park Service made the important decision to add a resident geologist to its permanent staff. That meant a change in style in the way in which research could be done. Instead of outsiders coming into the area for a few days or a summer field season, it was possible to mount an intensive and long-range program. This work initially had as its objective the delineation of the entire drainage basin by means of one of the largest water-tracing investigations ever undertaken. It was then possible to determine the interrelationships of the active drainage from surface streams, to cave streams, to the springs discharging on Green River. The fruits of this voluminous effort are summarized in J. F. Quinlan's groundwater basin map (Plate 1). Meanwhile, exploration in the cave system continued and much new insight was gained into the organization of the conduit system.

Another decade beyond the IHD has passed, and another plateau has been reached. The physical situation, the hydrogeologic setting of the karst aquifer has been well mapped out. Future investigations will require continuous monitoring of discharge and geochemical parameters through periods of flood and low flow to tackle new details of the hydrology. With a large number of the exploratory studies completed, and with the research paused on the threshold of an entirely new style of investigation, it seems an appropriate time to draw together the knowledge of this amazing karst system into a single volume.

All of the contributors to this volume have worked extensively in the south-central Kentucky karst, and the material presented in the chapters is based on first-hand personal experience. We have attempted, in constructing this book, to review what is known about the Mammoth Cave area and to cast it specifically into hydrologic rather than geomorphologic or speleologic terms. Much of the book, however, represents original contributions, which are published here for the first time. Although the book is a collaborative effort, and the authors have had the opportunity to review each other's chapters, we would not be so rash as to claim that we all agree on all points. We are offering herein a progress report; the work continues.

Because the conduit system plays such a critical role in carbonate aquifers, it seemed appropriate to recast the descriptions of the big cave systems, which are usually written from the perspective of the cave explorer, in terms of their hydraulic geometry and their hydrologic role in the drainage system. Standard groundwater theory "works" because the flow field is the solution of a linear differential equation. Conduits are an unpleasant nonlinearity that cannot, in a fundamental way, be incorporated into the theory. We have chosen to make them our centerpiece.

Units of measure have proved an insoluble problem. To the extent possible, units for length, velocity of flow, volume of flow, and related quantities have been given in metric form; however, the topographic and geologic base maps are available only in English units. Much of the discussion centers on elevations and distances that cannot be related to the base maps in other than English units. We have ended with a poor compromise in which both units are given in most cases, and some English units remain lurking on maps and in the text.

Much of the work reported in this volume was supported by the Cave Research Foundation and draws on its ongoing cartographic program. Much of the field work was based at the Foundation's Flint Ridge research facility. We also thank the Foundation for permission to use its large cave map (Plate 2) and CRF Treasurer Roger McClure for arranging the printing.

We—the collective "we"—are grateful to the National Park Service for permission to carry out the studies and for the support and encouragement of the superintendent and staff.

We—the editorial "we"—are grateful to the authors for producing their material under a tight deadline and for responding cheerfully to a lot of last-minute requests and discussion. We would also like to thank Mr. Charles Hutchinson of the former Hutchinson Ross Publishing Company for encouraging us to keep going during the long gestation period of this volume.

CONTRIBUTORS

Thomas A. Brucker
937 Forest Acres Court
Nashville, TN 37220

George H. Deike, III
P.O. Box 108
Cass, WV 24927

Ralph O. Ewers
Department of Geology
Eastern Kentucky University
Richmond, KY 40475

Angelo I. George
Consulting Hydrogeologist
1869 Trevillian Way
Louisville, KY 40205

John W. Hess
Water Resources Center
Desert Research Institute
2505 Chandler Avenue
Suite No. 1
Las Vegas, NV 89120

Arthur N. Palmer
Department of Earth Sciences
State University College
Oneonta, NY 13820-1380

James F. Quinlan
National Park Service
P.O. Box 8
Mammoth Cave, KY 42259

Stephen G. Wells
Department of Geology
University of New Mexico
Albuquerque, NM 87131

Elizabeth L. White
Department of Civil Engineering
The Pennsylvania State University
University Park, PA 16802

William B. White
Department of Geosciences and
Materials Research Laboratory
The Pennsylvania State University
University Park, PA 16802

1

INTRODUCTION TO THE KARST HYDROLOGY OF THE MAMMOTH CAVE AREA

William B. White

LIMESTONE AQUIFERS

Karst in Carbonate Terrains

Scattered widely over the Earth is a rather peculiar landscape known as "karst." Karst landscapes are often pocked and pitted lands of sinkholes, limestone towers and steep-sided hills, underground drainage, and caves. Most karst is formed on carbonate rocks such as limestones or dolomites, sometimes on gypsum, and more rarely on rocks of other lithologies. Limestone karst is the most extensively developed, has the broadest regional extent, and has the most elaborate and highly integrated underground drainage and cavern systems.

Karst landscapes have attracted human attention since the dawn of recorded history. Many karst lands are difficult to put into agriculture; sinkholes are not easy to plow. In contrast, some of the richest farmlands on Earth are thick, mildly alkaline limestone soils on landscapes that are also called karst. However, soils are easily lost to the subsurface through solution cavities, and some of the most inhospitable terrains on Earth are located in karst. The Burren of western Ireland is a bleak rocky plain, barren of trees, bushes, or shrubs, with only sparse blades of grass on a shingly rubble of solution-carved limestone rock. In the Mediterranean, overgrazing thousands of years ago destroyed the soil-holding plants, which allowed soils to be flushed down solution cavities; a land that might once have been rich and fertile is now essentially a desert. In the tropics (Cuba, Jamaica, Puerto Rico, Borneo, New Guinea, and South China) massive limestones have been carved by solution into residual towers, conical hills, and intermediate deep bowl-shaped depressions.

Karst landscapes can include underground rivers. Streams that flow from island interiors; streams that flow from borderlands on clastic rocks; streams that flow down the mountains of the Rockies, the Alleghenies, the Cumberlands, or the Ozarks sink into open cave entrances or into swallow holes at the contact between the clastic rocks and the underlying limestones. The sinking streams then reappear many kilometers away as huge limestone springs. Underground water in karst, unlike groundwater in nonkarstic rocks, is concentrated in natural pipes dissolved from the solid rock. When water tables are lowered, these water-filled conduits drain, dry out, and fill with air, and some develop entrances that become the caves accessible to human exploration.

Cave exploration has been a booming pastime in North America since the 1950s. With improved equipment, new types of rope and devices for descending deep pits, and wet suits and scuba equipment for exploring water-filled passages, the well-trained, motivated, and dedicated explorers of today are pushing far beyond the limits of what would have been possible only a few years ago. Indeed, the decade of the 1980s might well be called the "golden age" of cave exploration in North America. This growth of exploration has added to our knowledge of the hydrology of karst aquifers.

Sometimes developed to dramatic extents, sometimes visible only to the trained professional eye, karst terrains make up, according to several estimates, about 15% of the Earth's land surface. Of these, few have attracted more attention than the doline karst and the great limestone caves that lie under and around Mammoth Cave National Park in south-central Kentucky.

Land-Use Hazards in Karst

Development of karst terrains runs into hazards that result from three unique characteristics of the karst landscape.

1. The soil-bedrock contact is often irregular. Deep solution along joints and fractures produces a cutter-and-pinnacle topography that can be completely masked by the soil cover.
2. The network of solution cavities extending downward from the surface permits the efficient transport of clastic material by groundwater infiltration. The phenomenon of soil piping is extremely efficient in transporting soils from the surface to the subsurface.
3. The presence of shallow solution cavities makes subsidence and bedrock collapse a very real possibility.

The irregular bedrock topography is mainly of concern in foundation engineering. Structures are usually designed on the assumption that the underlying basement material will compact uniformly. This assumption is certainly not true on a karst surface where one corner of a structure might be supported by a bedrock pinnacle while another corner might extend over a soil-filled crevice. The weight of the structure causes differential compaction, which in turn causes differential settling and distortion of the structure. There could be cracks in floors and walls; doorways and windows might be twisted and no longer close properly. The problems generally increase in severity with increase in the size of the structure. Although one might successfully put a single family home on a karst surface, a good deal of preconstruction investigation and remedial action would be needed to build office buildings, apartment buildings, shopping malls, and factories in karst terrains.

The transport movement of soils into the subsurface gives rise to a class of sinkhole features known as cover-collapse and cover-subsidence sinks. These are particularly endemic where soils are thick and infiltration rates high. Florida seems to be the classic locality for this kind of problem (Beck, 1984; Beck and Wilson, 1987).

Soil piping is also a land-use hazard in developed regions built on karst. Any action that changes the infiltration characteristics of the soils will change the rate of soil transfer to the subsurface. Parking lots, street pavement, roofs, and driveways change the land surface from one of uniform infiltration capacity to one of alternate areas of essentially zero infiltration interspersed with small areas in which all of the runoff is concentrated. The point-source infiltration localizes the soil piping, which in turn undermines pavements, basements, and parking lots, which can then collapse.

Some structures have been lost because of solution cavity collapse. This risk is unnecessary because a program of test drilling before construction should reveal cavities having roof spans large enough to weaken and collapse under the weight of the intended structure.

Groundwater Supply and Groundwater Pollution

The most characteristic feature of karst terrain is the concentration of water flow in underground solution conduits. This means that there are two or more parts of the aquifer with very different response times. In the conduit portion, infiltration and runoff are flushed through very rapidly giving a system in which the karst springs have very flashy responses to storms and in which the base flow is exceptionally low because the open conduits drain completely. Because of the very low flow resistance, the karsted carbonate aquifer is unable to support much hydrostatic head, and, as a result, the water levels in the conduit drop very close to local base level. In the diffuse flow part of the aquifer, there is storage composed of joints, fractures, and primary porosity. The diffuse flow system drains laterally into the conduit system with a long response time and provides base flow. There is also deep storage in both flooded conduits and in joints and fractures below the elevation of local base level.

Because of the heterogeneous character of karst aquifers, evaluation of water resources is difficult. The usual procedure of drilling exploratory wells, test pumping the wells, and calculating the transmissibility and water yield of the aquifer from time-drawdown curves is almost meaningless when used in karst terrain. The diameter of the well bore is very small compared with the heterogeneities of the aquifer, and changing well location by only a few meters can make dramatic differences in the field-determined aquifer parameters. Water balance measurements have some special peculiarities because of the rapid movement of water in the conduits. Such calculations applied specifically to the central Kentucky karst are the subject of Chapter 4.

Because much of the input to the karst groundwater system is through sinkholes and sinking streams, and because of the open character of the aquifer and lack of thick soil covers, karst systems are susceptible to pollution. Sources of pollutants include industrial and hydrocarbon wastes; sinking polluted surface streams; sinkhole dumps; agriculturally derived nitrates; herbicide and pesticide residues; highway spills; and leaking sewer lines, pipelines, and storage tanks. Almost any imaginable source of pollution can be transmitted rapidly to the subsurface and into the groundwater system.

Again because of the highly anisotropic character of the aquifer, point-source pollutants do not form a plume as they do in most porous-media aquifers. Instead, if the pollutants make their way into an open conduit, they can be transmitted for very long distances with relatively little dilution and very little dispersion. In the Ozark karst of Missouri, Aley (1972) traced the source of a polluted spring to a municipal sinkhole dump 25 km away.

The Occurrence of Karst Aquifers

In his major compendium on groundwater resources in the United States, McGuinness (1963) listed 15 states in which carbonate aquifers supply a significant amount of groundwater for public consumption. The carbonate rocks throughout the United States are often used as water supplies either as domestic wells for homes and farms or as municipal wells for entire towns and industries. The latter consumers have often been supplied by limestone springs. As concern for groundwater pollution has grown, the use of limestone springs has gradually been replaced by production wells drilled in the vicinity of the springs. Although the use of drilled wells avoids contamination problems at the mouth of a spring, the same aquifer system provides the water, and contaminated conduit flow will affect the wells as much as it would the springs.

Table 1.1 lists some of the most important carbonate aquifers in the United States. A wide variety of these occur. Many of the most valuable are fracture aquifers where solution-modified joints and bedding-plane partings are the primary storage volume and pathways for groundwater movement. Many are conduit aquifers with large solution cavities and air-filled caves above the zone of saturation. A number of the aquifers—including the Floridan aquifer, the Edwards aquifer, and the Roswell aquifer—are artesian, and there is deep flow of the groundwater before it discharges from the springs.

GEOGRAPHIC SETTING

The south-central Kentucky karst, the subject of this volume, is a world-class example of a shallow, intensely karsted, carbonate aquifer unit. The particular area is less significant for its water supply than it is as an extreme example of an aquifer system in which much of the groundwater flow is through open conduits. The area has the additional advantage of having been investigated by many individuals for a period of 150 years and is the subject of a number of current research programs. Although this volume might seem to have a regional caste, it is really a case study of a common type of karst aquifer from which much can be learned that will apply to many other limestone aquifers in the United States and elsewhere.

The papers presented as chapters in this volume draw on the 30-year program of the Cave Research Foundation (CRF), on the survey and cartography of the giant cave systems, and on many individual hydrologic and geomorphologic studies. It also includes the program of the National Park Service carried out by the National Park geologist.

While the detailed geologic framework of the area is described in Chapter 2, Figure 1.1 gives a regional perspective. The area is underlain by relatively thin—several hundred meters—Mississippian limestone, which has the gentle dip of a regional structural basin. In this regard, the Mammoth Cave area is also a model for many smaller and less highly integrated carbonate conduit aquifers found in many other parts of the Appalachian Mountains and interior lowlands of eastern United States.

REVIEW OF THE LITERATURE

Early Science

The point of origin for discussion of the south-central Kentucky karst is Mammoth Cave. In its historical setting, Mammoth Cave was accessed

Table 1.1 Some Important Carbonate Aquifers in the United States

Name	*Location*	*Rock Formation*
Appalachian Valleys	Central Pa. through Va., W. Va., Tenn., Ga., to Ala.	Cambrian and Ordovician limestone and dolomite
Appalachian Plateaus	W. Va., Tenn., Ky., and northern Ala.	Mississippian Greenbrier/Bangor/Monteagle/ St. Louis limestones
Floridan	Coastal Plain of Fla., Ga., and S.C.	Eocene through Miocene limestones
Niagara	Northern Ill. and along Great Lakes	Silurian Niagara Dolomite
Illinois Basin	Southern Ind., central and western Ky., southern Ill.	Mississippian Girkin, Ste. Genevieve, and St. Louis limestones
Inner Bluegrass	North-central Ky.	Ordovician Lexington Limestone
Highland Rim	Central Tenn. and northern Ala.	Mississippian limestones
Nashville Dome	Central Tenn.	Ordovician limestones
Ozark Dome	Southern Mo. northern Ark.	Mississippian limestone and Ordovician limestone and dolomite
Edwards	Central Tex.	Cretaceous Edwards, Glen Rose and related limestones
El Capitan	Southeastern N. Mex. and west Tex.	Permian El Capitan Limestone
Roswell	Eastern N. Mex.	Permian San Andres Limestone and Artesia Group
Pahasapa	Western S. Dak.	Mississippian Pahasapa Limestone
Madison Aquifer	Wyo., Colo., Mont., and Utah	Mississippian Madison Limestone
Nevada Deep Aquifer	Nev.	Sequence of Cambrian and Ordovician carbonates

Hydrogeology	*Reference*
Many local aquifers in folded and faulted carbonates; both fracture and conduit aquifers common	Parizek et al. (1971)
Many local aquifers; highly karstified; shallow conduit systems common	Piper (1932) Theis (1936)
Mainly artesian; much primary porosity; drowned conduits	Stringfield (1966)
Low relief fracture aquifer	Zeizel et al. (1962)
Many local aquifers; shallow conduit systems common	
Shallow mixed fracture aquifers	Thrailkill (1982) Palmquist and Hall (1961)
Mainly localized conduit aquifers	Smith (1962)
Conduits of moderate depth	Moore et al. (1969) Burchett and Moore (1971)
Many and diverse aquifers	
Deep aquifer feeding large springs plus smaller shallow systems	Abbott (1975)
Deep system with some artesian parts	
Artesian system with some influence of gypsum beds	Havenor (1968)
Deep artesian aquifer	Swenson (1968)
Many local aquifers, many in alpine settings	Miller (1976)
Deep groundwater flow system	Hess and Mifflin (1978)

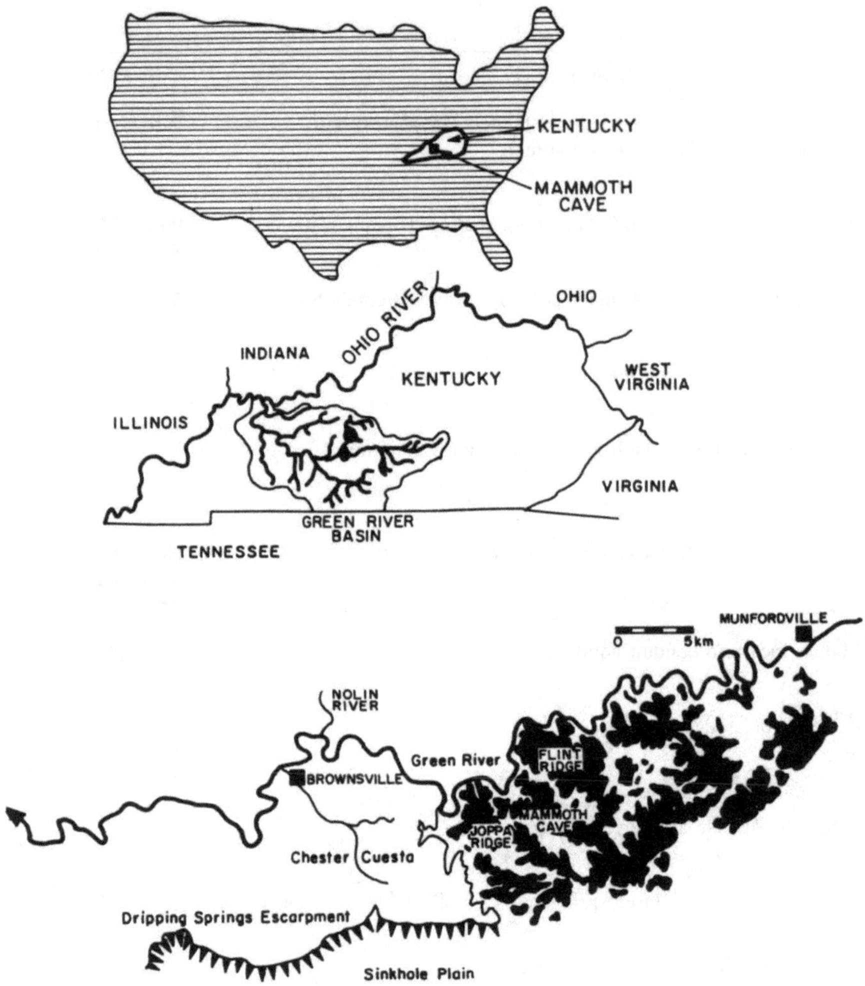

Figure 1.1 Location map for the Mammoth Cave area and the south-central Kentucky karst.

through one of the rare natural entrances found in the region. The entrance gave immediate access to a complex of large, high-level trunk passages that were mined for saltpeter in the early 1800s (DePaepe and Hill, 1981). Mammoth Cave is one of the oldest tourist caves in the United States; visitors were shown through as early as the saltpeter mining days. The cave was purchased by Franklin Gorin in 1838. Gorin built a hotel and assigned as guide his slave, Stephen Bishop, who was to become the pioneering explorer of Mammoth Cave. The fame of the cave soon spread throughout the United States and Europe. Many scientists and naturalists of the day found occasion to tour the cave and a few recorded their observations. One of the earliest scientific papers was written by Benjamin Silliman (1851). Silliman, one of the founding fathers of geology in the United States, was preoccupied with the winds and air currents that blow through the cave and apparently gave little notice to the hydrologic implications of what he deemed to be vast dry riverbeds in the upper-level passages. Others visited the caves at various times through the nineteenth century. John Locke (1841) gave the first description of secondary gypsum decoration, which is the most common mineralization in the cave.

Most important of the early research in the Mammoth Cave area were the classic cave-origin papers by Davis (1930), Swinnerton (1932), Gardner (1935), and Bretz (1942). These were attempts to provide a universal theory for cave development and all of them lean heavily on the geologic situation of the Mammoth Cave area. However, inadequate cave maps and misinterpretation of vertical shafts were, in large part, responsible for incorrect interpretation of the groundwater movement and its relationship to cave origin (Watson and White, 1985). Indeed, there has been a curious closing of the circle with regard to the hydrologic interpretation of big caves.

To the early investigators, it was completely self-evident that the large-diameter trunk conduits in Mammoth Cave represented the dry beds of former underground streams. William Morris Davis (1930) advocated cave development by deep circulating groundwater flow following the new groundwater theory that was coming into prominence at the time. After decades of more or less continuous and sometimes acrimonious debate, we have returned to the idea that cave systems are old stream channels, but with some important differences from the earlier ideas. Although cave passages are analogs of surface drainage channels, they are not the same phenomenon. In the underground conduit system there is the possibility of both pipe flow and open-channel flow, a situation that is absent in surface streams. The similarities and differences are of considerable interest.

Although the Mammoth Cave area has long been known as one of the great classic karst regions, and reproductions of segments of the topographic maps often appear in elementary geology and in geomorphology textbooks, there is, in fact, relatively little literature on the surface landforms. The recharge area to the southeast of Mammoth Cave National Park is a classic sinkhole plain that has received surprisingly little attention from karst geomorphologists.

The modern era of scientific research in the Mammoth Cave area began in the late 1950s with a number of independent but concurrent events. The U.S. Geological Survey's mapping program, which was released as the GQ series of geologic quadrangle maps, provided a much more complete geologic database. Investigations by USGS hydrologists produced a number of important water-resource investigation reports. The scientific investigation of the cave systems was initiated by the Cave Research Foundation in 1957 (see Watson, 1981). Current literature is cited as appropriate in the following chapters.

Exploration

The exploration of the large cave systems of the central Kentucky karst has also had a spotted career. The guides, notably Stephen Bishop, were early explorers. Bishop, the black slave who discovered much of the classic part of Mammoth Cave from 1839 through the 1840s, extended the cave from the upper-level trunks down to the level of active groundwater circulation at Echo River and beyond to additional high-level trunks in the middle part of Mammoth Cave Ridge. Hovey (1896) recounted a description of the cave and its exploration by Bishop and the other guides. The survey of Mammoth Cave by Max Kaemper, assisted by guide Ed Bishop, in 1908 was a milestone. Kaemper's precise compass and pace topography produced the best map of Mammoth Cave available until the current CRF cave surveys.

Exploration through the first half of the twentieth century was a sporadic and sometime thing. Pete Hanson and Leo Hunt found New Discovery in Mammoth Cave in 1939 (Lix, 1955), but there was little exploration and even fewer papers published because of intense local rivalry between competing commercial caves. The era was known as the "cave wars." The New Entrance to Mammoth Cave was found in 1921 by George Morrison, and for a while the two ends of Mammoth Cave were operated in competition with each other.

The modern systematic survey of the cave system began in 1955 in the then privately owned Floyd Collins Crystal Cave. Following the 1953 National Speleological Society Expedition to Crystal Cave (Lawrence and Brucker, 1955), the CRF was founded to continue the exploration and survey and to maintain continuity in the collection of scientific and survey data. The mapping of the Flint Ridge system (Smith, 1957, 1964) very quickly extended beyond the boundaries of the Crystal Cave property, and when this property was included within the National Park system in 1961, agreements with the National Park Service allowed the surveys to be extended to the entire cave system with the first collection of maps published in 1964 (Brucker and Burns, 1964). Surveying in Mammoth Cave began in 1964, the first new exploration in Mammoth Cave for many decades. Continued exploration in the Flint Ridge Cave system produced a connection between the two systems in 1972 to form what was unequivocally the longest cave in the world with a mapped length greater than 250 km (Brucker and Watson, 1976). The link-up of caves to form larger and larger cave systems included the tying of Mammoth Cave to Proctor Cave through the newly discovered

Hawkins River in 1979 (Brucker and Lindsley, 1979; Coons and Engler, 1980). The connection of the eastern part of Mammoth Cave to Roppel Cave brought the system close to its present length of more than 500 km. The Fisher Ridge Cave system, still farther east, might very well be part of the system although a connection has not been made as of this writing.

It is the tremendous concentration of detailed cave surveys that makes possible the sort of analysis presented in this volume. In few other karst areas of the United States are so many cave passages, representing such a variety of hydrologic functions, available on such a regional scale.

REFERENCES

Abbott, P. L., 1975, On the hydrology of the Edwards limestone, south-central Texas, *Jour. Hydrology* **24:**251–269.

Aley, T., 1972, Groundwater contamination from sinkhole dumps, *Caves and Karst* **14:**17–23.

Beck, B. F., ed., 1984, *Sinkholes: Their Geology, Engineering, and Environmental Impact,* Rotterdam: A. A. Balkema, 429p.

Beck, B. F., and W. L. Wilson, 1987, *Karst Hydrogeology: Engineering and Environmental Applications,* Rotterdam: A. A. Balkema, 467p.

Bretz, J H., 1942, Vadose and phreatic features of limestone caves, *Jour. Geology* **50:**675–811.

Brucker, R. W., and D. P. Burns, 1964, *The Flint Ridge Cave System,* Washington, D.C.: Cave Research Foundation, Map Folio.

Brucker, R. W., and P. Lindsley, 1979, New Kentucky junction, *Natl. Speleol. Soc. NSS News* **37:**231–236.

Brucker, R. W., and R. A. Watson, 1976, *The Longest Cave*, New York: Knopf, 327p.

Burchett, C. R., and G. K. Moore, 1971, Water resources in the upper Stones River Basin, central Tennessee, *Tennessee Div. Water Resources, Water Resources Ser. No. 8,* 62p.

Coons, D., and S. Engler, 1980, In Morrison's footsteps, *Natl. Speleol. Soc. NSS News* **38:**127–132.

Davis, W. M., 1930, Origin of limestone caverns, *Geol. Soc. America Bull.* **41:**475–628.

DePaepe, D., and C. A. Hill, 1981, Historical geography of United States saltpeter caves, *Natl. Speleol. Soc. Bull.* **43:**88–93.

Gardner, J. H., 1935, Origin and development of limestone caverns, *Geol. Soc. America Bull.* **46:**1255–1274.

Havenor, K. C., 1968, Structure, stratigraphy, and hydrogeology of the northern Roswell artesian basin, Chaves County, New Mexico, *New Mexico Bur. Mines and Mineral Resources Circ. 93,* 30p.

Hess, J. W., and M. D. Mifflin, 1978, A feasibility study of water production from deep carbonate aquifers in Nevada, *Desert Research Inst., Water Resources Center Pub. 41054,* 125p.

Hovey, H. C., 1896, *Celebrated American Caverns,* Cincinnati: Robert Clarke & Co., 228p. (Reprinted by Johnson Reprint Corp., New York, 1970.)

Lawrence, J., Jr., and R. W. Brucker, 1955, *The Caves Beyond,* New York: Funk & Wagnalls, 283p. (Reprinted by Zephyrus Press, Teaneck, N. J., 1975.)

Lix, H. W., 1955, Mammoth Cave's underground wilderness, in *Celebrated American Caves*, C. E. Mohr and H. N. Sloane, eds., New Brunswick, N. J.: Rutgers University Press, pp. 105–115.

Locke, J., 1841, Alabaster in the Mammoth Cave of Kentucky, *Amer. Jour. Sci.* **42:**206–207.

McGuinness, C. L., 1963, The role of ground water in the national water situation, *U.S. Geol. Survey Water-Supply Paper 1800,* 1121p.

Miller, W. R., 1976, Water in carbonate rocks of the Madison group in southeastern Montana—A preliminary evaluation, *U.S. Geol. Survey Water-Supply Paper 2043,* 51p.

Moore, G. K., C. R. Burchett, and R. H. Bingham, 1969, Limestone hydrology in the upper Stones River Basin, central Tennessee, *Tennessee Div. Water Resources,* 58p.

Palmquist, W. N., Jr., and F. R. Hall, 1961, Reconnaissance of groundwater resources in the Blue Grass region, Kentucky, *U.S. Geol. Survey Water-Supply Paper 1533,* 39p.

Parizek, R. R., W. B. White, and D. Langmuir, 1971, Hydrogeology and geochemistry of folded and faulted rocks of the central Appalachian type and related land use problems, *The Pennsylvania State Univ. Coll. Earth and Mineral Sci. Expt. Sta. Circ. 82,* 210p.

Piper, A. M., 1932, Ground water in north-central Tennessee, *U.S. Geol. Survey Water-Supply Paper 640,* 238p.

Silliman, B., Jr., 1851, On the Mammoth Cave of Kentucky, *Amer. Jour. Sci. Arts* **11:**332–339.

Smith, O., Jr., 1962, Ground-water resources and municipal water supplies of the Highland Rim in Tennessee, *Tennessee Div. Water Resources, Water Resources Ser. No. 3,* 257p.

Smith, P. M., 1957, Discovery in Flint Ridge, 1954–1957, *Natl. Speleol. Soc. Bull.* **19:**1–10.

Smith, P. M., 1964, The Flint Ridge Cave System: 1957–1962, *Natl. Speleol. Soc. Bull.* **26:**17–27.

Stringfield, V. T., 1966, Artesian water in Tertiary limestone in the southeastern United States, *U.S. Geol. Survey Prof. Paper 517,* 226p.

Swenson, F. A., 1968, New theory of recharge to the artesian basin of the Dakotas, *Geol. Soc. America Bull.* **79:**163–182.

Swinnerton, A. C., 1932, Origin of limestone caverns, *Geol. Soc. America Bull.* **43:**663–694.

Theis, C. V., 1936, Ground water in south-central Tennessee, *U.S. Geol. Survey Water-Supply Paper 677,* 182p.

Thrailkill, J., 1982, Groundwater in the inner Bluegrass karst region, Kentucky, *Kentucky Univ. Water Resources Research Inst., Research Rept. No. 136,* 108p.

Watson, R. A., 1981, *The Cave Research Foundation: Origins and the First Twelve Years, 1957–1968,* Mammoth Cave, Ky.: Cave Research Foundation, 494p.

Watson, R. A., and W. B. White, 1985, The history of American theories of cave origin, *Geol. Soc. America Centennial Special Volume* **1**:109–123.

Zeizel, A. J., W. C. Walton, R. T. Sasman, and T. A. Prickett, 1962, Ground-water resources of DuPage County, Illinois, *Illinois Water Survey Coop. Ground-Water Rept. 2*, 103p.

2

HYDROGEOLOGY OF THE SOUTH-CENTRAL KENTUCKY KARST

John W. Hess, Stephen G. Wells, James F. Quinlan, and William B. White

HYDROGEOLOGIC SETTING

The hydrogeologic setting provides the physical constraints determining the flow characteristics of an aquifer. Stratigraphy, lithology, and structure determine the type of aquifer, the location of recharge and discharge areas, and the nature of the porosity and permeability. If groundwater flow and the evolution of the conduit system are described by rate equations, the hydrogeologic setting defines the boundary conditions.

Considering the historic significance of the Mammoth Cave area as a classic karst region, the literature on the hydrogeology of the area is surprisingly sparse. Pohl (1936) early recognized the role of caves as abandoned conduits, and his ideas were developed in later Cave Research Foundation (CRF) work (Watson, 1966). Much important work has been carried out as part of the water-resource investigations of the U.S. Geological Survey (USGS) and appears in a series of reports (Brown and Lambert, 1963; Cushman et al., 1965; Brown, 1966; Cushman, 1968; Lambert, 1976).

Surface Drainage

The surface drainage in south-central Kentucky is shown in Figure 2.1. The karst lies entirely within the drainage basin of Green River, the trunk of which flows in a meandering channel from east to west across the upper portion of the map. Green River has two main tributaries, the Nolin River

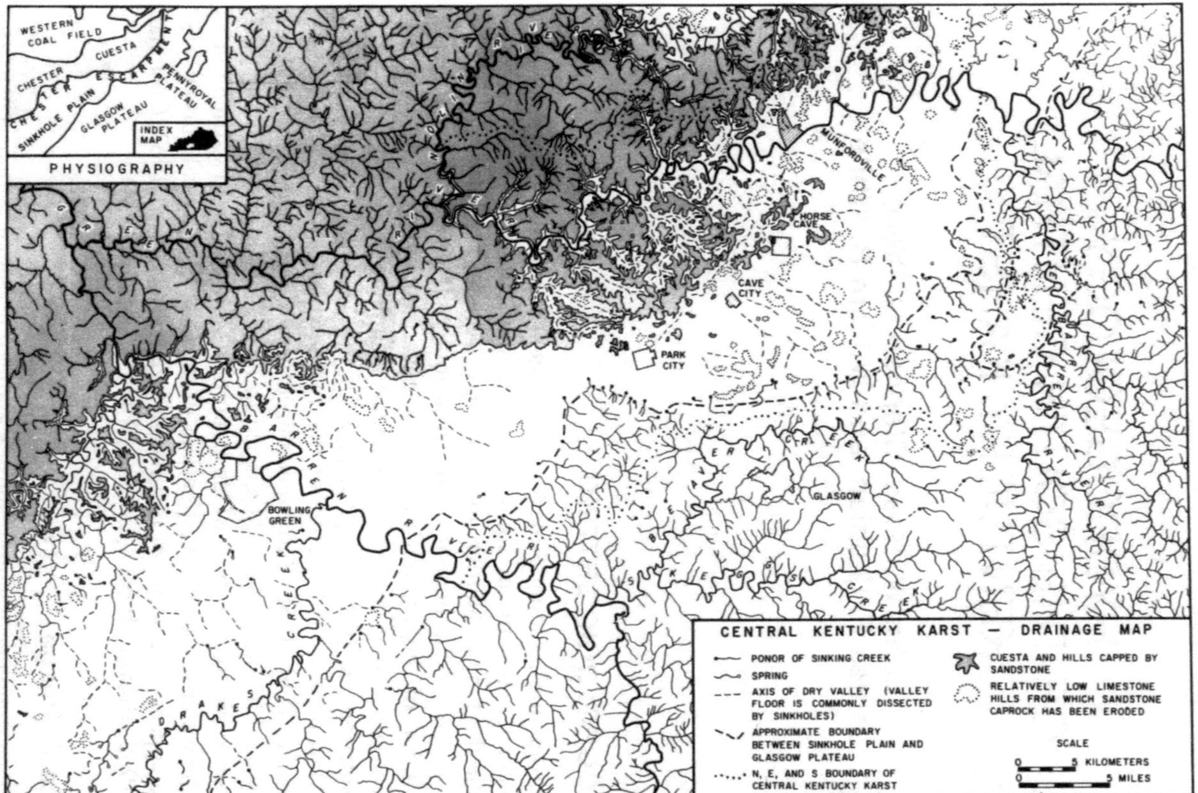

Figure 2.1 Surface drainage patterns in south-central Kentucky. (*Adapted from Quinlan, 1970, and Quinlan and Ewers, 1981)*

and the Barren River. Nolin River flows from the north and east and enters Green River in the north-central portion of the map. Because the Nolin River flows mainly on the clastic rocks above the cavernous limestones, it will play a relatively minor role in the discussion. The other major tributary, the Barren River, flows from the south and east and drains much of the southern portion of the region.

Three physiographic subprovinces cross the area. To the north, the stippled area on Figure 2.1 is the Chester Cuesta, underlain mainly with Mississippian clastic rocks that dip gently to the northwest beneath Pennsylvanian rocks of the Western Coal Field (Illinois Basin). To the south is the Glasgow Upland, a part of the Pennyroyal Plateau, underlain with lower Mississippian argillaceous limestones and shales. Between the two is the Sinkhole Plain, clearly visible on Figure 2.1 as a band extending from northeast to southwest along which there is no surface drainage except where the plain is crossed by major streams. The Sinkhole Plain is bounded on the north by the Chester or Dripping Springs Escarpment and on the south by the line of sinking creeks that separate it from the Glasgow Upland.

The overall drainage basin of the south-central Kentucky karst is defined by a series of drainage divides. On the north, where the area is underlain by clastic rocks, the boundary is the surface drainage divide between the Green River and its tributary, the Nolin River. There is a distinct drainage divide along the southern margin of the Sinkhole Plain, south of which are surface tributaries to Beaver Creek and Little Barren River on the Salem, Harrodsburg, and Fort Payne formations. North of this divide there is surface drainage consisting of short segments of streams that flow northward toward the Sinkhole Plain and then sink more or less at the contact between the Salem and St. Louis limestones. The drainage divide is south of the physiographic boundary (which is mapped by a geologic boundary) between the Sinkhole Plain and the Glasgow Upland. An eastern boundary is the Little Barren River, which drains north into the Green River from the Glasgow Upland. An equivalent western boundary is the Barren River, although some observations of interest have been made along Drakes Creek and Jennings Creek west of the Barren River. The karsted Mississippian limestones extend both to the northeast and the west so that the delineation of the Mammoth Cave area or south-central Kentucky karst is arbitrary to some extent.

A second major divide separates the Green River drainage from the Barren River drainage. This divide crosses the region north-south just west of Park City. East of this boundary, subsurface drainage beneath the Sinkhole Plain is north to the Green River. West of the divide, subsurface drainage is west to the Barren River. The highly dissected portion of the Chester Cuesta north of Park City is the location of Mammoth Cave National Park and the large cave systems. In this area, Green River has bisected the Cuesta down to the level of the cavernous limestones making possible an outlet for the subsurface drainage. West of the divide, Green River is floored with clastic rocks above the limestones, and the Cuesta is drained only by surface streams.

A characteristic feature of karst aquifers is the existence of well-defined subsurface basins, which also delineate catchment areas that feed into a drainage system through the active conduit system. The Sinkhole Plain and Cuesta south of Green River have been divided into subsurface basins by groundwater tracing and by mapping of the water table. The substance of these investigations and the properties of the subsurface basins are covered in Chapter 3.

Stratigraphy

The rocks underlying the south-central Kentucky karst belong to the Meramecian and Chesterian series of Mississippian age except for the Pennsylvanian Pottsville (Caseyville) Sandstone that crops out on the tops of the ridges. Geologic mapping has been completed for the entire karst area by the USGS and is available through the USGS geologic quadrangle (GQ) series. Figure 2.2 shows the overall stratigraphy. A. N. Palmer, who examined the lithologic character of the cavernous limestone units in considerable detail (Fig. 2.3), found that lithologic character plays a role both in the gross behavior of the drainage system and more subtly in the placement and morphology of individual conduits (see Chap. 11).

The oldest rocks, the Salem and Harrodsburg formations, which crop out along the southern and eastern margins of the area, are sufficiently argillaceous that surface karst and underground conduit systems are very poorly developed. Drainage tends to remain on the surface of these formations and their aquifer systems are more like porous media than karstic aquifers.

The St. Louis Limestone, which underlies much of the southern Sinkhole Plain, tends to have much interbedded chert. At the top of the formation is the Lost River chert, which controls both sinking streams and the development of sinkholes on the Sinkhole Plain. Much of the irregular pattern of sinkhole development has been ascribed to the outcrop pattern of the Lost River chert (Howard, 1968). Cave streams tend to be either trapped under the chert or to flow down the top of the bed.

The Ste. Genevieve Limestone underlies that portion of the Sinkhole Plain nearest to the escarpment and also underlies the karst valleys that dissect much of the Chester Cuesta. Most of the lower levels of Mammoth Cave and the other cave systems are in the Ste. Genevieve Limestone. The Ste. Genevieve Limestone is 35–40 m thick and consists of interbedded limestone and dolomite in the lower half and limestone interspersed with

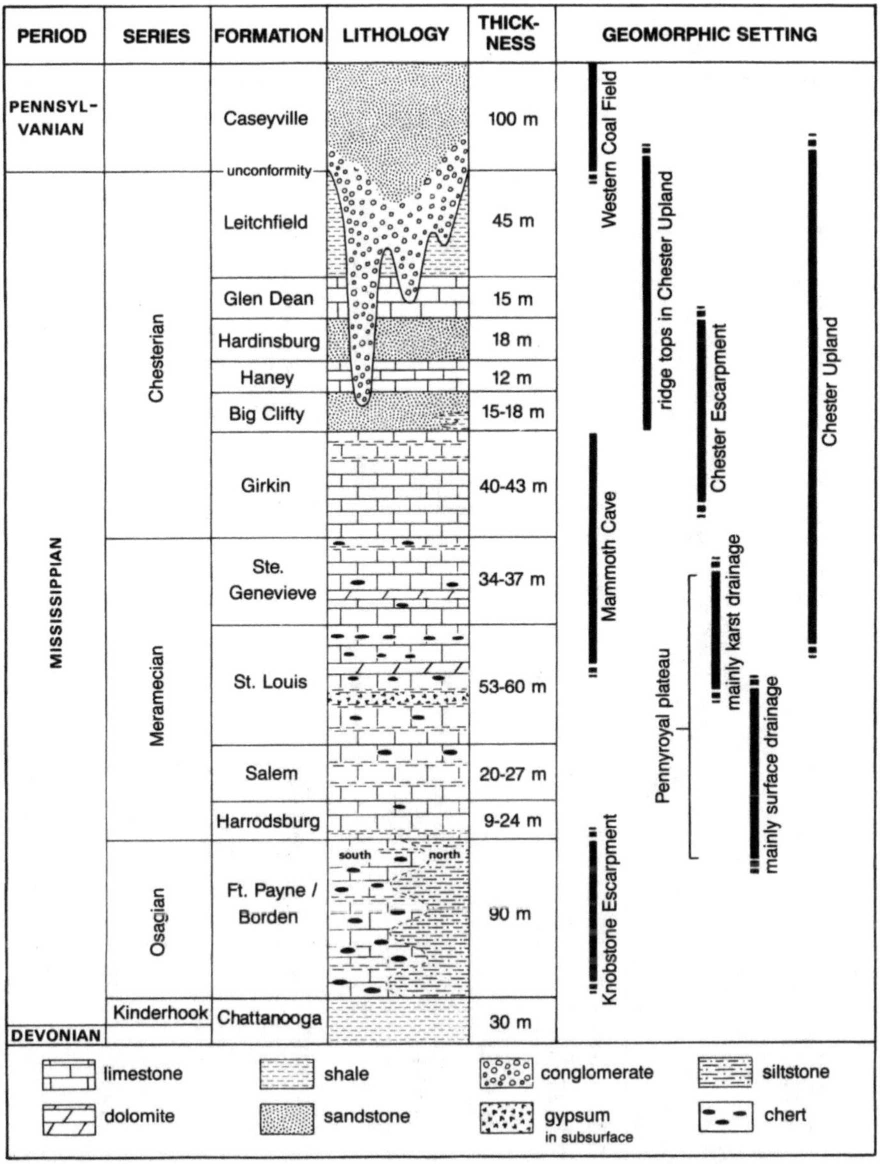

Figure 2.2 Stratigraphic section for the Mississippian and lower Pennsylvanian rocks of south-central Kentucky. *(After Pohl, 1970; from Palmer et al., 1981)*

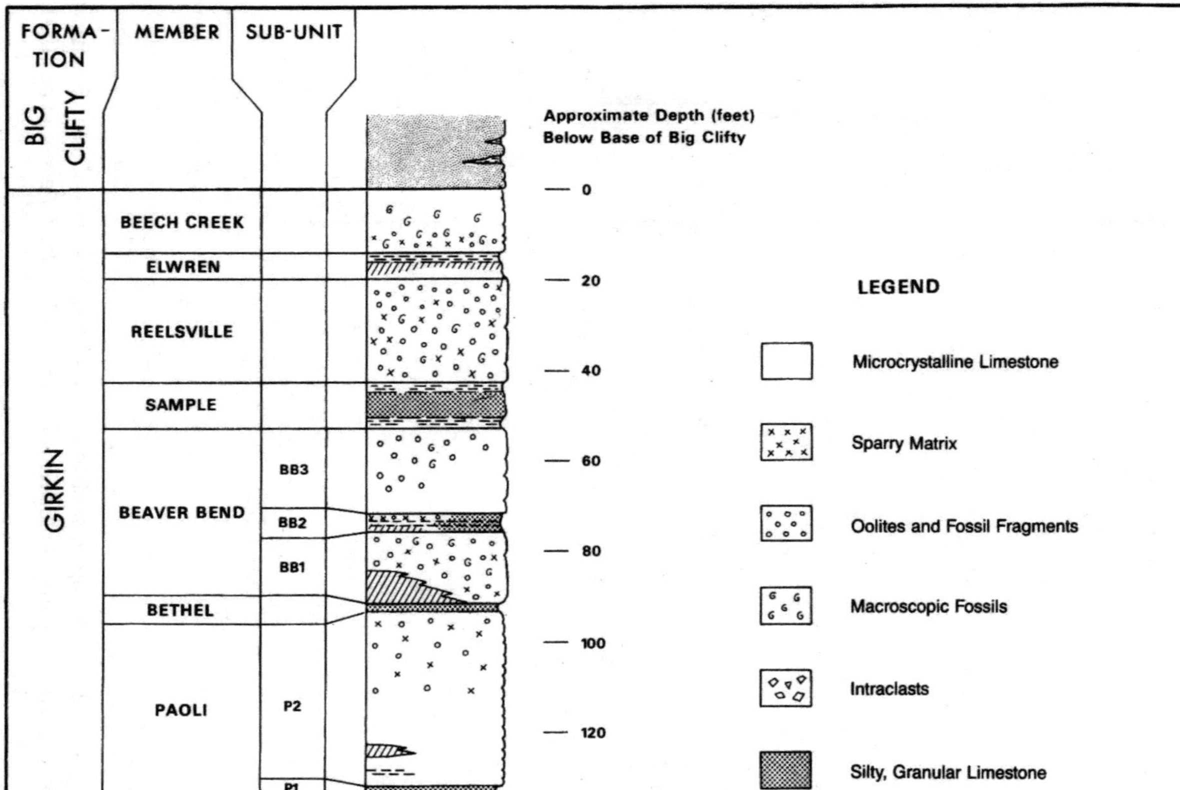
FORMATION
MEMBER
SUB-UNIT
BIG CLIFTY
GIRKIN
BEECH CREEK
ELWREN
REELSVILLE
SAMPLE
BEAVER BEND
BETHEL
PAOLI
BB3
BB2
BB1
P2
P1
Approximate Depth (feet)
Below Base of Big Clifty
0
20
40
60
80
100
120
LEGEND
Microcrystalline Limestone
Sparry Matrix
Oolites and Fossil Fragments
Macroscopic Fossils
Intraclasts
Silty, Granular Limestone

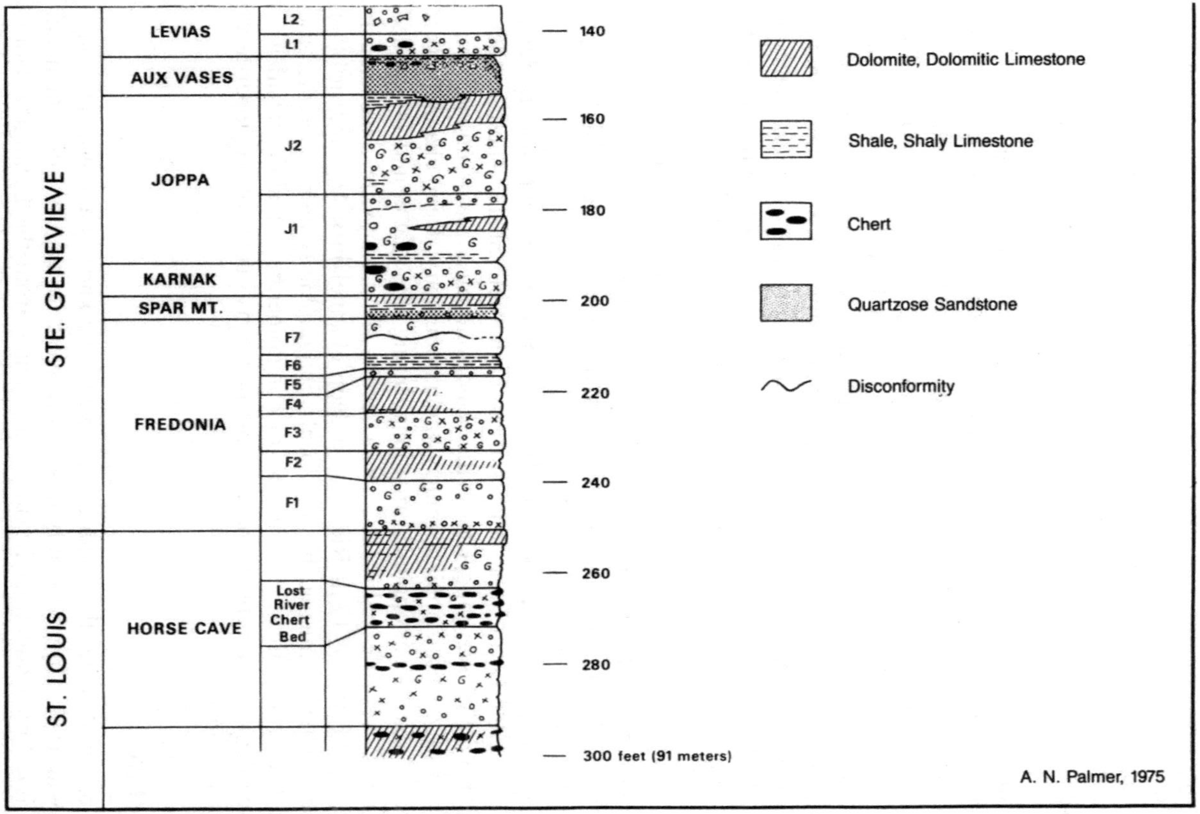

Figure 2.3 Detailed stratigraphy of the cavernous limestones. *(After Palmer et al., 1981)*

thin beds of competent, silty limestone in the upper half. Dolomite tends to dissolve less rapidly than limestone, and the dolomite bands and lenses stand out in relief in the cave passages.

The uppermost carbonate unit of the main aquifer sequence is the Girkin Limestone, which has a considerable number of thin, interbedded shales and siltstones. The clastic units are less than 1 m thick and have only a local influence on the development of underground drainage channels.

A most important hydrogeologic role is played by the Big Clifty Sandstone. It acts as a protective caprock that retards the erosion and dissection of the Chester Cuesta and is responsible for the steep south- and east-facing escarpment that marks the edges of the plateau. At the base of the Big Clifty are a few meters of discontinuous black shale. Where the shale is present, vertical percolation of groundwater is essentially nonexistent. Where the shale is absent, some percolation apparently occurs although most of the vertical infiltration is at the margin of the caprock. Near the top of the Big Clifty is an organic-rich zone containing quantities of pyrite, which might be the source of sulfate minerals found in the cave systems below.

The Haney Limestone forms a minor aquifer perched well above the main cavernous units. The Haney is a well-jointed crystalline limestone about 12 m thick. It crops out on Flint Ridge and in the dissected Hilly Country north of Green River.

Other formations underlying the Hilly Country are the Hardinsburg Sandstone, the Glen Dean Limestone, and the Leitchfield Shale. These are separated from the overlying Caseyville Sandstone by a major unconformity that marks the Mississippian–Pennsylvanian boundary. Deep channels were eroded in the Mississippian formations during this hiatus and later filled with the Caseyville sediments. As a result, the Caseyville Sandstone is found on the top of Flint Ridge as a deep channel in-filling and in other locations where it mismatches the local stratigraphic sequence. At the base of the Caseyville are several meters of conglomerate containing characteristic rounded quartz pebbles that are common constituents of the upper-level cave sediments.

Structure

On a regional scale, the south-central Kentucky karst is a structural monocline with rocks dipping gently to the northwest. To the east, the Cincinnati Arch is a broad anticlinal flexure that extends from the Lexington Dome in the north to the Nashville Dome in the south. To the northwest is the Illinois basin. North of the Mammoth Cave area, the outcrop line of Mississippian rocks extends parallel to the Cincinnati Arch continuously into southern Indiana. West of the Mammoth Cave area, the outcrop line of

Mississippian rocks parallels the Illinois basin westward to the Mississippi River embayment.

On a local scale there is much fine detail to the structure. Dips range widely from 1:1000 to 1:50. The structural contour map (Fig. 2.4) shows the general northwest-trending dip broken by minor flexures with axes oriented perpendicular to the regional strike and many minor structural highs and structural troughs. A sudden steepening of the dip occurs in an east-west pattern.

Faulting is most pronounced in the northern part of the region where normal faults with throws of tens of meters have been found. There are also many smaller faults, most unmapped, with throws of a few meters or less scattered throughout the region.

Conceptual Framework for the Hydrogeology

The Green River, 15–30 m wide and usually less than 8 m deep, extends some 64 km in a narrow meandering channel through the Mammoth Cave area. It is the hydrologic base level for the region and receives discharge from some 81 springs. There are five major sources that supply the water discharging from these base-level springs.

1. On the southeastern edge of the Sinkhole Plain, the sinking streams contribute water injected into the aquifer at specific swallow holes.
2. Between the sinking streams and the Chester Escarpment, thousands of sinkholes on the Sinkhole Plain conduct precipitation directly underground.
3. Precipitation that falls on the caprock-protected ridges of the Chester Cuesta recharges the perched Haney Limestone and Big Clifty Sandstone aquifers. This water is released at the edge of the caprock as a series of springs and seeps and is then conducted downward through vertical shafts.
4. Precipitation that falls on the karst valleys flows into the aquifer through many swallow holes and small sinkholes on the floors of the larger depressions.
5. Water from the Green River backfloods into the base-level aquifer.

The Sinkhole Plain catchment makes up 47% of the total area and 60% of the area south of Green River. The Chester Cuesta catchments make up 53% of the total area, 100% of the area north of the river, and 40% of the area south of the river.

A conceptual model for the groundwater system and overall hydrologic arrangement for the south-central Kentucky karst is illustrated in Figure 2.5. It contains elements common to many carbonate aquifers of eastern

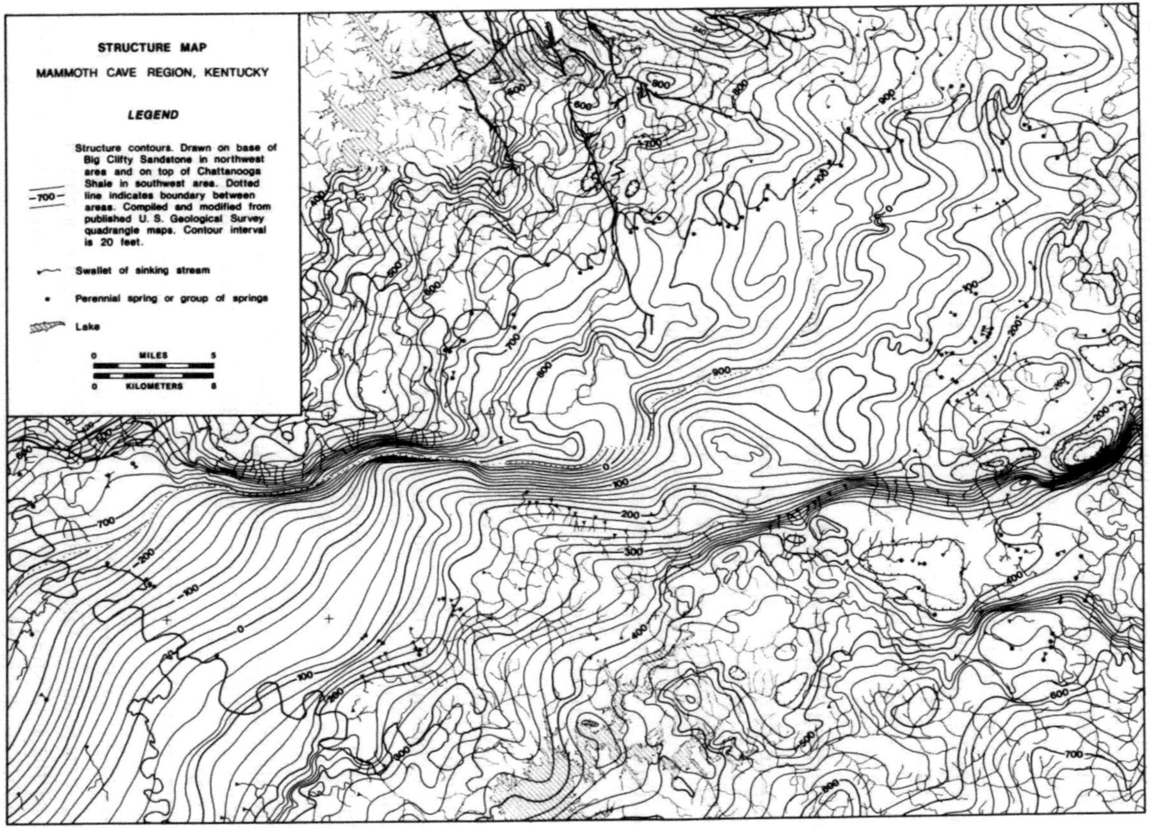

Figure 2.4 Structure of the Mammoth Cave area. (*From Quinlan and Ewers, 1981*)

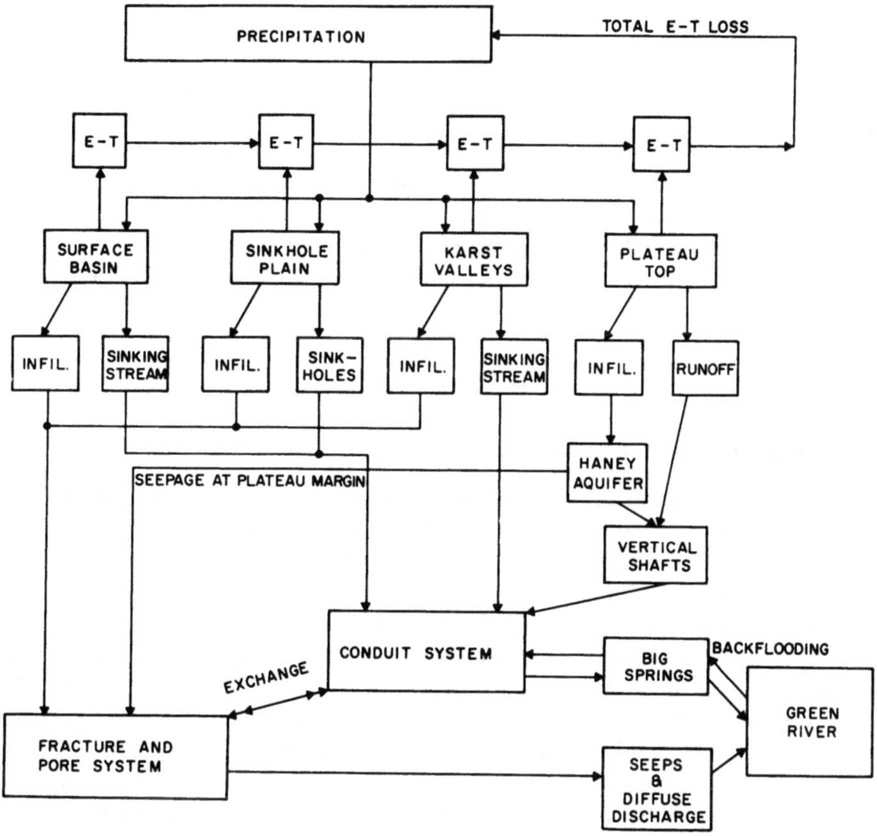

Figure 2.5 Conceptual model for groundwater and surface-water flow systems in the south-central Kentucky karst.

United States. Many of the features of the conceptual model appear on the schematic cross section of Figure 3.2.

Precipitation collected on each of the catchment areas forms the input to the system. Evapotranspiration (E-T) losses discharge some of the water back into the atmosphere, and E-T is a loss term for each type of catchment. The E-T loss is controlled largely by climate and plant cover. The various types of catchment areas and the several parts of the aquifer are described in the sections that follow. One can view Figure 2.5 either as a schematic of the physical system or the parts of the diagram can be related to mass balance terms in an overall water balance.

Each spring can be regarded as an outlet point for an underground drainage system of some degree of complexity. The regional springs are fed by large conduits carrying water from the collector systems of the surface basins and sinkhole drainages many kilometers away. Closer to the river, valley drains from the karst valleys feed into the main system and shaft drain systems feed into the valley drains. Some valley drains lead to the river directly as springs of intermediate size, while shaft drain systems in small re-entrant valleys near the river might also drain directly to the river through small springs and seeps. Thus we can account for the large number of springs of diverse flow characteristics. The discharge of the fracture and pore system is shown in Figure 2.5 as a separate discharge to the river but little is known about it.

GROUNDWATER RECHARGE

Sinking Stream Catchments

The sinking stream catchments are made up of an aggregate of many small surface catchments ranging up to a few square kilometers in area. Sinking Creek, Little Sinking Creek, and Blue Spring Creek have several surface tributaries, but many of the sinking creeks are short first-order streams. The gradients of these streams range from 3.8 m/km to 8.0 m/km with a mean gradient of 6.1 m/km. The distributions of stream lengths and number of tributaries as a function of stream order for the sinking streams along the southern boundary between Sinking Creek and Blue Spring Creek are shown in Figure 2.6. The sinking streams are incised up to 35 m below the general level of the Sinkhole Plain and most of their extension out onto the plain is in incised blind valleys. The blind valleys are alluviated and the streams flow in irregular channels cut into the alluvial blanket. At the lower end of the sinking streams, the streams' channels branch out in a distributary pattern with each channel ending at a swallow hole clogged with logs and other flood debris. The swallow hole farthest upstream diverts surface drainage into the groundwater system. As the stream's discharge increases, surface water is diverted from the upstream swallow hole to swallow holes farther down the main channel. The terminal swallow hole usually is never reached by surface drainage during periods of low flow. As the flow increases, more and more of the swallow holes are utilized until, under flood conditions, the entire blind valley could be flooded. In no case, however, is there a surface overflow channel downstream from the terminal swallow hole. Only rarely can access to an open cave system be gained through the swallow holes of the sinking streams; almost without exception, the openings in the channel are blocked with fallen trees, other debris, and alluvium.

Infiltration from sinking streams into the groundwater system is the most rapid form of recharge for carbonate aquifers. Figure 2.7 shows two hydrographs for a sinking stream (Little Sinking Creek) and the groundwater level

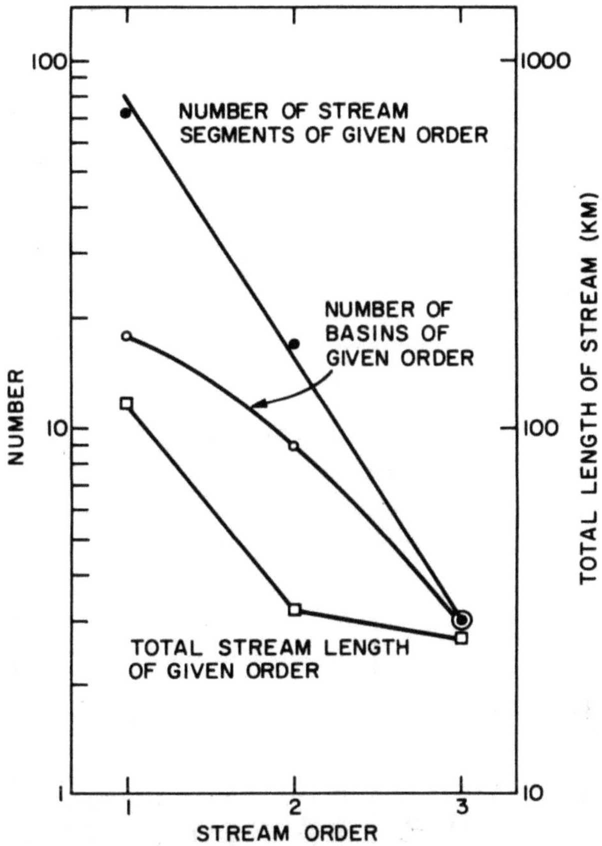

Figure 2.6 Relationships between number of stream segments, order of basin, and stream length as a function of stream order for 30 sinking creek basins between Sinking Creek and Blue Spring Creek along the southern boundary of the Sinkhole Plain.

in a well approximately 600 m from the terminal swallow hole of the sinking stream. During six flood events in 1961 and 1962, the groundwater level in the well rose a minimum of 0.4 m in two days following the peak discharge of the sinking stream. The minimum stage for the stream during these flood events was 2.8 m. The largest increase in the level of the groundwater was 3.1 m for a peak stage of the sinking stream equal to 7.3 m.

The recovery time for the sinking stream is usually 3–5 days. However, for groundwater in the carbonate aquifer approximately 600 m from the stream, the recovery time varies from 10 days to 20 days (Fig. 2.7). White and Reich (1970) found that the mean of the annual series of floods for basins developed in thick carbonates is low compared to basins developed

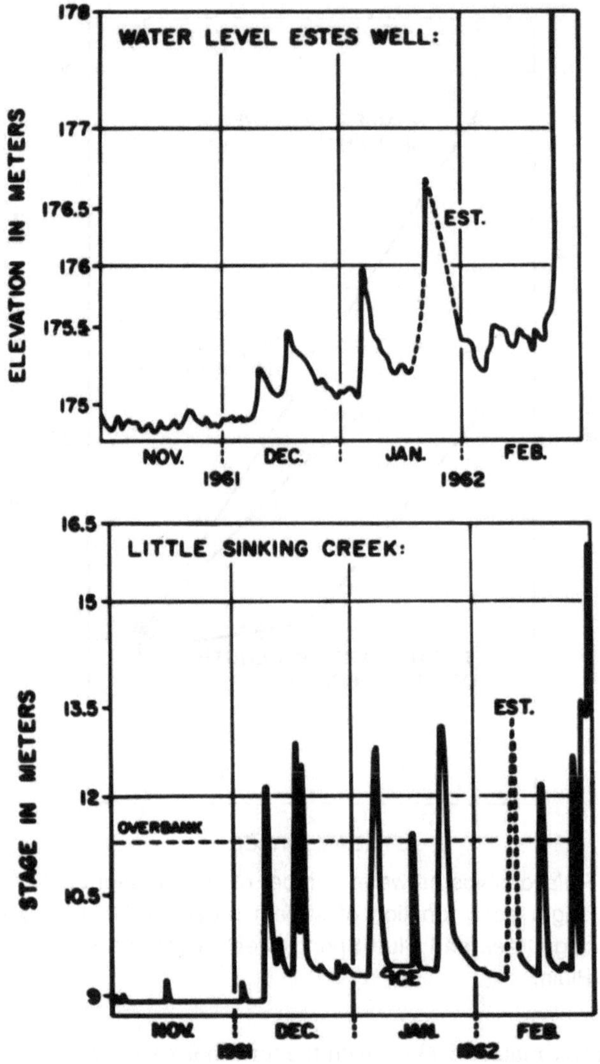

Figure 2.7 Hydrographs of flood events on Little Sinking Creek and corresponding changes in groundwater levels near the terminus of the sinking stream. Data based on USGS gaging stations.

on clastic rocks. They suggested that this phenomenon was due to floodwater storage in the aquifer with slow resurgence over a long period of time. These data support the groundwater storage idea offered by White and Reich (1970). Additionally, the long rate of decay of the flood curves for groundwater in carbonate aquifers has been attributed to ponding in the epiphreas, the upper portion of the saturated zone (Ashton, 1966). Flood waters cannot escape quickly from the bottom of the vadose section of the aquifer and back up, creating an increase in head. Finally, the rate of flow increases to restore equilibrium to the system.

In the case of Little Sinking Creek, groundwater appears to be ponded for several days. This storage of flood waters in the aquifer produces the asymmetric curves for the groundwater hydrographs. Two reasons for groundwater ponding might be related to the degree of integration of the solutional conduits and the degree of segmentation of the conduits (which is related to collapse of the overlying bedrock). The latter process could be important with respect to the Little Sinking Creek drainage because at least two major collapse sinks penetrate the conduit. Mill Hole and Cedar Sink are impressive collapse sinkholes that have intercepted the main subsurface drainage line for this sub-basin, and during flood events they block the flow of groundwater in the system. This results in an increase in head and flooding of the sinkhole floors. Groundwater in the Little Sinking Creek catchment cannot escape quickly, which results in a long recovery time for the aquifer.

The storage of flood waters in the aquifer and subsequent increase in hydraulic head affect the hydrologic behavior of the sinking streams. Sinking streams often attain overbank stage because of the increase in hydraulic head and the inability of the swallow holes to pass large volumes of water. During a 4-month period in 1961 and 1962 (high discharge period), Little Sinking Creek was overbank 10 times. Ephemeral lakes form in the terminal portions of the sinking creek watersheds during floods when surface runoff is high and infiltration into the aquifer is low (Fig. 2.8). Ephemeral lake development is characteristic of most sinking streams during flood events. Although it may seem strange at first thought, sinking streams and sinkhole terrains are flood-prone areas. Sinkhole flooding is an important environmental problem in the city of Bowling Green at the western end of the south-central Kentucky karst (Crawford, 1984).

The watersheds of each sinking stream vary considerably in size, but they have a similar characteristic morphology. In the Glasgow Upland the watersheds exhibit few solutional features; rather, they are characterized by high wide interfluves and narrow flat floodplains. Northward, the sinking streams exit the Glasgow Upland and drain into the more soluble carbonate units of the Sinkhole Plain. In this region the sinkhole density is large and the interfluves change from the wide high divides to a complex of saddles separating individual closed depressions. At the contact of the Sinkhole Plain and Glasgow Upland, the floodplains of the sinking streams become very extensive, creating broad flat regions. The development of the broad alluvial flats between the two physiographic units results from several interacting processes.

Figure 2.8 Blind valley of Little Sinking Creek during dry weather (top) and after heavy rains (bottom).

Water from sinking streams is lost through rapid infiltration into the groundwater system, which has developed in the more soluble carbonate units of the Sinkhole Plain. As the discharge is reduced through infiltration, the suspended and bedload material transported by sinking streams is deposited. An analogy might be that of ephemeral washes on alluvial fans where recharge occurs through transmission losses and sediments are subsequently deposited.

As illustrated previously, sinking streams frequently reach overbank stage and form ephemeral lakes. During these periods, suspended sediments are deposited in these blind valleys. Up to one-half inch of fine-grained sediment has been deposited on the sinking streams' floodplains as a result of a single ephemeral lake event.

The termini of the sinking streams are not necessarily stationary, but shift through time and possibly with increasing discharge. This can be seen by the numerous abandoned blind valleys associated with the terminal portions of the sinking streams. Deposition in these blind valleys, resulting from the processes stated above, shifts along the contact between the uplands and Sinkhole Plain and further extends these broad alluvial flats.

Soils representative of the south-central Kentucky karst are described in the Soil Conservation Service surveys for Barren County (Latham, 1969) and Warren County (Barton, 1981). Soil surveys for Hart County and Edmonson County are being compiled. The published maps of soil associations in Barren and Warren counties do not correlate with one another and there are text errors in the description of some of them. In brief, however, the principal soils are fine, mixed mesic Typic Hapludalfs and Paleudalfs and Rhodudultic Paleudults that range from 1 m to 20 m thick. A foot or more of loess can be locally found where plowing has disturbed the profile minimally.

Sinkhole Plain Catchments

The Sinkhole Plain consists of a rolling topography pocked with thousands of dolines. There are no surface streams and little evidence for channels or distinguishable valley development. All overland flow is diverted into sinkholes. The relief within short distances on the Sinkhole Plain varies from 10 m to 40 m with individual depressions usually from 8 m to 10 m deep. Exceptions are the occasional occurrence of knobs, outliers of the Chester Cuesta, which rise 70–100 m above the plain (Fig. 2.9). One can sub-divide the Sinkhole Plain catchment into a large number of catchment cells by drawing drainage divides along the high ground between individual sinkholes. Within each catchment cell there is some diffuse infiltration

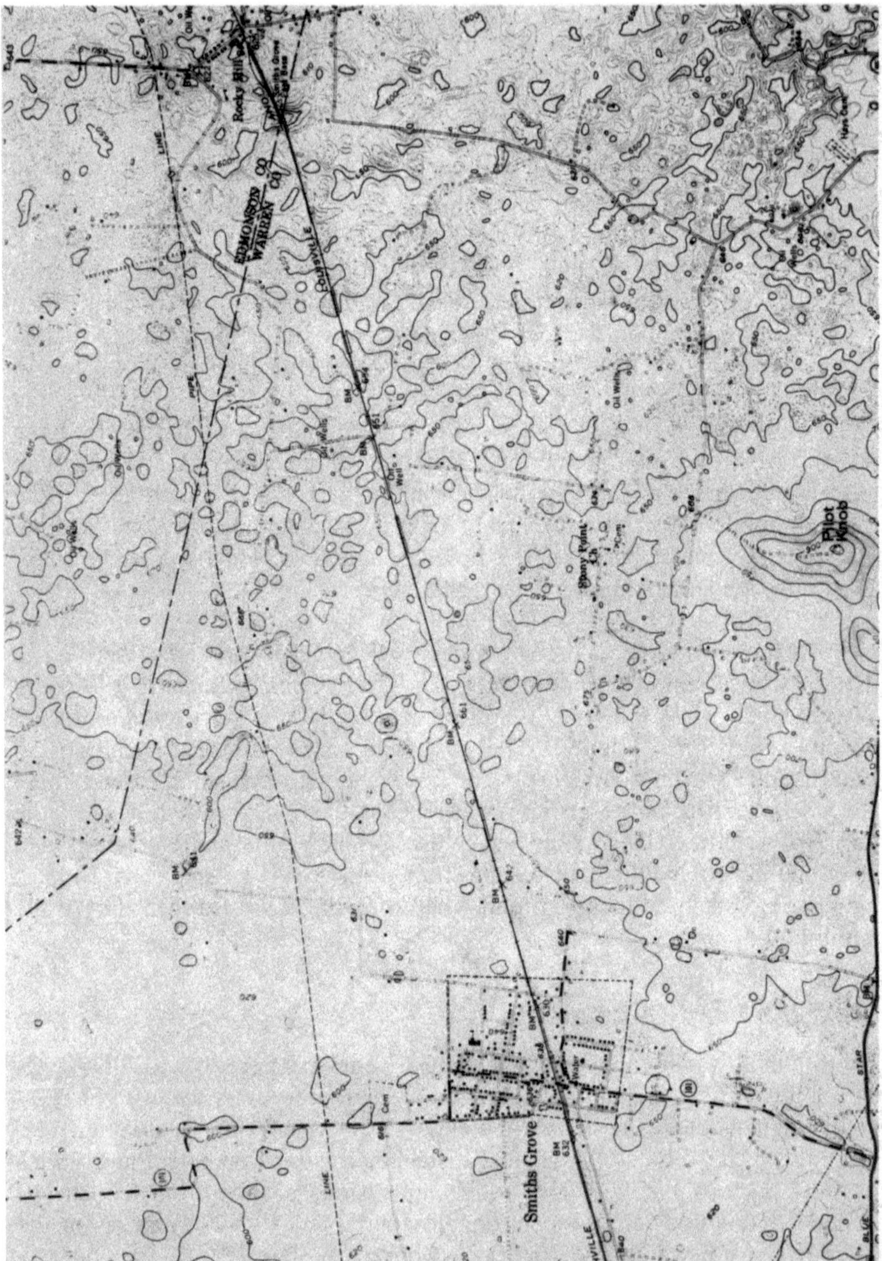
EDMONSON CO
WARREN CO
Rocky Hill
Pilot Knob
Smiths Grove

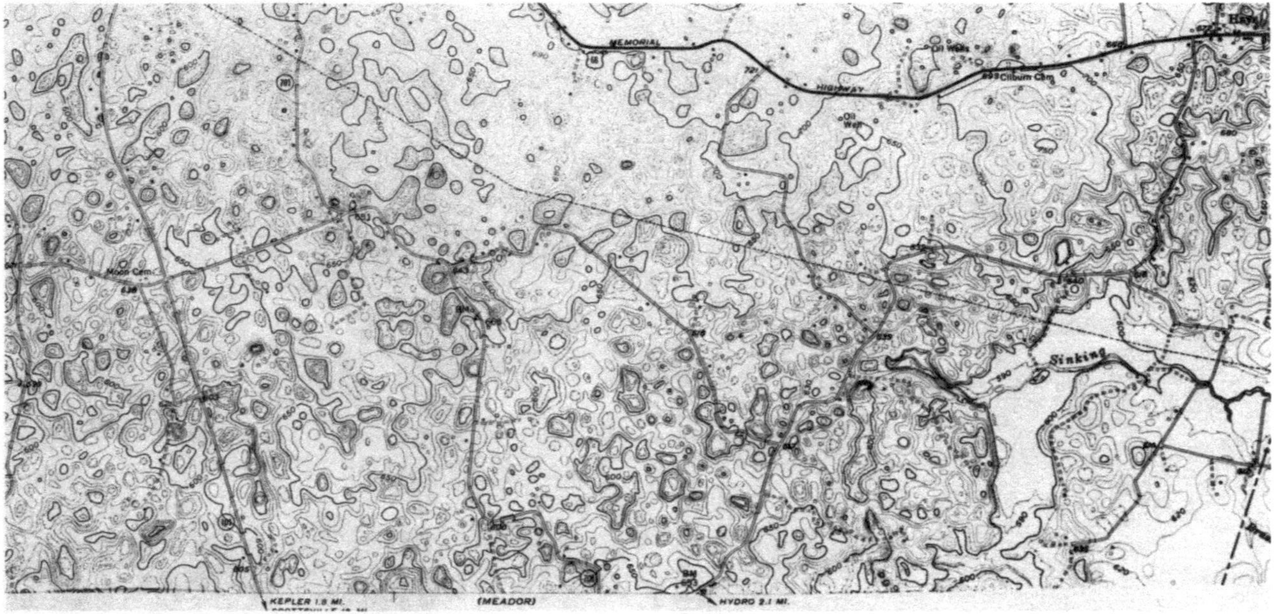

Figure 2.9 A segment of Smiths Grove quadrangle illustrates most of the topographic features of the Sinkhole Plain. The Ste. Genevieve Limestone crops out in the northwest corner of the map and the sinkhole-free area near Smiths Grove is in the region of the Lost River chert at the Ste. Genevieve–St. Louis contact. The sinkhole-pitted terrane that makes up most of the map is underlain by the St. Louis Limestone. Near the center of the map is Pilot Knob, a residual hill of Ste. Genevieve Limestone with a smaller remnant of Girkin Limestone on its summit. The blind valley and floodplain of Sinking Creek can be seen in the southeast corner.

through the soil into the bedrock below and there is some overland flow or flow at the soil-bedrock contact into the bottom of the closed depression where it enters the subsurface through the sinkhole drain.

Although sinkholes are of diverse origins, all have three common components: the bedrock depression; the cover materials of soil, alluvium (glacial drift in some northern areas), and other unconsolidated material; and the drain. Three processes are also at work in sinkhole enlargement and development: localized dissolution at the soil-bedrock contact; stoping and collapse of bedrock into pre-existing subsurface solution openings; and piping of soil and other unconsolidated material into the subsurface. Depending on the dominant process, at least four distinct types of sinkhole are found in the Mammoth Cave area.

Solution sinkholes are basinlike depressions in the bedrock. The solution sinks of the south-central Kentucky karst are usually draped with soil so that the bedrock surface is not seen. The closed depressions have gentle slopes and flat floors with thick layers of unconsolidated material. Studies in Indiana sinkholes (Ruhe, 1975) suggest that the fill material there is not an in-situ weathering phenomenon. The sediment layers are sufficiently impermeable to support sinkhole ponds (Fig. 2.10), but some limited pollen measurements (Wright et al., 1966) suggest that the sinkhole ponds are

Figure 2.10 Compound sink showing sinkhole ponds and linearly aligned depressions below the level of the Sinkhole Plain. This feature appears near the top center of the topographic map shown in Figure 2.9.

Figure 2.11 Resurgence (Elk Spring) in karst window on the Sinkhole Plain serves as domestic water supply.

ephemeral and contain only recent pollen and sediments. The drains from solution sinks are likewise seldom accessible to observation, but most seem to be solutionally widened joints and fracture sets.

A second type of sinkhole is the collapse sink, which often has steep walls with bare bedrock exposed. Collapse sinks form by roof collapse of shallow cave passages. Sometimes the cave can be entered from the bottom of the sinkhole. Other collapse sinks are blocked by fallen rubble and draped with soils so that the surface expression of the collapse sink is indistinguishable from a solution sinkhole. When collapse sinkholes intersect the subsurface drainage, they are called karst windows. In karst windows, the subsurface drainage resurges and flows a short distance over an alluviated floor and sinks again on the downstream side. Residents of the Sinkhole Plain use the karst windows as a form of domestic water supply (Fig. 2.11).

A variant of the collapse sinkhole common in the south-central Kentucky karst is the collapsed vertical shaft. Vertical shafts form an integral part of the vertical drainage on the plateau margin. As the plateau retreats, the shafts collapse, fill with rubble, and the lower parts of the shaft remain as sinkholes. Filled shaft structures can be seen in road cuts along Route 255 where the highway climbs the escarpment, suggesting that they are a very common feature.

Where soils are thick, cover collapse sinks can form (Fig. 2.12). Unlike the other types of sinkholes, cover collapse sinks do not involve solution

Figure 2.12 Cover collapse sinkhole in thick soil near Park City. Note sharp edge marking shear zone at the top of the soil arch.

or collapse of the bedrock, only the movement of unconsolidated cover material. Figure 2.13 illustrates the essential features of cover collapse sinkholes. Infiltrating water carries soil down the drain system forming a small void between the soil and the bedrock surface. As more and more soil is lost to the subsurface, the enlarging void forms an arched roof that provides mechanical stability. With continued enlargement the void will reach a size where the roof will no longer support its own weight. There is an abrupt collapse that forms a sinkhole with vertical sides entirely in the soil. These form rapidly and also decay gradually as the sides slump. Unlike solution and bedrock collapse sinks, which are more or less stable features of the landscape on human time scales, cover collapse sinkholes are a major threat to property in karst areas.

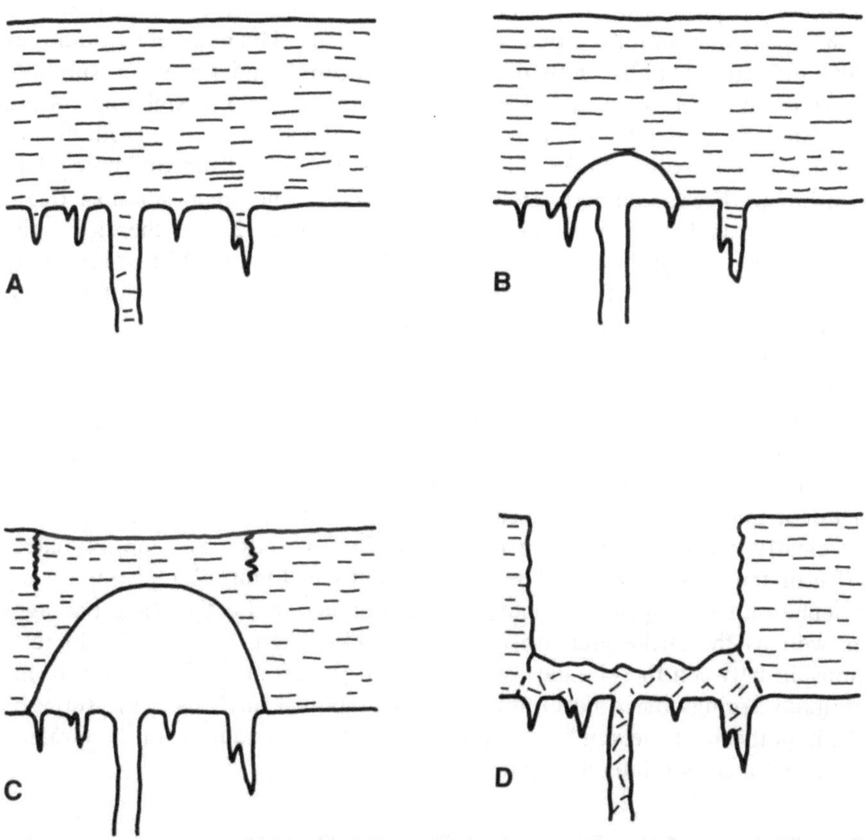

Figure 2.13 Sketches showing the evolution of a cover collapse sinkhole. *A*, Initiation of the arch by soil piping through solutionally widened joints. *B*, Enlargement of the arch. *C*, Destabilization of the cover. Shear cracks and some sagging might be visible at the surface during this phase. *D*, Collapse of the arch. Drain and initial void space are concealed by in-fallen soil.

The distribution of sinkholes over the Sinkhole Plain is quite irregular, and sink densities vary considerably from place to place. Some of the variation is probably of structural origin, related to the outcrop pattern of the chert horizons with the lower Ste. Genevieve and St. Louis limestones. The chert serves as a local capping bed and localizes the groundwater infiltration (Howard, 1968). Some zones of deep depressions are related to major groundwater flow paths such as the trunk drainage of the Graham Springs system and perhaps the line of sinks along Waterloo Valley.

A sample study of sinkhole depth and frequency on the Sinkhole Plain in the Graham Springs basin was conducted to determine whether the major conduits of the carbonate aquifer were segmented as a function of sinkhole collapse. The depth of 830 sinkholes was measured from USGS topographic maps (although sinkhole depths determined from topographic maps do not accurately represent the depth of penetration of sinkholes, they provide the best approximation for regional studies; Fig. 2.14). The number of sinkholes falls off exponentially with depth as has been found for other sinkhole populations (Troester et al., 1984). Sixty percent, or 498 sinkholes, were 3 m or less in depth, whereas the average elevation of the groundwater surface is approximately 38–43 m below the surface of the Sinkhole Plain (Cushman, 1968; Lambert, 1976; Wells, 1973). The mean sinkhole depth was 5.5 m and only one sinkhole was deeper than 27 m; therefore, the results suggest that the major conduits within the aquifer could be relatively free from collapse in Graham Springs basin. There is nothing in the depth distribution to distinguish sinkholes of different origins.

The other aspect of sinkhole morphology that might suggest a relationship with conduits in the carbonate aquifer is the orientation of elongate sinkholes. That is, elongate sinkholes that collapse into cave passages should follow the trend of the cave passage. In this study an elongate sinkhole is defined as any sinkhole having a major axis twice the length of a minor axis. Using this criterion, the total number of sinkholes that can be clearly defined as elongate in Graham Springs basin are approximately 10% of the total sampled. The rose diagrams in Figure 2.15 compare the orientation of elongate sinkholes with abandoned and active cave passages as well as the strike and dip of the bedrock. There is little visible relationship between the elongate sinkholes and the groundwater system in the Graham Springs basin based on a purely graphical analysis. This supports the hypothesis of the sinkhole-depth study in that the aquifer for the Graham Springs basin is relatively open.

The Ridges of the Dissected Chester Cuesta

Green River flows at an angle to the structural grain of the south-central Kentucky karst and thus crosses physiographic boundaries. Using Green River as a reference line, this results in a transition from the western region

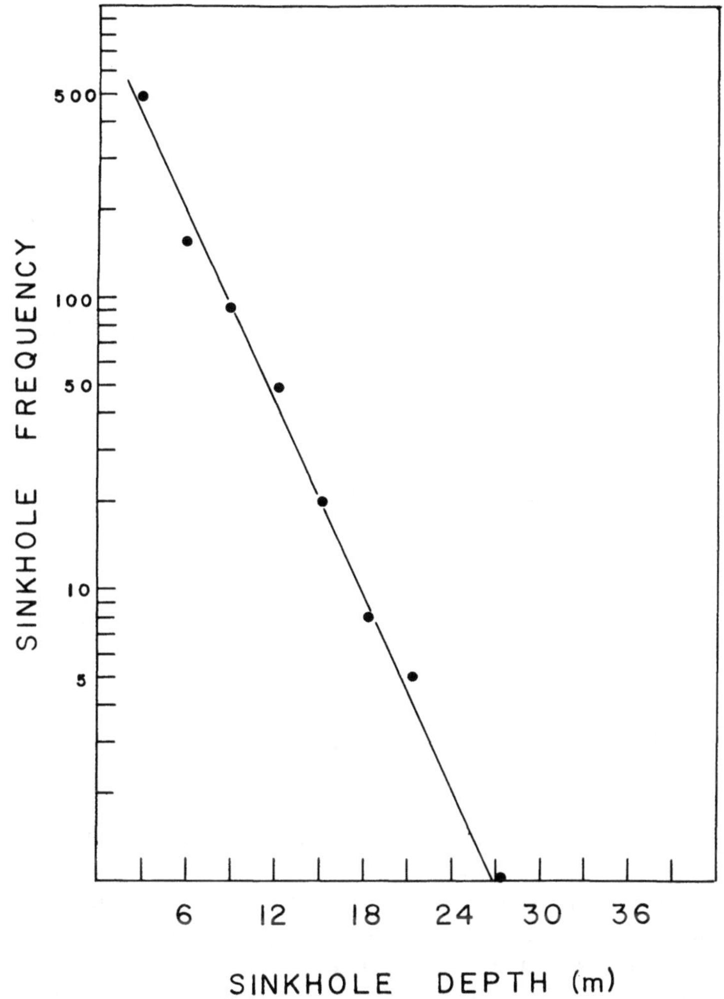

Figure 2.14 Sinkhole depth and frequency distribution for the Sinkhole Plain.

of the Cuesta with unbreached clastic rocks, through a transitional region where the sandstone-capped Cuesta has been dissected into isolated ridges separated by karst valleys, to the eastern region where the caprock has been completely removed south of Green River and only isolated knobs remain. The principal ridges and intermediate valleys are shown in Figure 2.1 in the central area of the map.

Precipitation falling on the sandstone-capped ridges recharges a perched aquifer in the Hardinsburg Sandstone, Haney Limestone, and Big Clifty Sandstone. This aquifer is tapped by shallow wells on the ridge tops. The aquifer discharges through a series of small springs. The water is then

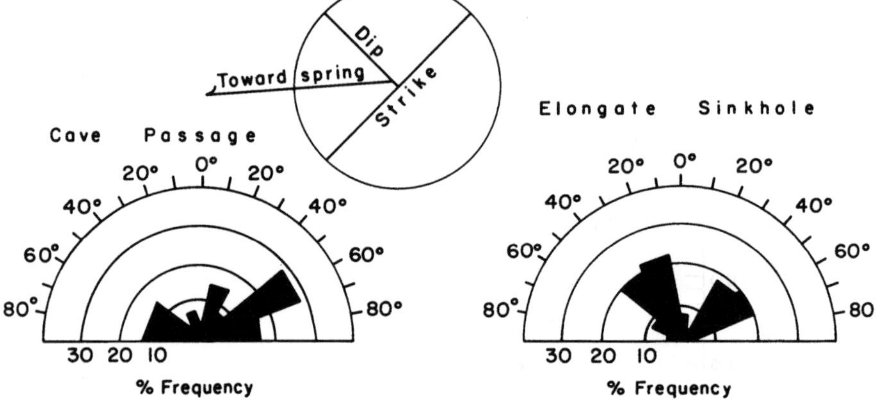

Figure 2.15 Rose diagrams of the orientation of cave passage segments and the orientation of elongate sinkholes in Graham Springs basin.

directed downward to the basal drainage system via vertical shafts. Where substantial areas of caprock occur, there can be enough accumulation of surface runoff to supply perennial streams, which usually sink when they cross from the Big Clifty Sandstone to the Girkin Limestone. In the western portion of the region, the surface streams persist all the way to Green River. Flat areas of the Cuesta sometimes support small perched bodies of water such as Sloan's Crossing Pond (Fig. 2.16).

Vertical shafts are cylindrical voids created by the dissolution of limestone by vertically moving water in the vadose zone. Isolated shafts often have a near circular cross section (Fig. 2.17) although complexes of overlapping shafts and shafts that become retreating canyons are also common. The walls of the shafts are usually perfectly vertical, a dissolution surface that truncates beds of greatly varying lithology. The tops of the shafts are formed some distance under the caprock so that, in general, shafts are not open to the surface. Shaft floors, where not blocked by rubble, are bedrock basins. Water exits the shafts through drains that are cave passages much smaller in volume than the shaft itself (Brucker et al., 1972). Pohl (1955) recognized that although shafts often connect to cave passages, they are much younger than most of the cave passages and form entirely in the unsaturated zone. Quinlan and Pohl (1967) argued that densities of shafts are very high along the margins of the retreating caprock and that vertical shaft development is responsible for the oversteepened slopes along the escarpment. Distribution of mapped shafts located from within the cave system (Fig. 2.18) are closely related to the present-day position of the caprock margin. The relationship is not perfect. Some abandoned shafts are found beneath valley walls and occasionally shafts are found beneath the caprock well away from the margin.

Figure 2.16 Sloan's Crossing Pond, a perched water body on the sandstone caprock near the entrance to Mammoth Cave National Park.

The Cuesta top has a cover of second growth forest, mainly hardwoods with some cedar in the confines of Mammoth Cave National Park. South and east of the Park boundary, large areas of the Cuesta are in farmland.

The soils and the underlying karsted limestone surface retard the infiltration of precipitation. Traditionally, infiltration through karst surfaces was assumed to be very rapid because of observations of rapid runoff into sinkhole drains (internal runoff). However, recent evidence (Williams, 1983, 1985) suggests that the subcutaneous zone (or the epikarstic zone of the French literature) holds infiltration water in storage for periods of weeks before releasing it into the solution channels in the underlying bedrock. Figure 2.19 shows the bedrock contact with solutionally widened joints (grikes) and other sculpturings of the limestone surface. These are soil-filled and appear to be the reservoir that smooths out infiltration pulses so that many observed drip points in the underlying caves continue to drip during dry seasons.

Karst Valley Catchments

Once the sandstone caprock has been breached the processes of solution can begin the construction of underground drainage routes. This process has gone to different degrees of completeness in different parts of the

Figure 2.17 Keller Well, an active vertical shaft in the Flint Ridge section of the Mammoth Cave system. Meter stick is 8 m below the observer. Depth to water is 17 m. *(Photo by Roger W. Brucker)*

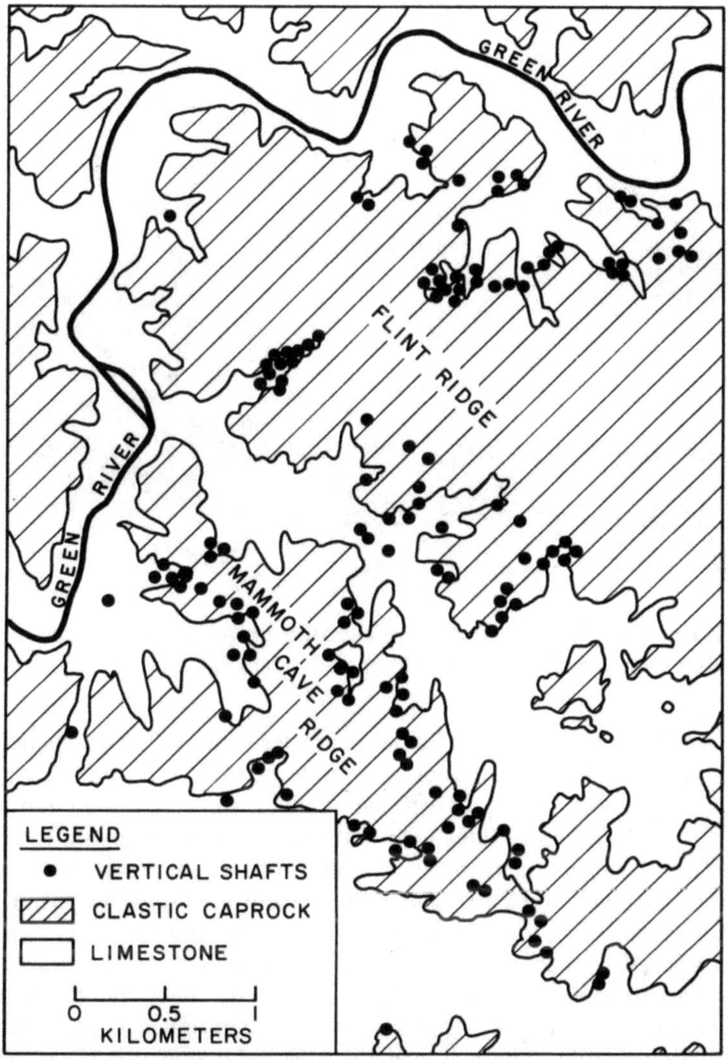

Figure 2.18 Distribution of vertical shafts at the caprock margins on Flint Ridge and Mammoth Cave Ridge. Although these points are representative, only a small fraction of the total number of known shafts are shown.

transition region. South of Green River as far downstream as Turnhole Bend, the valleys are completely underdrained. Although short reaches of surface channel, used by flood runoff, appear in many of the valleys, these channels are discontinuous and rarely reach Green River. The valley profiles are discontinuous and many parts of the valleys are transitional between fluvial valleys and large dolines. The upstream reaches of the

Figure 2.19 The subcutaneous zone exposed on a road cut on the Chester Escarpment between Cave City and Mammoth Cave National Park.

valleys maintain much of their fluvial character, but the lower reaches give way to sequences of closed depressions. Indeed, there is a limestone barrier wall that completely blocks Doyel Valley, indicating that diversion of the drainage to underground routes occurred very early in the evolution of the valley. This is substantiated by the valley profile, which shows the outlet to Doyel Valley hanging 70 m above the present level of Green River.

Farther to the east, the dissection of the cuesta and the replacement of fluvial by karst processes is so nearly complete that the original valley patterns are difficult to reconstruct. To the west, however, the tributary valleys to Green River from the south are underdrained, but maintain a normal valley profile. There is usually a well-developed surface channel extending from the sandstone caprock to the river, although the channel is often dry except during periods of high runoff. North of Green River the valleys are also underdrained, but maintain a normal valley profile. The geomorphology of the area north of the river is discussed in more detail in Chapter 8.

The valley drain system, which carries the integrated underground runoff from entire tributary valleys, is one of the principal concepts for understanding the hydrology of the south-central Kentucky karst. The valley drain systems can develop independently of each other; north of the river only valley drain systems can exist. South of the river the drains from the karst valleys are complicated by the other drainage coming from the Sinkhole Plain resulting in larger and more complex cave systems.

The karst valley catchment areas are underlain by two soil types: the Weikert and the Caneyville. They are described as shallow to moderately deep, well to somewhat excessively drained, sloping to steep, rocky or stony soils. They range up to 1 m in thickness with moderately rapid to moderately slow permeabilities. The available water capacities are approximately 7 cm.

Base-level Backflooding

An unusual source of recharge for the karst aquifer is backflooding from the Green River. The various sources of recharge drain to the river through low gradient open conduits. The stage of Green River fluctuates through more than 20 m in its natural regime (moderated considerably since the construction of a dam at Greensburg, upstream from the karst area). Sudden rises in the river resulting from storms in the upstream reaches of the Green River basin reverse the hydraulic gradients and cause water to flow into the springs. Because gradients are low, backflooding can force water into the aquifer for distances of several kilometers from the river. Base-level backflooding can be considered a form of bank storage of an extreme sort.

Evidence for base-level backflooding was provided by wildcat oil drillers, who, in the late 1950s, contaminated the upper Green River basin with large quantities of brine. High chloride levels were found in Echo and Styx Rivers in Mammoth Cave, thus proving that river water flowed at least 1–2 km into the system when the river stage increased faster than inputs from the Sinkhole Plain (Krieger and Hendrickson, 1960; Hendrickson, 1961).

THE HANEY AQUIFER

General Setting

The geology and topography of the Chester Cuesta combine to create conditions for the formation of several perched aquifers above the basal water table. These zones are as follows:

1. A generally continuous groundwater body in the Haney Limestone perched on shale layers in the lower part of the Haney.
2. A generally continuous body of groundwater in the Big Clifty Sandstone perched on a shale at the base of the Big Clifty.
3. Discontinuous bodies of perched water above local shale beds near the top of the Girkin Limestone (Cushman et al., 1965).

The perched groundwater in the Haney Limestone is of interest here. The Haney and the overlying Hardinsburg Sandstone and the underlying

Big Clifty Sandstone make up the caprock on the ridges south of the river. The Haney is about 12 m thick on Flint Ridge and is exposed above the Big Clifty at elevations of about 225–232 m (740–760 ft). Its areal extent on Flint Ridge is approximately 10.9 km^2. The caprock is broken by numerous joints. The openings along joints in the Hardinsburg and Big Clifty are small, but the openings along the joints in the Haney are enlarged by solution and the limestone is cavernous.

Most of the precipitation infiltrates or flows along the relatively level surface of Flint Ridge and sinks at the edge of the caprock. Water penetrating the permeable surface rock moves downward until it reaches the relatively impermeable shale at the base of the Haney Limestone where it then moves laterally in solution channels to emerge as seeps and springs along the margin of the ridge.

Haney Springs

Concentrated discharge from the Haney aquifer on Flint Ridge occurs at eight known springs located typically at the heads of deep re-entrants on the sides of the ridge (Figs. 2.20 and 2.21). The discharge from these springs crosses the Big Clifty Sandstone and sinks. The USGS has monitored the flow of seven of the springs: Three Springs, Bransford, Blair, Adwell, Cooper, Collins, and CCC No. 1 Springs. The combined average flow is 5.4 l/sec with the base flow being 4.5 l/sec (Cushman et al., 1965). These figures yield an average base-flow runoff for the seven Haney springs of 0.42 l/sec/km^2, which is 26% of the base flow per unit area of the central Kentucky karst as a unit (1.64 l/sec/km^2). This indicates that most of the water is discharged as small seeps and as springs that are not obvious or is leaked downward into the underlying perched Big Clifty Sandstone aquifer.

THE CAVERNOUS LIMESTONE AQUIFER

The main groundwater body in the south-central Kentucky karst occurs in the cavernous St. Louis, Ste. Genevieve, and Girkin limestones. Most of this volume is devoted to interpretation of the cavernous limestone aquifer; this section outlines only a few of its geologic characteristics.

Porosity and Permeability

The overall permeability of carbonate rock is the sum of three contributions:

1. Primary porosity and permeability that is due to the presence of communicating pore spaces.

2. Permeability that is due to the three-dimensional network of joints, fractures, and bedding-plane partings.
3. Permeability due to cavernous openings (Table 2.1).

Primary porosity of the Ste. Genevieve Limestone is 3.3%, and the coefficient of permeability is 0.0016 l/day/m^2, as determined from core samples (Brown and Lambert, 1963). Specific capacities of wells drilled in the St. Louis Limestone range from 70 to 8700 l/min/m of drawdown.

It is the interior plumbing of carbonate aquifers that separates them from other aquifers. In addition to the pore spaces that serve as storage volume for water and joints, fractures, and bedding-plane partings that form more

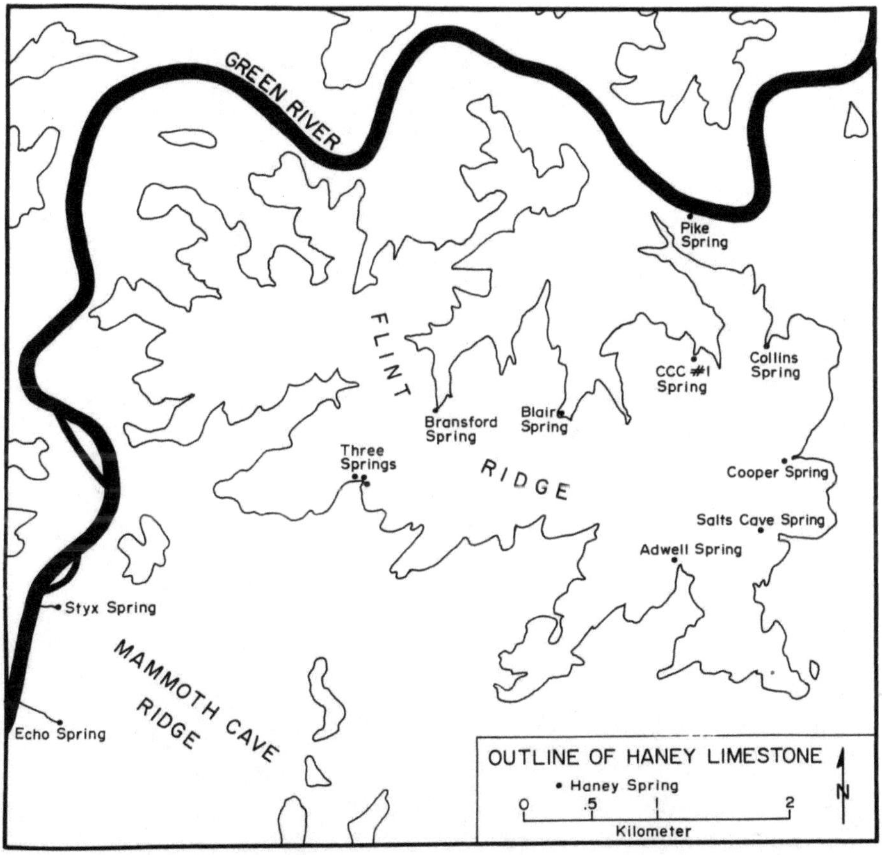

Figure 2.20 Map of the Haney Limestone aquifer on Flint Ridge, showing the locations of eight Haney springs and their relationship to the Haney–Big Clifty contact.

Figure 2.21 Collins Spring, a typical spring in the Haney Limestone on Flint Ridge. The larger Haney Springs have been capped and attached to a collector system to provide a water supply for the visitor center of Mammoth Cave National Park.

Table 2.1 Types of Porosity in Carbonate Aquifers

	Primary Porosity	*Fracture Porosity*	*Conduit Porosity*
Physical description	Intergranular pores and vugs	Concentrations of joints, fractures, and bedding-plane partings	Open pipes and channels of various sizes
	Isotropic	Might be isotropic in vertical plane or statistically isotropic over large volume of aquifer	Anisotropic
Regime	Laminar	Laminar	Turbulent
	Darcy flow	Might be deviations from Darcy flow	Non-Darcy
Water table	Well-defined surface	Irregular	Poorly defined with deep troughs and other irregularities
Response to short-term events	Slow	Moderate	Rapid

permeable pathways through the rock, the karsted carbonate aquifers contain a network of integrated solution cavities that act as pipes for the rapid transport of underground water. The usual model of groundwater flow in porous media has been extended to fracture aquifers by considering the rocks to be statistically isotropic if one examines a sufficiently large volume. The extension to conduit aquifers does not seem possible, partly because of the difficulty in averaging over small numbers of conduits in large volumes of rock and because the flow in the conduits is turbulent and of high velocity. Flow in the porous medium part of the aquifer could be meters per year while the flow velocity in the conduits could be meters per second. Because flow in conduits is so localized and because high flow velocities make possible rapid response to short-term events, there is an advantage in considering this water to be more like surface water than groundwater.

Stratigraphic and Structural Controls

The cave passages are a series of low-gradient branching conduits with steeper cutoffs and vertical shafts connecting them. They follow lines of least resistance through the rock along the hydraulic gradient, with various relationships to dip and strike, folds, and fractures, depending on the local conditions. The effects of fractures and joints on the locations of cave passages are less important than bedding-plane partings; less than 50% of the length of passages are joint-controlled (see Chap. 10).

Palmer (see Chap. 11) has concluded that the variations in stratigraphy and structure influence the trend and gradient of cave passages, but not their elevations. Most of the major passages are concordant to local structure because of the fact that bedding-plane partings represent the most efficient paths of flow at any given horizon within the limestone. The bedding planes appear to have greater influence on passage orientation than do variations in lithology.

Stratigraphy and structure have influenced the regional groundwater-flow pattern. The shales in the lower St. Louis and Upper Salem formations have acted as groundwater dams and are the primary reason that the southern drainage divide is close to Barren River. Flow is generally down structure to the north toward the Green River. On the western end of the Sinkhole Plain, the dip changes to the west. Here the shales are below the Barren River and groundwater flow is westward to Graham Springs.

Diffuse-Flow and Conduit-Flow

It has long proved valuable to describe karst aquifers as *conduit-flow* or *diffuse-flow* types. Diffuse-flow aquifers are those in which groundwater movement takes place only (or mainly) through the primary permeability and the fracture permeability (Shuster and White, 1971). Conduit-flow aquifers are those that have a well-developed conduit permeability. Further classification is possible depending on the type of recharge and the presence of capping or perching beds in the stratigraphic sequence (White, 1969, 1977). While this sort of classification is helpful in labeling aquifers and in discussing their properties, it is less useful in describing the actual groundwater-flow fields. All conduit aquifers, including the south-central Kentucky karst aquifer, have both diffuse-flow and conduit-flow components as indicated in Figure 2.5.

The type of recharge provided to a carbonate aquifer also has much to do with its hydrologic response and with the subsequent evolution of the conduit portion of the permeability (Palmer, 1984). Sinking streams and the drains from large closed depressions act as point sources that inject large quantities of water at very localized positions in the aquifer. Storm runoff mounds the groundwater near these injection points, developing large hydrostatic heads, which in turn are the driving force for conduit development. The combination of high input rates and high heads produces conduit systems with branchwork patterns. The sinking streams or closed depression drains are the infeeders.

Smart and Hobbs (1986) extended these concepts with a three-coordinate model for carbonate aquifers (Fig. 2.22). Flow type, recharge source, and storage are taken to be independent variables that range, for example, from pure diffuse flow (such as a chalk or fractured dolomite aquifer) to pure conduit flow (difficult to realize as a pure end member, but a conduit that

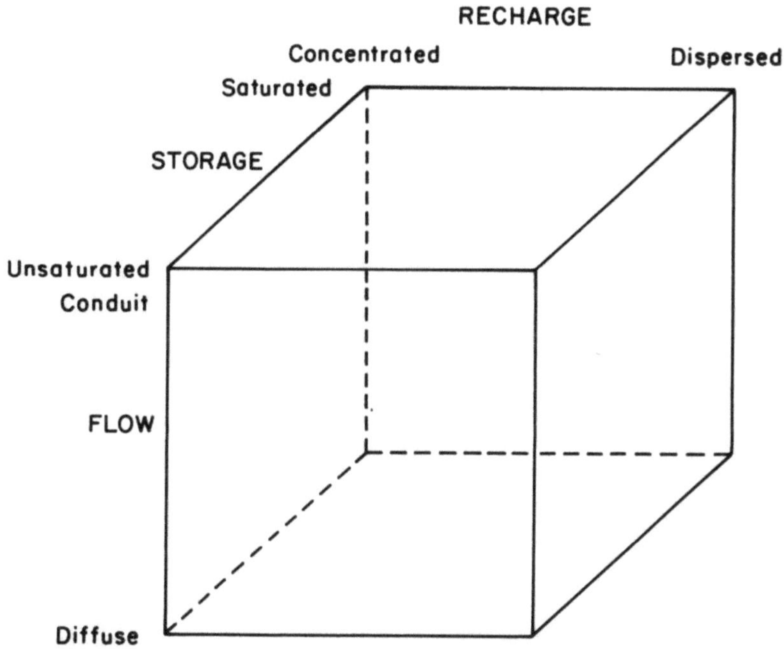

Figure 2.22 Classification of karst aquifers according to Smart and Hobbs (1986).

acts as a storm drain for some surface catchment is an example). Smart and Hobbs' model is more descriptive and more flexible than earlier attempts, but likewise does not take into account the variability within a given aquifer.

Certainly much of the water moving through the south-central Kentucky karst aquifer is localized in conduits. Because of infilling of the Green River channel with clastic sediment in the late Pleistocene, the lowest level conduits are perennially flooded. They are low-gradient pipes 5–20 m below base level. Conduits beneath the Sinkhole Plain, where the flow is often down dip along resistant chert beds, carry free-surface streams. In Parker Cave, both surface and subsurface tributaries of a branchwork drainage system have been identified (discussed in more detail in Chapter 3).

Cave Systems as Abandoned Conduits

Although karstic aquifers—that is to say carbonate aquifers with extensive conduit development—have given hydrologists great difficulties because of the breakdown of the usual mathematical methods for aquifer analysis, they compensate with one extraordinary advantage. Conduit systems evolve through time and, as base levels and water tables are lowered, the upper

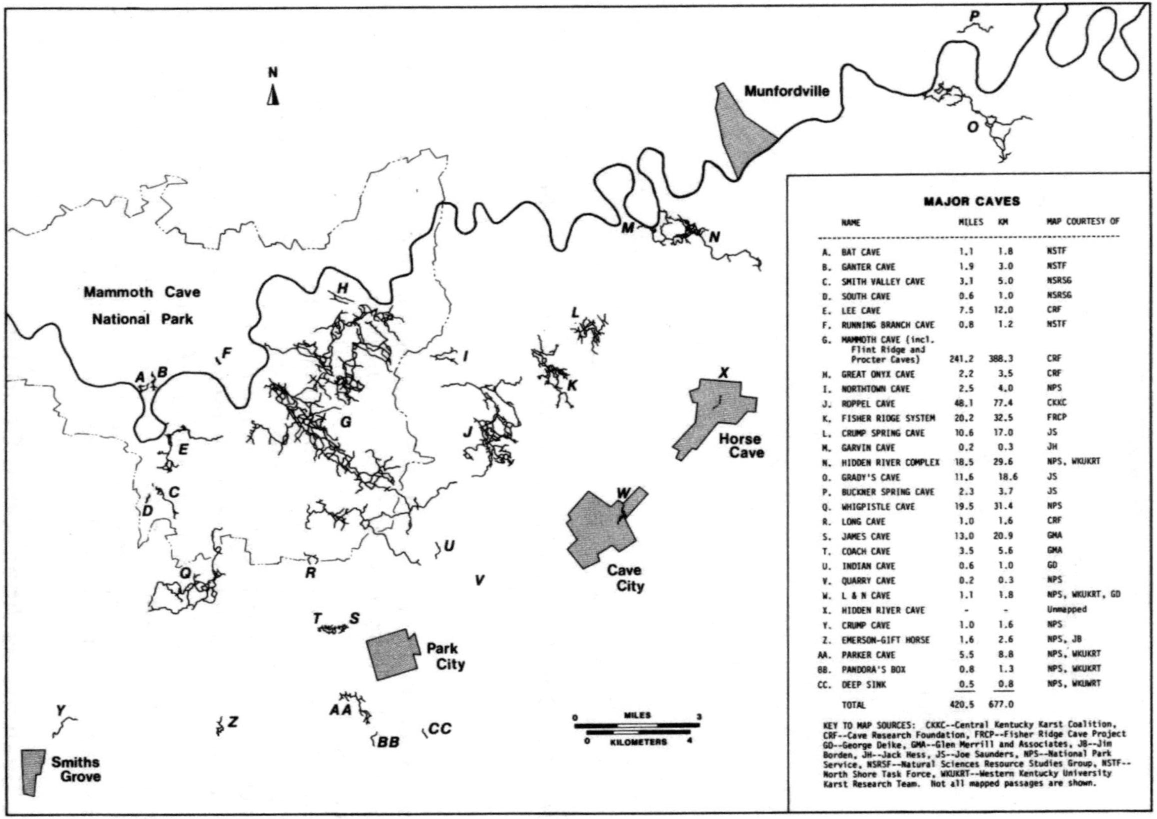

MAJOR CAVES

	NAME	MILES	KM	MAP COURTESY OF
A.	BAT CAVE	1.1	1.8	NSTF
B.	GANTER CAVE	1.9	3.0	NSTF
C.	SMITH VALLEY CAVE	3.1	5.0	NSRSG
D.	SOUTH CAVE	0.6	1.0	NSRSG
E.	LEE CAVE	7.5	12.0	CRF
F.	RUNNING BRANCH CAVE	0.8	1.2	NSTF
G.	MAMMOTH CAVE (incl. Flint Ridge and Procter Caves)	241.2	388.3	CRF
H.	GREAT ONYX CAVE	2.2	3.5	CRF
I.	NORTHTOWN CAVE	2.5	4.0	NPS
J.	ROPPEL CAVE	48.1	77.4	CKKC
K.	FISHER RIDGE SYSTEM	20.2	32.5	FRCP
L.	CRUMP SPRING CAVE	10.6	17.0	JS
M.	GARVIN CAVE	0.2	0.3	JH
N.	HIDDEN RIVER COMPLEX	18.5	29.6	NPS, WKUKRT
O.	GRADY'S CAVE	11.6	18.6	JS
P.	BUCKNER SPRING CAVE	2.3	3.7	JS
Q.	WHIGPISTLE CAVE	19.5	31.4	NPS
R.	LONG CAVE	1.0	1.6	CRF
S.	JAMES CAVE	13.0	20.9	GMA
T.	COACH CAVE	3.5	5.6	GMA
U.	INDIAN CAVE	0.6	1.0	GD
V.	QUARRY CAVE	0.2	0.3	NPS
W.	L & N CAVE	1.1	1.8	NPS, WKUKRT, GD
X.	HIDDEN RIVER CAVE	-	-	Unmapped
Y.	CRUMP CAVE	1.0	1.6	NPS
Z.	EMERSON-GIFT HORSE	1.6	2.6	NPS, JB
AA.	PARKER CAVE	5.5	8.8	NPS, WKUKRT
BB.	PANDORA'S BOX	0.8	1.3	NPS, WKUKRT
CC.	DEEP SINK	0.5	0.8	NPS, WKUWRT
	TOTAL	420.5	677.0	

KEY TO MAP SOURCES: CKKC--Central Kentucky Karst Coalition, CRF--Cave Research Foundation, FRCP--Fisher Ridge Cave Project GD--George Deike, GMA--Glen Merrill and Associates, JB--Jim Borden, JH--Jack Hess, JS--Joe Saunders, NPS--National Park Service, NSRSF--Natural Sciences Resource Studies Group, NSTF--North Shore Task Force, WKUKRT--Western Kentucky University Karst Research Team. Not all mapped passages are shown.

Figure 2.23 Caves in part of the Mammoth Cave region. *(From Quinlan and Ewers, 1981)* parts of the system are drained making abandoned segments of conduit accessible to human exploration. It is thus possible to map the conduit system in some detail and to get direct information on past and present pathways for the conduit portion of the groundwater flow. In spite of the wealth of information that has been gleaned by thousands of hours of tedious exploration and mapping, the known caves (Fig. 2.23) offer only fragments of the overall drainage system. For example, Parker Cave, which provides the best example of an upstream branchwork, is many kilometers away from other parts of the Turnhole Spring system to which it drains.

Beneath the Chester Cuesta are many long caves. Cave length is a human concept. Cave passages that are of human size can be explored and surveyed and are thus part of the "length," but there are many solutionally widened fractures and bedding-plane partings that act hydraulically as conduits, which are not of human dimensions. The caves of the Chester Cuesta are more highly integrated than cave systems found elsewhere. The collection of passages provides records of master trunk drains, valley drains, shaft drains, shafts, and various other passage types that were left behind as the Green River deepened its valley. The Mammoth Cave system is very old. Paleomagnetic investigations (Schmidt, 1982) indicate that the sediments in the highest levels are of early Pleistocene or late Pliocene age (see Chap. 12).

Caves on the Sinkhole Plain are smaller and more isolated but more of them are part of the active drainage system. Two types of caves were found in the Graham Springs basin: (1) active, large, segmented trunk passages graded to the Barren River and serving as master conduits for the aquifer (Fig. 2.24*A*); and (2) abandoned, dry, upper-level trunk passages, which once served as pre-existing flow paths for the Sinkhole Plain groundwater (Fig. 2.24*B*). The active caves are Mill Cave (Graham Springs), Wolf Sink Cave, and Grant-Palmore Cave, all developed in the St. Louis Limestone. Entrance to these caves is gained through large collapse sinkholes more than 30 m in depth. Mean passage width is 13 m, and the maximum passage width is 26 m for Mill Cave. The caves are oriented across strike and dip of the local bedrock. Wolf Sink and Graham Springs caves are serving as the active groundwater-flow path for the Graham Springs basin.

The abandoned cave, Smiths Grove Cave (Crump Cave), was first described by Fowke (1922). This cave is also developed in the St. Louis Limestone but differs from the active caves in that it is a dry upper-level trunk passage with no major active stream. It is perched approximately 20 m above the present level of the groundwater. The mean passage width is 10 m. Two types of passage morphology can be distinguished: (1) the pre-abandonment shape as a stream passage whose walls contain solutional features (scallops), which indicate a flow direction to Barren River; and (2) breakdown, or collapse, modified shape. The vertical shafts that modify the

Figure 2.24 *A,* Mill Cave (Graham Springs) in the Graham Springs basin serves as an active base-level drain for the groundwater system. *B,* Smiths Grove Cave represents a previously higher stand of the water table.

passage shape are more frequent on the east side of the trunk passages, or on the updip side.

A leveling survey was conducted in the southern section of Smiths Grove Cave where little collapse has occurred to determine the direction of passage slope. Leveling substantiated the evidence given by the solutional features and indicated a pre-existing southwestward flow of the groundwater when base level was 20 m higher.

A new method applied to karst terrain was devised to determine the geomorphic history of the karst basins in the Pennyroyal Plateau (Wells, 1974). Using the equation of best fit curves, longitudinal profiles of the sinking streams were projected from the adjacent uplands beyond the streams' terminal swallow holes. The general form of the equation is

$$Y = Ae^{-KX}$$

where Y is elevation, A is the elevation of the stream's head, K is a coefficient that varies with basin size, and X is the horizontal distance. The longitudinal profiles of Little Sinking Creek and Sinking Branch were fit with these curves (Table 2.2). The good regression coefficients suggest that the stream profiles can be extrapolated for some distance beyond the swallow holes to predict the elevation of the underground portion of the drainage trunk.

The longitudinal profiles of these two sinking streams were projected over 3 km across the groundwater divide between Graham Springs basin and Turnhole Bend basin. For that distance, the computed elevation was approximately 165 m (540 ft), which is the elevation of Smiths Grove Cave. Additionally, the approximate distance from the terminal portions of the sinking streams to Smiths Grove Cave is 3 km. This suggests that these two sinking streams drained into the ancestral Barren River drainage and were subsequently pirated to the base-level conduit that now drains to

Table 2.2 Fitting Coefficients for Sinking Stream Profiles

Karst Basin	*Sinking Creek*	*A* *(m)*	*K* *(km^{-1})*	*Regression Coefficient* *(r)*
Graham Springs	Sinking Creek	221	0.03	0.984
	Pondsville Creek	206.5	0.06	0.999
	Doty Creek	204.5	0.05	0.982
	Molser Creek	189	0.09	0.972
	Three Forks Creek	177	0.10	0.993
Turnhole Bend Spring	Sinking Branch	215	0.07	0.955
	Little Sinking Creek	222.5	0.03	0.985
	Gardner Creek	211	0.06	0.992

Green River. The surface profiles of these two sinking streams are out of equilibrium with the present hydrologic setting.

Although exponential fitting functions have proved useful in the interpretation of karst areas in the Appalachian Mountains (White and Hess, 1982; White and White, 1983) there is some skepticism about the extrapolation over such distances in the low relief Sinkhole Plain (see Chap. 3, Discussion, for a contrary point of view).

Additional evidence for the occurrence of subterranean piracy and northward diversion of Barren River drainage can be illustrated by the relationship between the location of Smith Grove Cave and the groundwater divide between Graham Springs basin and Turnhole Bend basin. Smiths Grove Cave terminates in breakdown approximately 1.5 km from the subsurface divide of the Barren and Green rivers drainages, and the cave represents a segment of groundwater-flow path that extended farther northeastward toward the Cuesta. The passage dimensions of Smiths Grove Cave show negligible reduction in size in the upstream direction toward its termination near the divide. The only terrane from which the abandoned groundwater-flow path could have been recharged is presently draining to Green River.

There has been capture of Barren River drainage and a decrease in its catchment area, whereas the catchment area of the Green River has increased at the expense of Graham Springs basin. Deike (1967) suggested that breaching of the sandstone caprock a few kilometers southwest of Brownsville on the Green River would provide another location for a subsurface drainage outlet. It can be speculated that eventually as the outlet develops at this point on the Green River, more Barren River drainage will be diverted northward into the Green River watershed, and the boundary between the two basins will shift to a still more westerly position.

GROUNDWATER DISCHARGE

The outlet points for the water collected in the karst aquifer are a series of springs along Green River and Barren River. Some springs are very large implying a large catchment; others are small suggesting discharge of water from local sources.

Springs on Green River

To obtain the location, description, and distribution of discharges for all springs on Green River, several careful inventories of springs between the Munfordville and Brownsville river gages have been conducted. The first method (Hess et al., 1974) was to float down the current along the bank of the river and to watch for evidence of springs while towing a thermistor and conductivity electrode beside the boat. The physical manifestation of springs ranged from small subtle notches under trees on the riverbank to resurgence streams up to 1000 m long. Some of the springs flowed from open cave mouths, while others had rise pools ranging from 1 m to 100 m in diameter. Morphological evidence was not adequate for many smaller springs. Small notches, for example, could be caused either by springs or by bank slumpage. In these cases thermal and electrical conductivity methods were of great value. The observations were made in the summer or early fall when the river water–spring water temperature contrast was at a maximum. The river temperature ranged between 18° C and 26° C and the springs ranged from 11° C to 20° C. The summer is also the low-water period for the river and the morphological evidence for the springs was most visible. Discharge was estimated, the temperature and specific conductance were recorded, and a brief description was written for each spring. Eighty-one springs were found in this fashion. The inventory has since been extended and refined. The most important springs and their drainage basins are shown on Plate 1, and many of them are discussed further in Chapter 3. The discharge distribution (Fig. 2.25) shows that most of the springs are quite small, some with no measurable discharge during the summer months.

The springs can be placed into a 2 x 2 classification. They are classified as alluviated or gravity springs based on their morphology at pool stage and into regional or local springs based on their discharge and specific conductance. A similar separation of alluviated (bluehole) and gravity springs has been proposed for other Kentucky springs (Van Couvering, 1962).

The alluviated springs are developed below river level and have rise pools that are fed at depth (Fig. 2.26). Pike Spring and Turnhole Spring, both alluviated springs, have been examined by scuba divers, and both are fed from large conduits approximately 10 m below pool stage of the river. Divers also entered Echo River Spring and were able to follow a submerged conduit more than 1000 m until they surfaced in Echo River in Mammoth Cave. Springs with unconfined outlets and free-flowing streams are the gravity-type springs. They have developed at or above the present pool stage of the river.

The regional springs have relatively high discharge and specific conductance. They receive their major recharge from the Sinkhole Plain, which is capable of providing the necessary catchment area to maintain the high discharge and provides the longer flow path and residence time necessary to account for the high conductance. The local springs have lower discharges and highly variable conductances. The discharge of these springs is smaller because of the limited catchment area, and the conductance is highly variable because of the variable flow path and residence time. Many local springs receive their main recharge from the Chester Cuesta and the valleys dissecting it. Their headwaters consist of vertical shaft complexes carrying the water from the caprock and valley floors down into the karst aquifer.

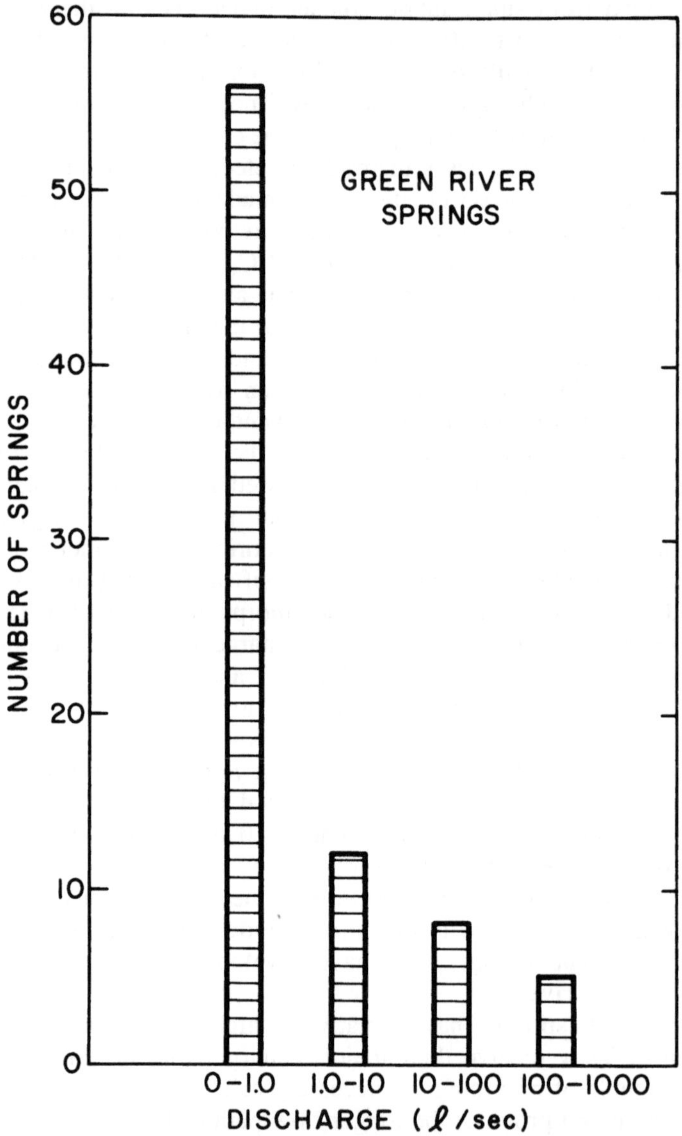

Figure 2.25 Distribution of spring discharges for 81 springs along Green River (north and south banks) between Munfordville and Brownsville river gages.

Each of the karst valleys dissecting the Chester Cuesta has its own, sometimes separate, drainage system. Surface water sinks near the clastic rock–limestone contact except in periods of very high runoff. This water then either becomes part of the main drainage or remains separate and discharges at its own spring along the river near the mouth of the valley. This accounts for many of the large number of small springs on either side of Green River.

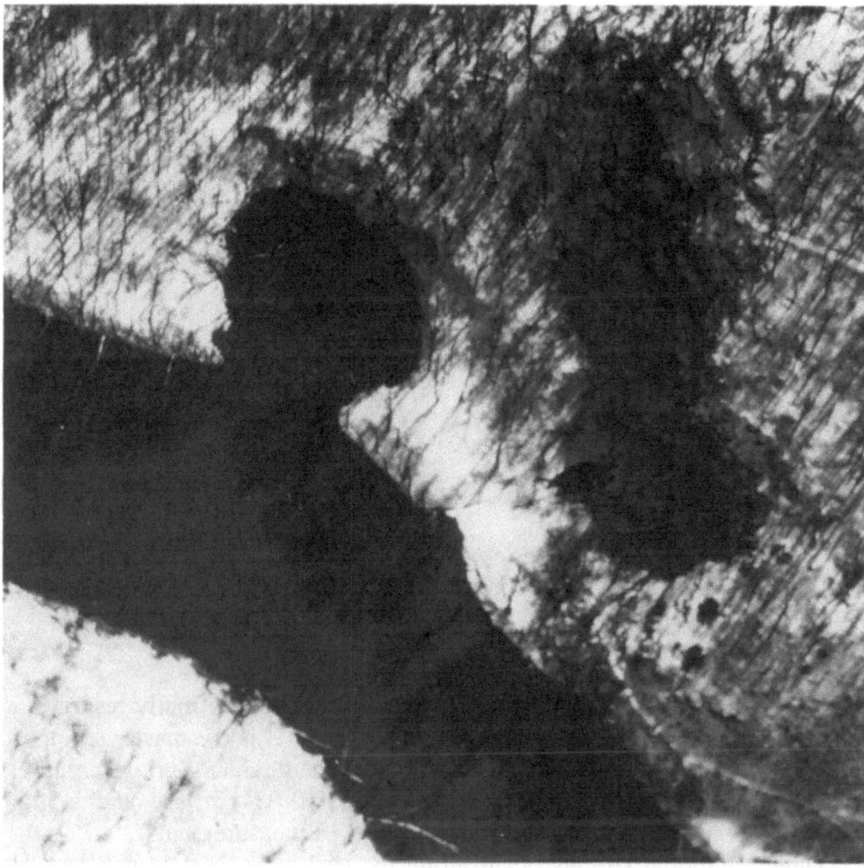

Figure 2.26 Infrared photograph in winter of the rise pool of Turnhole Spring, one of the largest alluviated springs in the south-central Kentucky karst. *(Photo by Gordon Smith)*

Springs on the Barren River

For that portion of the Pennyroyal Plateau east of Barren River, but not draining to Green River, 67 km^2 are drained by approximately seven springs with a total low-flow discharge of 0.12 m^3/sec. The remaining 360 km^2 of area drains to four alluviated outlets at one point known as the Graham Springs complex. The total low-flow discharge for this spring complex is approximately 0.55 m^3/sec. The Graham Springs complex discharges into a common valley forming a resurgence river approximately 360 m long.

The hydrology of these springs is complex. Various outlets function at certain discharges or river stages. For example, outlets 1 and 2 (Fig. 2.27) at the base of the bedrock bluff do not flow during low stages. However, with

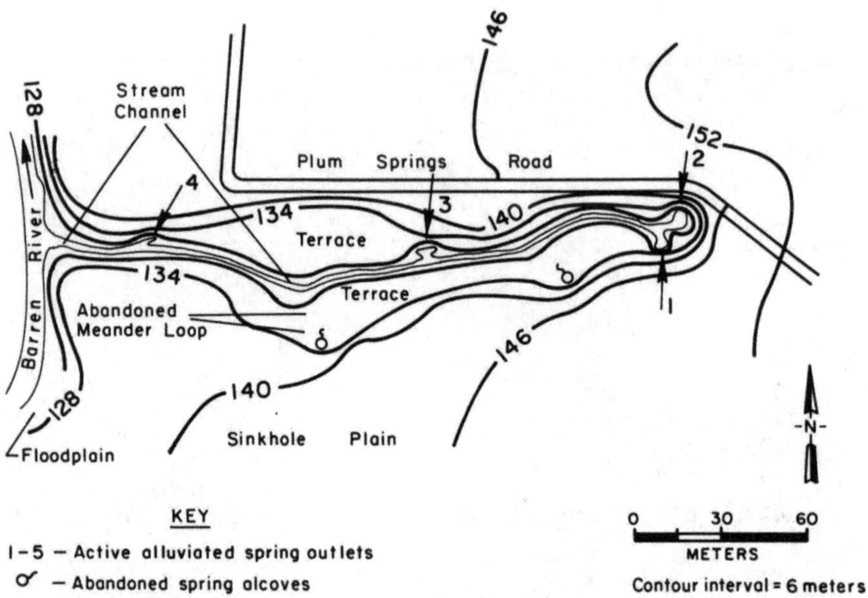

Figure 2.27 Features of the Graham Springs groundwater discharge zone on Barren River near Bowling Green, Kentucky.

increasing discharge, outlet 1 drains into outlet 2, which finally resurges at outlet 3 halfway down the resurgence river. Outlet 3 is the master spring; it and outlet 4 flow perennially. Discharge from the Graham Springs complex is variable, with a low-flow measurement of 0.034 m^3/sec and a high discharge measurement of 0.765 m^3/sec (USGS measurements).

Three additional springs occur along the east bank of Barren River, 50 m, 200 m, and 330 m downstream from where the Graham Springs stream joins the river. The second and third of these springs are part of the Graham Springs distributary, as described by Quinlan and Ewers (1981, p. 488). The first of these three springs, at 50 m, is not. It is Waterworks Spring, the only perennial diffuse-flow spring known in the entire Mammoth Cave area (Quinlan and Ewers, 1981, pp. 488, 490).

Dishman Mill Rise is located on Jennings Creek, a tributary to Barren River on its west bank. Dishman Mill Rise is the major outlet for the Lost River basin (George, 1973). The main outlet of the Dishman Mill Rise is approximately 19.8 m wide with a small 2.4-m high bedrock bluff. George divided the spring into two segments: a main resurgence and a ponded flood route. The main outlet during low flow is open and it is questionable as to whether this outlet would be classified as an alluviated spring. The Lost River drainage has been the object of intensive investigations by N. C. Crawford and his students at the Center for Cave and Karst Studies, Western Kentucky University.

REFERENCES

Ashton, K., 1966, The analysis of flow data from karst drainage systems, *Cave Research Group Great Britain Trans.* **7**:161–208.

Barton, A. J., 1981, *Soil Survey of Warren County, Kentucky*, Washington, D.C.: U.S. Department of Agriculture, Soil Conservation Service, 114p.

Brown, R. F., 1966, Hydrology of the cavernous limestones of the Mammoth Cave area, Kentucky, *U.S. Geol. Survey Water-Supply Paper 1837*, 64p.

Brown, R. F., and T. W. Lambert, 1963, Reconnaissance of ground water resources in the Mississippian Plateau region, Kentucky, *U.S. Geol. Survey Water-Supply Paper 1603*, 58p.

Brucker, R. W., J. W. Hess, and W. B. White, 1972, Role of vertical shafts in the movement of groundwater in carbonate aquifers, *Ground Water* **10**(6):5–13.

Crawford, N. C., 1984, Sinkhole flooding associated with urban development upon karst terrain: Bowling Green, Kentucky, in *Sinkholes: Their Geology, Engineering, and Environmental Impact*, B. F. Beck, ed., Rotterdam: A. A. Balkema, pp. 283–292.

Cushman, R. V., 1968, Recent developments in hydrogeologic investigations in the karst area of central Kentucky, *Internat. Assoc. Hydrogeol. Mem.* **8**:236–248.

Cushman, R. V., R. A. Krieger, and J. A. McCabe, 1965, Present and future water supply for Mammoth Cave National Park, Kentucky, *U.S. Geol. Survey Water-Supply Paper 1475–Q*, pp. 601–647.

Deike, G. H., III, 1967, The development of caverns of the Mammoth Cave region, The Pennsylvania State University, Ph.D. dissertation, 235p.

Fowke, G., 1922, Archeological investigations, *Bur. Am. Ethnology Bull. 76*, 204p.

George, A. I., 1973, *Guidebook to the Kentucky Speleo-fest*, Louisville, Ky.: Speleopress, 70p.

Hendrickson, G. E., 1961, Sources of water in Styx and Echo Rivers, Mammoth Cave, Kentucky, *U.S. Geol. Survey Prof. Paper 424–D*, pp. 41–44.

Hess, J. W., S. G. Wells, and T. A. Brucker, 1974, A survey of springs along the Green and Barren Rivers, central Kentucky karst, *Natl. Speleol. Soc. Bull.* **36**(3):1–7.

Howard, A. D., 1968, Stratigraphic and structural controls on landform development in the central Kentucky karst, *Natl. Speleol. Soc. Bull.* **30**:95–114.

Krieger, R. A., and G. E. Hendrickson, 1960, Effects of Greensburg Oilfield brines on the streams, wells, and springs of the upper Green River Basin, Kentucky, *Kentucky Geol. Survey Rept. Inv. No. 2*, 36p.

Lambert, T. W., 1976, Water in a limestone terrane in the Bowling Green area, Warren County, Kentucky, *Kentucky Geol. Survey Rept. Inv. No. 17*, 43p.

Latham, E. E., 1969, *Soil Survey of Barren County, Kentucky*, Washington, D.C.: U.S. Dept. Agriculture, Soil Conservation Service, 85p.

Palmer, A. N., 1984, Geomorphic interpretation of karst features, in *Groundwater as a Geomorphic Agent*, R. G. LaFleur, ed., Boston: Allen and Unwin, pp. 173–209.

Palmer, A. N., M. V. Palmer, and W. B. White, 1981, *A Guide to the Historic Section of Mammoth Cave*, Bowling Green, Ky.: 8th International Congress of Speleology, 59p.

Pohl, E. R., 1936, Geologic investigations at Mammoth Cave, Kentucky, *Am. Geophys. Union Trans.* **17**:332–334.
Pohl, E. R., 1955, Vertical shafts in limestone caves, *Natl. Speleol. Soc. Occasional Papers 2*, 24p.
Pohl, E. R., 1970, Upper Mississippian deposits of south-central Kentucky, *Kentucky Acad. Sci. Trans.* **31**:1–15.
Quinlan, J. F., 1970, Central Kentucky karst, *Études et Travaux de Méditerranée* **7**:235–253.
Quinlan, J. F., and R. O. Ewers, 1981, Hydrogeology of the Mammoth Cave region, Kentucky, in *GSA Cincinnati '81 Field Trip Guidebooks*, vol. 3, T. G. Roberts, ed., Falls Church, Va.: American Geological Institute, pp. 457–506.
Quinlan, J. F., and E. R. Pohl, 1967, Vertical shafts actively promote slope retreat and dissection of the solution escarpment and the Chester Cuesta in the central Kentucky karst, *Natl. Speleol. Soc. Bull.* **29**:109–110 (abstract).
Ruhe, R. V., 1975, Geohydrology of karst terrain, Lost River Watershed, southern Indiana, *Indiana Univ. Water Resources Research Center, Rept. Inv. 7*, 91p.
Schmidt, V. A., 1982, Magnetostratigraphy of sediments in Mammoth Cave, Kentucky, *Science* **217**:827–829.
Shuster, E. T., and W. B. White, 1971, Seasonal fluctuations in the chemistry of limestone springs: A possible means for characterizing carbonate aquifers, *Jour. Hydrology* **14**:93–128.
Smart, P. L., and S. L. Hobbs, 1986, Characterization of carbonate aquifers: a conceptual base, in *Proceedings of the Environmental Problems in Karst Terranes and Their Solutions Conference*, Dublin, Ohio: National Water Well Association, pp. 1–14.
Troester, J. W., E. L. White, and W. B. White, 1984, A comparison of sinkhole depth frequency distributions in temperate and tropic karst regions, in *Sinkholes: Their Geology, Engineering, and Environmental Impact*, B. F. Beck, ed., Rotterdam: A. A. Balkema, pp. 65–73.
Van Couvering, J. A., 1962, Characteristics of large springs in Kentucky, *Kentucky Geol. Survey, Information Circ. 8*, 37p.
Watson, R. A., 1966, Central Kentucky karst hydrology, *Natl. Speleol. Soc. Bull.* **28**:159–166.
Wells, S. G., 1973, Geomorphology of the Sinkhole Plain in the Pennyroyal Plateau of the central Kentucky karst, University of Cincinnati, M.S. thesis, 122p.
Wells, S. G., 1974, Drainage basin morphology in the Sinkhole Plain of the central Kentucky karst, in *Proceedings of the 4th Conference on Karst Geology and Hydrology*, H. W. Rauch and E. Werner, eds., Morgantown, W. Va.: West Virginia Geological Survey, p. 91 (abstract).
White, E. L., and B. M. Reich, 1970, Behavior of annual floods in limestone basins in Pennsylvania, *Jour. Hydrology* **10**:193–198.
White, E. L., and W. B. White, 1983, Karst landforms and drainage basin evolution in the Obey River Basin, north-central Tennessee, *Jour. Hydrology* **61**:69–82.
White, W. B., 1969, Conceptual models for limestone aquifers, *Ground Water* **7**(3):15–21.
White, W. B., 1977, Conceptual models for carbonate aquifers: revisited, in

Hydrologic Problems in Karst Terrains, R. R. Dilamarter and S. C. Csallany, eds., Bowling Green, Ky.: Western Kentucky University, pp. 176–187.

White, W. B., and J. W. Hess, 1982, Geomorphology of Burnsville Cove and the geology of the Butler Cave–Sinking Creek System, *Natl. Speleol. Soc. Bull.* **44:**67–77.

Williams, P. W., 1983, The role of the subcutaneous zone in karst hydrology, *Jour. Hydrology* **61:**45–67.

Williams, P. W., 1985, Subcutaneous hydrology and the development of doline and cockpit karst, *Zeitschr. Geomorphologie* **29:**463–482.

Wright, H. E., Jr., B. Spross, and R. A. Watson, 1966, Pollen analyses of the sediment from sinkhole ponds in the central Kentucky karst, *Natl. Speleol. Soc. Bull.* **28:**185–188.

3

SUBSURFACE DRAINAGE IN THE MAMMOTH CAVE AREA

James F. Quinlan and Ralph O. Ewers

The research described in this chapter was begun in 1973—after most of the hydrologic and geologic studies described in many of the other chapters were completed. No other karst area in the United States has been as intensively studied by dye-tracing, cave mapping, potentiometric surface mapping, and continuous monitoring of water quality and stage as the Mammoth Cave area (described in this chapter) and the adjacent Lost River groundwater basin (Crawford, 1985, 1986). Indeed, in no other karst where these techniques have been applied have so much relevant data been available or able to be generated. More than 500 traces were run in the Mammoth Cave area from 1975 through 1987; most are summarized on Plate 1. As many as nine dye-traces have been under way in various basins at the same time.

The general strategy of investigation—as dictated by a balancing of research needs, topics of interest and ignorance, funding limitations, and work already done—has been to do dye-tracing, drafting, and data reduction during the rainy season, and cave mapping and potentiometric surface mapping during the relatively dry season. The fate of the heavy-metal-rich sewage effluent injected into the ground at the town of Horse Cave was recognized at once as being a problem, and a long-term study of heavy metals in springs, wells, cave streams, and the Green River was begun (Quinlan and Rowe, 1977). Indeed, interpretation of the chemistry of these waters led to a prediction of the existence of the cave system behind Hicks Spring and its distributary nature (Quinlan et al., 1977). We have gone on from this initial work to continuous monitoring for stage and water quality (Meiman et al., 1988).

METHODS OF STUDY

Techniques of water tracing with dyes have been reviewed by Quinlan (1981, 1985) and Aley et al. (1989). The techniques used for cave mapping and potentiometric surface mapping in the Mammoth Cave area were described by Quinlan (1981). Groundwater divides located by mapping of the potentiometric surface were tested and verified by dye-tracing. This procedure maximized the efficiency of the tracing program and served as a check on the reliability of interpretation of the potentiometric data.

The dyes used by the senior author for groundwater tracing were: Fluorescein (colour index [CI] Acid Yellow 73), Rhodamine WT (CI Acid Red 388), CI Direct Yellow 96, and Optical Brighteners (CI Fluorescent Whitening Agents 22, 28, and 351). Techniques for continuous monitoring of stage and water quality have been developed by R. O. Ewers and are summarized by Ewers et al. (1986) and Meiman et al. (1988).

Magnetic induction equipment, a so-called "cave radio," which transmits a low-frequency (3500 Hz) signal to the surface from a water-filled or air-filled cave passage (Reid, 1984), has been used to locate numerous sites for observation wells and for the construction of cave entrances.

Although extensive dye-tracing and detailed mapping of the potentiometric surface has been done both north and south of the Green River by Quinlan and National Park Service staff, only the hydrology of the area south of the river is discussed in this chapter; the dye-tracing north of the river is only about 70% complete.

GROUNDWATER BASINS

Although all surface water flows in a discrete surface-water basin, not all groundwater flows in a discrete groundwater basin (Thrailkill, 1985). Nevertheless, most groundwater flows in a groundwater basin in which the divides between adjacent basins do not coincide with surface-water divides and in which each groundwater basin is bounded by others.

In the Mammoth Cave area the groundwater basins discharge primarily at a spring or group of springs at or near a base-level stream. The springs are fed by a system of dendritic or trellised conduits that increase in size and order as they decrease in number in the downstream direction. There might be, and probably is, some seepage discharge at rivers, but it has not been detected or studied (see Chap. 4). Undoubtedly seepage discharge and recharge occur adjacent to the surface streams in the Glasgow Upland and Sinkhole Plain.

The general hydrogeology of the Mammoth Cave area is discussed in

Chapter 2, and the hydrogeology of the Bowling Green area has been described by Crawford (1986). Accordingly, the following subsections will discuss only selected aspects of the three larger basins: Turnhole Spring, Bear Wallow, and Graham Springs (217, 486, and 307 km^2 or 85, 190, and 120 mi^2, respectively). But first it is necessary to discuss the map showing all the groundwater basins that have been delineated. Plate 1 (Quinlan and Ray, 1981) depicts the groundwater basins, flow routes, potentiometric surface, wells, caves, and surface drainage of the Mammoth Cave area.

Explanation of Groundwater Basin Map (Plate 1)

Plate 1 shows the flow directions of groundwater in nearly all of the major karst area south of Green River between the Barren and Little Barren rivers and north of Beaver Creek. Flow direction is indicated by the curved red arrows; the wider the red lines, the greater the amount of water that is flowing to springs (blue circles) via cave streams (wiggly black lines). The red flow lines are drawn perpendicularly to the potentiometric surface, which is shown by green contours. The dashed red lines indicate flow routes above local base level that function only during flooding after rains. Each red line represents one or more dye-traces.

As shown in the box in the upper left corner of the map, the boundaries of 28 different groundwater basins and 7 sub-basins have been determined. The basin boundaries are indicated by dashed black lines; sub-basin boundaries are indicated by dotted black lines. Generally speaking, all surface and subsurface water within the confines of the boundary of a groundwater basin flows to the same spring or group of springs. Accordingly, the map can be used to identify the areas that might adversely affect the water quality of all of the springs and cave streams shown. It can also be used to predict the dispersal route of any hazardous material that might be discharged into the ground or spilled accidentally, and to predict what water supplies might be adversely affected by discharges or spills. For example, the map shows that Mammoth Cave and the connected Flint Ridge Cave system can be affected by anything that pollutes groundwater in basins 3 (including 3A, 3B, 3B1, and 3B2), 6, and 7.

Twenty-two of the 26 U.S. Geological Survey 7.5-min topographic quadrangle maps used to construct this map are identified by number and name in the northwestern corner of the box in the upper left corner of the map. The corners of these quadrangles are shown by black crosses, about 0.6 cm wide, on the big map of Plate 1.

The green lines are water-level contours. Their elevations are expressed as feet above sea level, and the contour interval is 20 ft. These lines indicate the elevation to which water would rise (and stand) in a well drilled at any locality. The flow of groundwater is downslope and generally in a direction that is perpendicular to the trend of these green lines. The depth from the ground surface to water in a well ranges from about 1 m to 120 m (from 3 ft to 400 ft).

The green circles, squares, and triangles indicate more than 1500 of the 4000+ water wells that are used for domestic water supplies in the map area. These wells are shown so that one can see the amount and location of data used to construct the potentiometric surface. The fine black squiggly lines are schematic depictions of mapped cave passages.

The gray tone in the southern and eastern areas of the map represents limestone with interbedded shale of the Salem–Warsaw Formation. These rocks prevent vertical circulation of groundwater; the shales and related silty rocks form an impermeable zone some 30 m thick that separates the overlying St. Louis, Ste. Genevieve, and Girkin formations from limestone that is below it.

This map is based on long-term field studies that include more than 500 dye-traces (approximately 100 of which were done north of Green River since 1981), 1800 water-level measurements, and mapping of more than 725 km (450 mi) of cave passage (480+ km by Cave Research Foundation teams; 80+ km by National Park Service teams; and 160+ km primarily by six other organizations; much of this exploration is shown schematically in Fig. 2.23). One of the more significant results of the dye-tracing studies was the discovery that most springs—indeed, those in all but 1 of the 21 larger groundwater basins—are part of a distributary system within these basins, one in which water (or pollutants) from a point source can be dispersed to as many as 53 springs over a 19-km (12-mi) reach of Green River, as happens in the Bear Wallow basin (Plate 1, basin 14). All other distributaries in the map area, however, are smaller.

Interpretation of the Potentiometric Surface

Study of the potentiometric surface shown on Plate 1, and comparison of it with Figures 2.1 and 2.4, lead to the conclusion that the potentiometric highs and most of the topographically highest recharge areas coincide with structural highs. Also, the steepness of the potentiometric slope south of Park City, but not more than about 3 km west of it, can be explained in terms of structure and stratigraphy. Groundwater south of Park City is perched on silty and shaly beds in the lower third of the St. Louis Formation. Assuming this situation, imagine that water is "cascading," so to speak, down the "steep" dip slope of 0.028–0.038 (150–200 ft/mi). Consider also that the filling of the valley of Green River by 15–18 m of Pleistocene alluvium has impounded the conduit systems that were graded (adjusted?) to a lower, now buried, level of the valley. The "groundwater lakes," thus created, would drain to one or more springs now discharging at a level higher than the floor of the buried valley. The upper surface of this discontinuous groundwater body is the potentiometric surface. Where the potentiometric surface of the "cascading" water intercepts the potentiometric surface of the "lake," there is an abrupt change in gradient. This abrupt change dies out about 3 km west of Park City because, as the monocline plunges westward into the subsurface, the relatively impermeable beds at the base of the aquifer are increasingly farther below the potentiometric surface that is here graded to the major discharge point on the Barren River (Graham Springs in basin 21).

A significant trough on the surface of a potentiometric map is most readily interpreted to be a locus of maximum flow of groundwater. Four of the larger troughs on Plate 1 are coincident with big caves that include major underground rivers (between Smiths Grove and Barren River, through Park City, 4.8 km west of Park City, and northeast of Cave City and Horse Cave). Conversely, the location of no known major cave river is inconsistent with the potentiometric data.

The potentiometric contours on Plate 1 for the Poorhouse Spring area (basin 19) are radically different from those shown by Lambert (1976, plate 3). His flow directions there are opposite to those on Plate 1. This is because Lambert misinterpreted water-level data from wells finished in perched water zones within the Ste. Genevieve Limestone. Such data from elevations above the potentiometric surface graded to the Barren River, shown as open squares on Plate 1, were judiciously ignored by Quinlan and Ray (1981). Their conclusions were verified by dye-tracing to and from the mapped caves shown and to Poorhouse Spring.

The potentiometric surface for the entire Mississippian aquifer, composed of the Girkin, Ste. Genevieve, and St. Louis limestones between the Ohio River and the Cumberland River, was mapped by Plebuch et al. (1985). Unfortunately, because most of their mapping was based on an insufficient number of water-level measurements per quadrangle, which were contoured at a 50-ft interval rather than a 20-ft interval, the utility and reliability of their map for spill response is highly questionable. A much more authoritative potentiometric map, done with more wells per km^2, with a 20-ft contour interval, and verified by dye-tracing, was compiled by Crawford (1985) for the adjacent Lost River area (which includes Bowling Green) immediately west of the Barren River.

Turnhole Spring Groundwater Basin

The Turnhole Spring groundwater basin (Figs. 3.1 and 3.2) and the adjacent Echo River, Pike Spring, and Sand Cave groundwater basins into which it discharges during moderate flow and flood-flow conditions (as shown by dashed lines) have been more thoroughly studied than any other in the Park

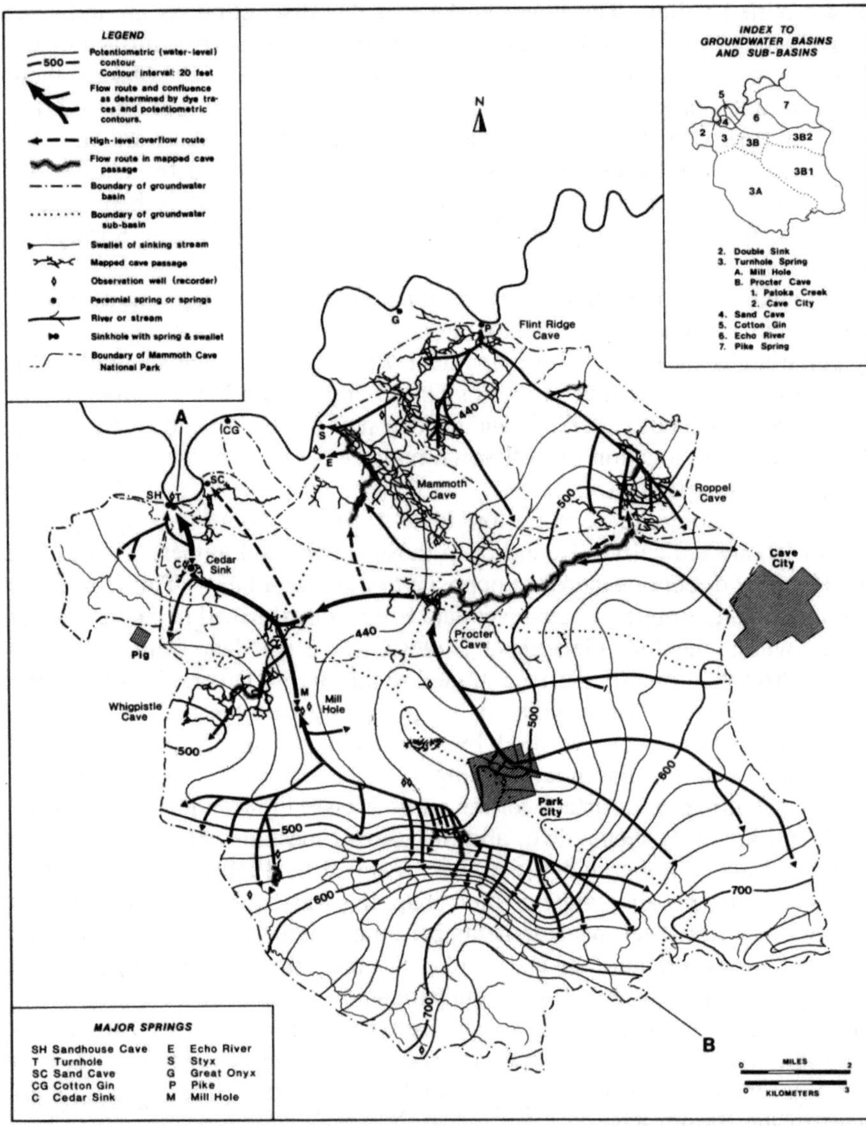

Figure 3.1 Map of the Turnhole Spring groundwater basin (1986 revision of a portion of Plate 1). This map shows the position of newly discovered west-flowing stream segment in Whigpistle Cave almost 3 km north of Mill Hole, changes in the discharge hydrology of this basin (see under Applications of Recent Hydrologic Research), streams in mapped cave passages, additional passage in Whigpistle Cave and in Roppel Cave, and the town of Pig. *(From Quinlan and Ewers, 1986b)*

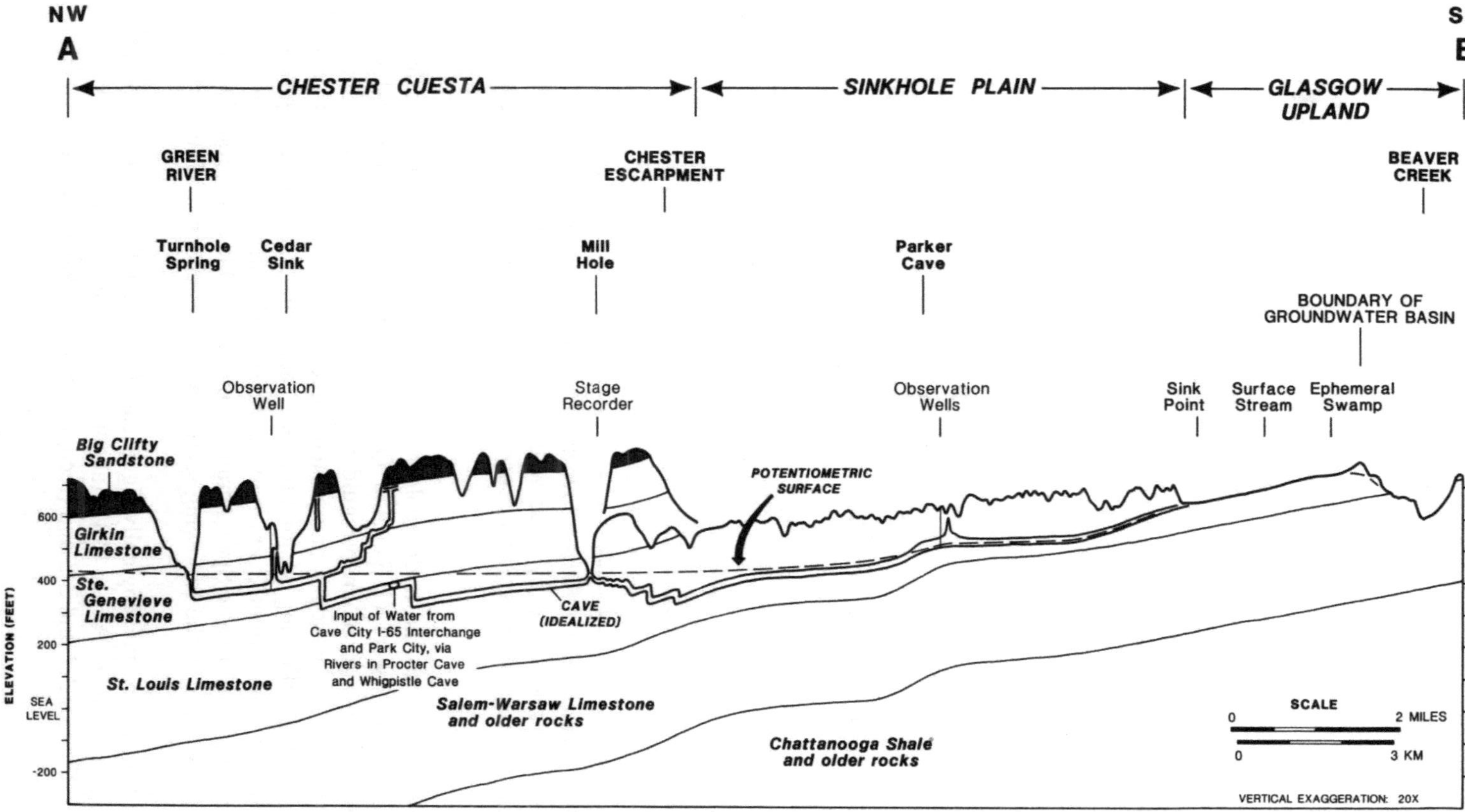

Figure 3.2 Cross section along the axis of major drainage in the Turnhole Spring groundwater basin depicting flow of water to the northwest as shown in Figure 3.1. The line of the section is oblique to the east-west strike of the beds. Also shown are stratigraphy, structure, caves, potentiometric surface, and physiography. (*From Quinlan et al., 1983*)

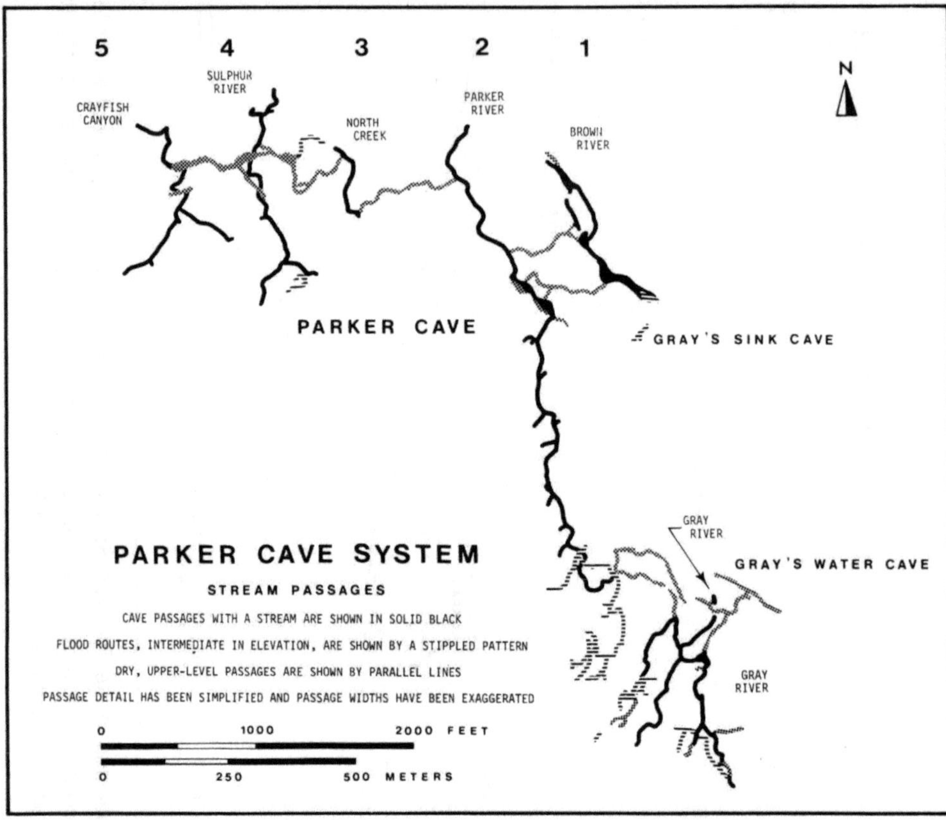

Figure 3.3 Sketch map of Parker Cave system, showing streams, intermediate-level overflow routes used by flood waters, and permanently dry upper-level passages. Approximately 9 km of passage are shown. *(From Quinlan and Ewers, 1981)*

area. Guidebooks to the hydrogeology of this basin have been published by Quinlan and Ewers (1981) and Quinlan et al. (1983, 1986), which complement the comprehensive guidebook to Mammoth Cave by Palmer (1981). Figure 3.1, which is an updated, annotated version of part of Plate 1, includes the following major revisions: addition of a stream segment in Whigpistle Cave about 3 km north-northeast of Mill Hole, and storm-related capture of Turnhole Spring drainage by the Double Sink groundwater basin (as discussed in the following section on recent changes in subsurface drainage).

One of the more hydrologically significant features of the Turnhole Spring groundwater basin is Parker Cave, about 1 mi southwest of Park City, as shown in Figures 3.2 and 3.3 and (recognizable on Plate 1 and Fig. 3.1) as discussed by Quinlan and Ewers (1981, pp. 472–474), Quinlan et al. (1983, pp. 16–21; 1986, pp. 16–21). Flow velocities between Parker Cave and Mill Hole, 6 km to the northwest, range from about 0.25 cm/sec (30 ft/hr) during base flow to more than 12.7 cm/sec (1500 ft/hr) during flood flow. This cave includes five subparallel, trellised streams, which are about 150 m apart. Interpretation of leveling data in it has demonstrated how intermediate-level overflow routes (streamless passages at a higher level that convey water only during and after rainstorms) can function to route water from one groundwater basin to another. The local complexity of groundwater flow in this basin is also suggested by examination of Figure 3.4 (and Fig. 3.14).

Transport of clastic sediment in Parker Cave and the development of surface and subsurface drainage in this basin have been studied in detail by Pickle (1985). His work complements the significant pioneering sedimentological studies by Davies and Chao (1959).

Bear Wallow Groundwater Basin

This is the largest groundwater basin in the Mammoth Cave area, about 500 km^2 (190 mi^2), and it is shown on Plate 1. Its most significant features are

- Hidden River groundwater sub-basin and associated cave system, near Horse Cave (Figs. 3.5, 3.6, and 3.7; Plate 1)
- Distributaries (Figs. 3.5, 3.6, and 3.7; Plate 1; discussed below)
- Shared headwaters (described briefly in the section on Research in Progress)

Heavy-metal-laden effluent from the Horse Cave Sewage Treatment Plant has been discharged into the ground for almost 20 years. It still is, but concentrations are now at permissible levels. Some of this effluent has included more than 10 mg/l chromium. (The maximum allowable chromium in drinking water is 0.05 mg/l.) Study of the dispersal of this effluent showed that it travels 1.6 km northeast to Hidden River Cave and then 6–8 km north to where it is discharged at as many as forty-six springs at 16 locations along an 8-km reach of Green River (Quinlan and Rowe, 1977). Interpretation of the chemistry of Hicks Spring inspired excavation at one of its high-level orifices and consequent discovery of more than 32 km of cave passage—the Hidden River Complex (Fig. 3.7).

Explanation of distributary flow is best done by reviewing some fundamentals of water movement in karst. A typical groundwater basin in maturely karsted rocks has three important elements:

1. Inputs from sinking streams and from water stored in the soil and in the epikarstic zone (Williams, 1983)

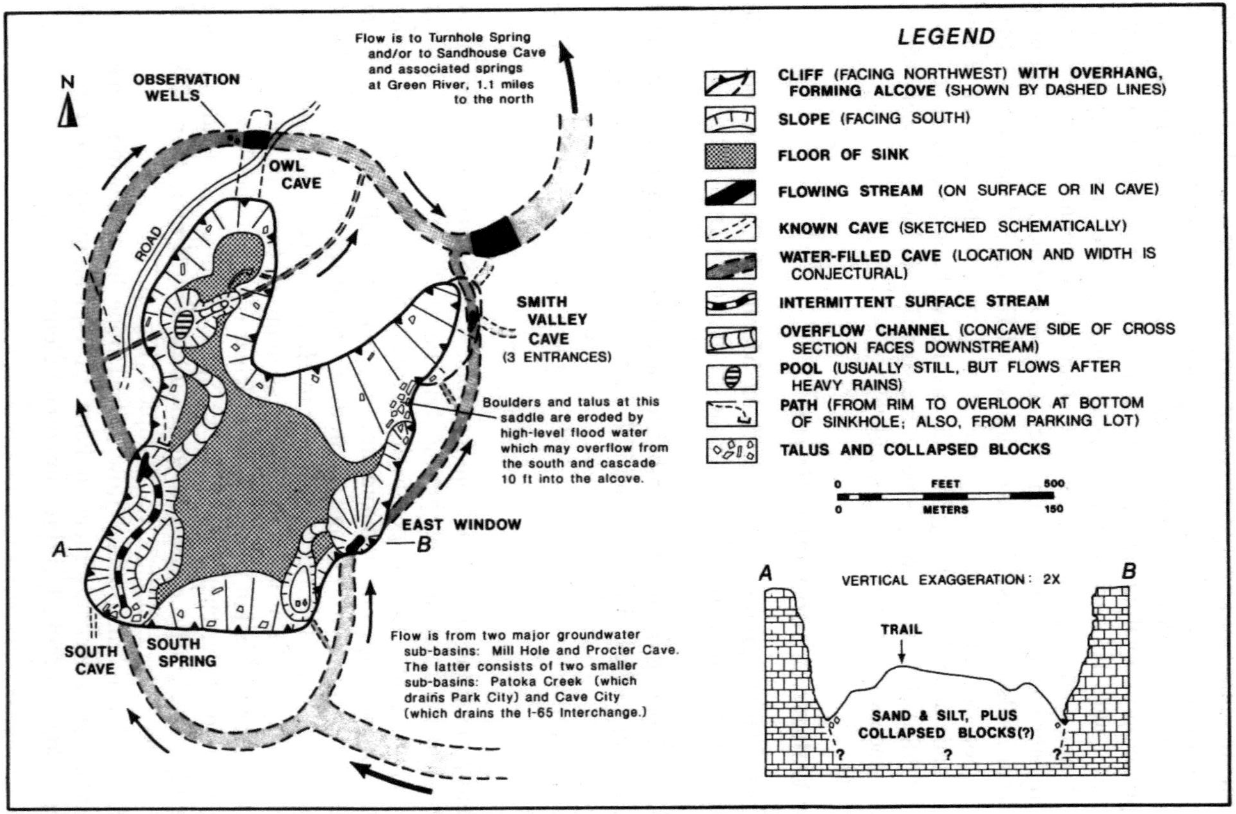

Figure 3.4 Sketch map showing Cedar Sink hydrology and geomorphology. *(From Quinlan and Ewers, 1986b)*

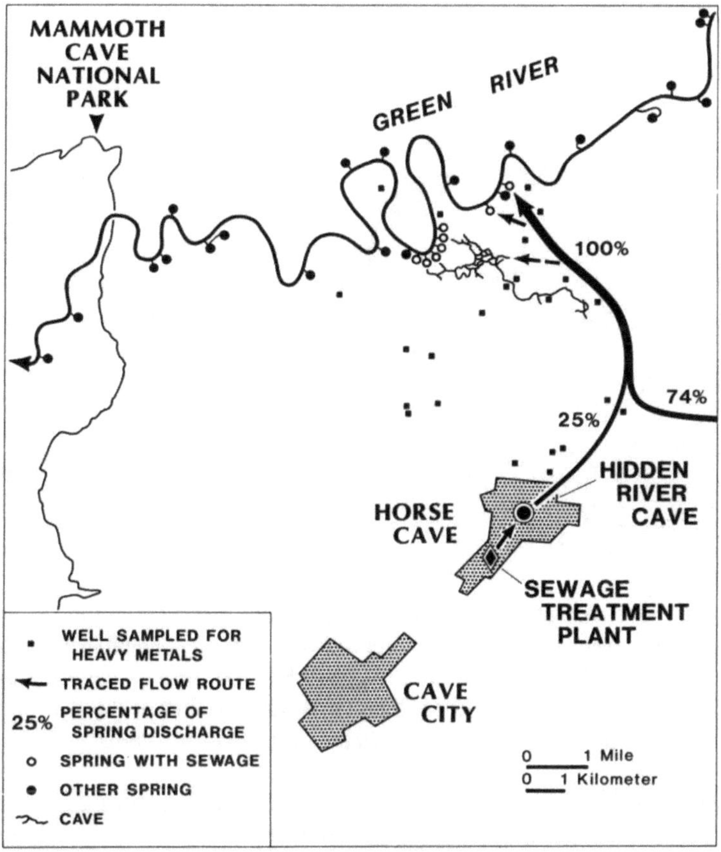

Figure 3.5 Dispersal of heavy-metal-laden effluent from a sewage treatment plant to Hidden River Cave and then to a distributary system ending at as many as 46 springs at 16 locations along a 8-km (5-mi) reach of Green River. Flow of effluent from the Cave City Sewage Treatment plant and of groundwater from the area south and east of the town of Horse Cave to Hidden River Cave has been omitted for clarity. Flow at Hidden River Cave is only 25% of the base-flow discharge of the springs. The two dashed arrows indicate distributary flow to springs that ceases discharging during lowest base-flow conditions; compare with Figures 3.6 and 3.7. *(From Quinlan and Ewers, 1985)*

2. Tributary conduits
3. A discharge point consisting of a spring or group of springs

Where carbonate rock is above the water table, infiltration water percolates vertically until it intercepts a cave stream in which most flow is nearly horizontal, in conduits, convergent, in the uppermost fringe of the phreatic

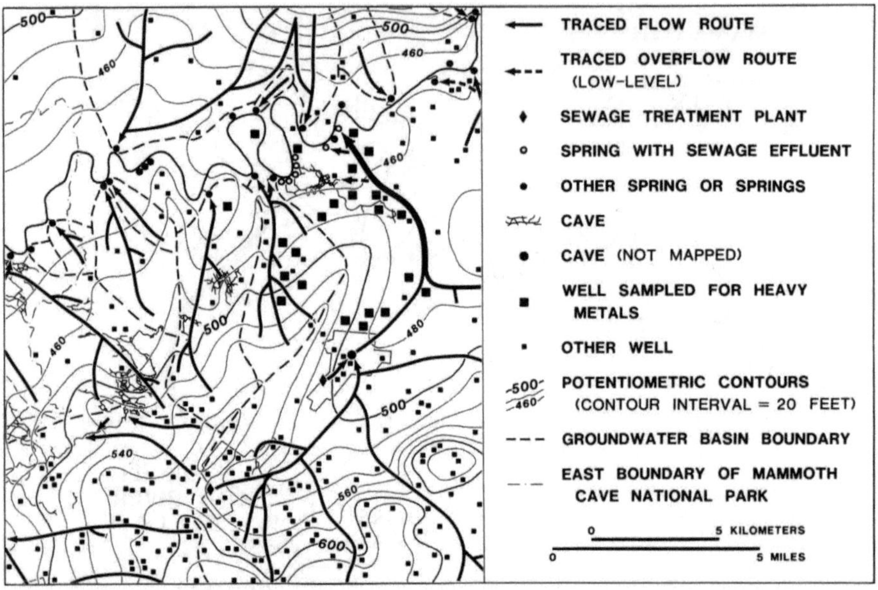

Figure 3.6 Groundwater-flow routes and pollutant dispersal in the Horse Cave and Cave City area; compare with Figures 3.5 and 3.7. Reliable monitoring of spills or pollutants here and in most maturely karsted terrains can only be done at springs and in wells drilled to cave streams that are known (by dye-tracing) to drain the site to be monitored. *(From Quinlan and Ewers, 1986b)*

zone, and to a spring (or group of springs) that drains an entire groundwater basin. Flow in a conduit system is described by pipe and channel-flow equations (Gale, 1984), but is in many ways analogous to flow in a surface river. Both are convergent, and each is fed by tributaries and by diffuse flow through the subjacent or surrounding bedrock or sediment. Both are characterized by rapid, turbulent flow over much of the network length. This is in sharp contrast to the dispersive, slow, laminar flow of most granular and fractured aquifers.

The discharge point of a groundwater basin in a karst terrain is usually a spring located at regional base level or at a stratigraphic or fault contact where insoluble rocks restrict the formation of dissolution conduits. Springs are ephemeral, transient features, and they migrate to lower elevations as the base level is lowered. Hydrodynamic and geochemical models suggest that the tributary pattern of the caves should give way to one of distributaries (Fig. 3.8), similar in function to those in the delta region of major surface rivers, but relatively larger and different in origin (Quinlan et al., 1978; Ewers et al., 1978; Ewers, 1982). Distributaries could be a result of one or more of the following:

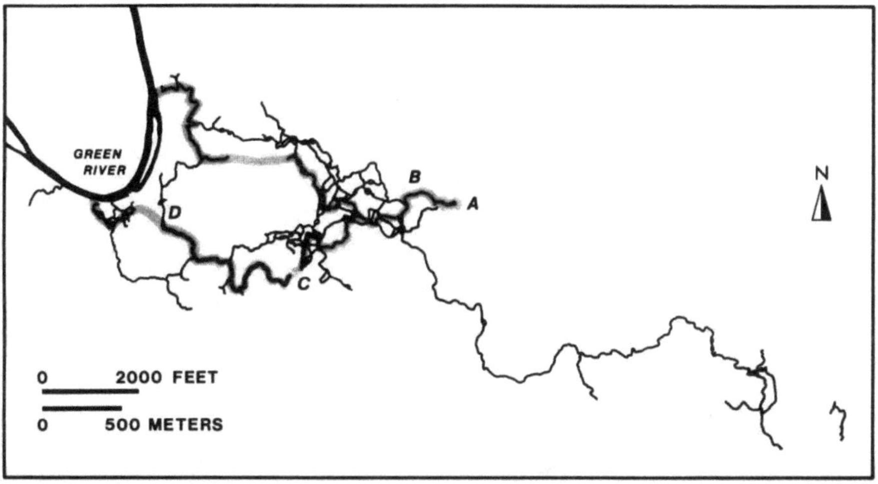

Figure 3.7 Map showing hydrology of the Hidden River Complex. Water enters the cave at *A,* diverges at base flow as shown by the gray pattern, and flows to numerous springs along Green River. More than 29 km of passage are represented. As shown in Figures 3.5 and 3.6 and Plate 1, heavy-metal-rich effluent flows from a metal-plating plant to Hidden River Cave, beneath the town of Horse Cave and 6.4 km south of *A,* and then to 46 springs at 16 locations along an 8-km reach of Green River. Most (and sometimes all) base flow is to Gorin Mill Spring. Water rises from a lift tube at *A* that is 17 m deep and 30 m wide. A magnetic-induction cave-radio and a compass and tape survey were used to locate an observation well at *B.* The well, 40 m deep, missed the center of the 8-m wide stream by 18 cm. From *C* to *D* is a 2.4-km stream segment that averages 12 m wide and 2.4 m deep. Its ceiling is 4 m high. *(From Quinlan and Ewers, 1981)*

1. Enlargement of small, pre-existing anastomoses by flow in response to large head-differentials between floodwater-filled passages and the river to which they discharge.
2. Collapse and blockage of a spring orifice and the development of alternative discharge routes.
3. Diversion (piracy) of cave streams to lower routes as base level and river stage are lowered.
4. Enlargement of vadose conduits that intersect anastomoses and other passages at the potentiometric surface or a water table that rises and falls in response to changes in the stage of a river. Backflooding forces relatively unsaturated water into and through available joints; falling stages are accompanied by a reversal of flow. Field observations and dye-trace results show that terminal distributaries and their related spring

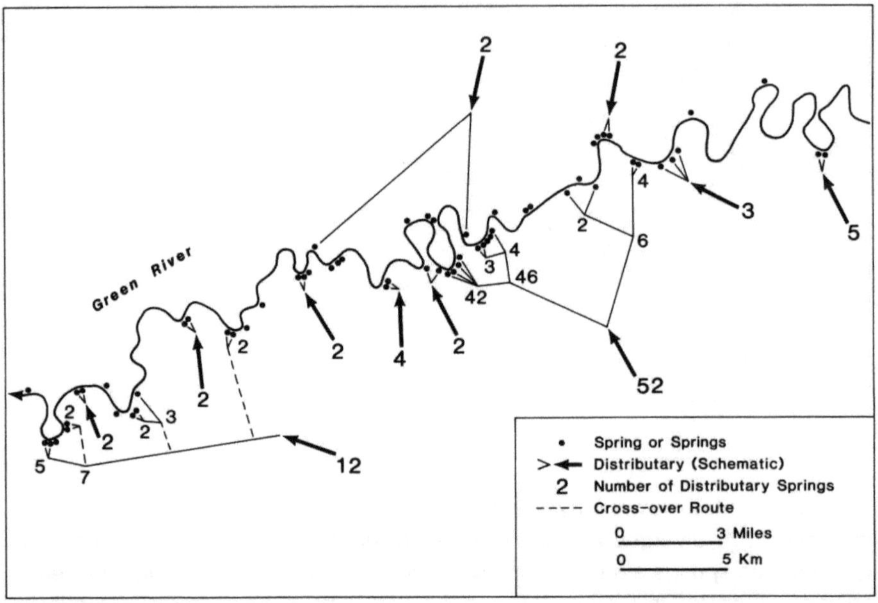

Figure 3.8 Distributary springs along Green River in the Mammoth Cave area. The numbers indicate the total number of springs in a given distributary system or subsystem. The positions and geometry of the underground bifurcations shown are schematic, but their existence is a certainty. Some of the bifurcations have been located and mapped. Pollutants at a point-source in the headwaters or mid-reaches of a groundwater basin or sub-basin will, depending on the stage, flow to all springs in its distributory system or subsystem. Each of the three crossover routes (shown with a dashed line) allows flood waters to travel from one groundwater basin to another. Data on the westernmost distributary system is shown in Figure 3.1. Not all springs between the extremities of a distributary system are necessarily a part of the system. *(From Quinlan and Ewers, 1985)*

clusters are a common feature of carbonate aquifers, especially in the Mammoth Cave area (see also Figs. 3.5 and 3.6).

Graham Springs Groundwater Basin

Perhaps one of the more interesting features of this basin is Graham Springs itself (a group of four springs also known as Plum Springs), which is shown in Figure 2.27, and an associated group of three springs downstream from the spring-fed 460-m (1500-ft) west-flowing stream, which is tributary to the Barren River. These groups are shown schematically in Figure 3.9 and with more detail in Figure 2.27 and Plate 1.

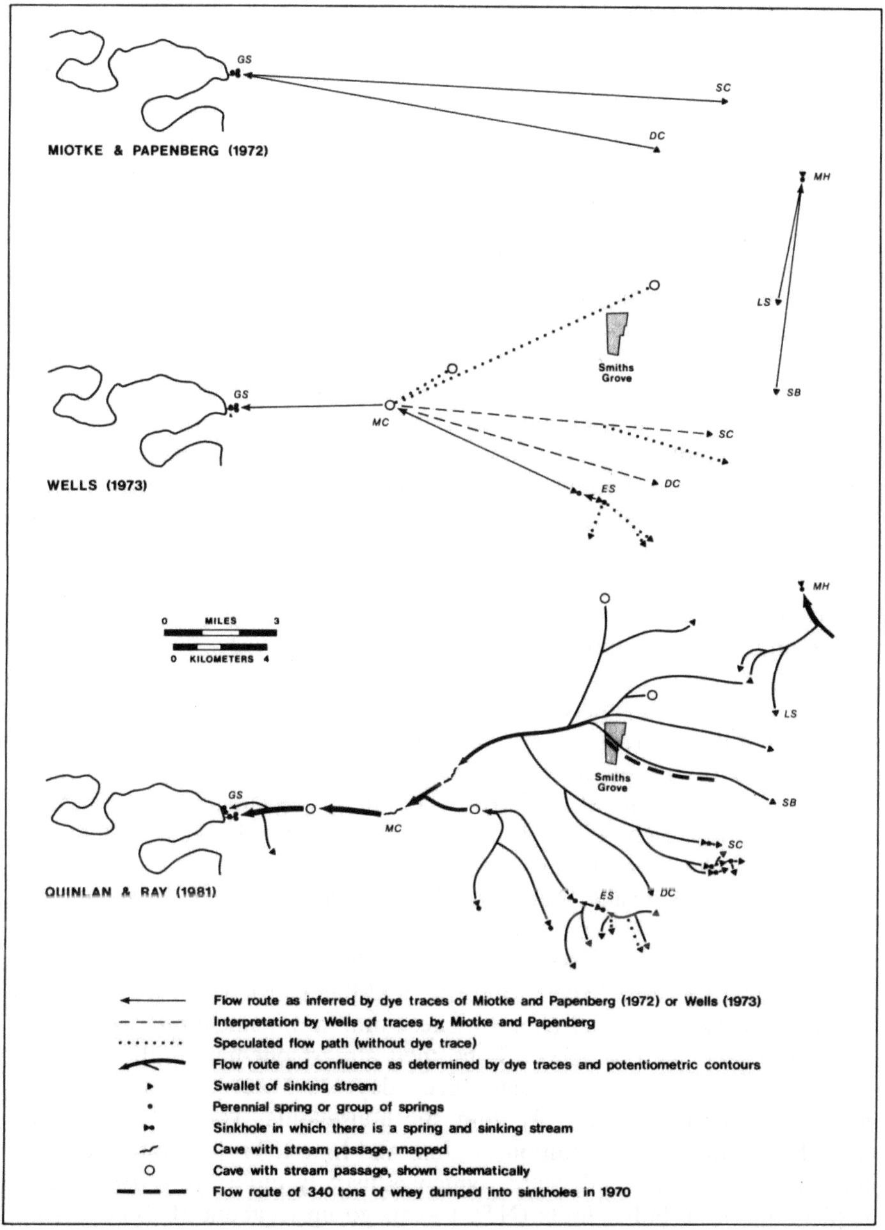

Figure 3.9 Maps showing the evolution of knowledge of the flow routes in Graham Springs groundwater basin. Key to selected sites shown: DC = Doty Creek; ES = Elk Spring; GS = Graham Springs; LS = Little Sinking Creek; MC = Mill Cave; MH = Mill Hole; SB = Sinking Branch; and SC = Sinking Creek. *(From Quinlan et al., 1983)*

The two easternmost springs closest to Plum Springs Road are high-level overflow routes, which do not discharge during base-flow conditions. Perennial discharge is from the two springs on the north bank of the spring channel (No. 3 and No. 4 in Fig. 2.27). Downstream from the junction of the spring channel with the Barren River are three more perennial springs, located approximately 45 m, 200 m, and 330 m from the junction. The springs at 210 m and the spring at 335 m are part of the Graham Springs distributary, which is 600 m wide and discharges from the second largest groundwater basin in the Mammoth Cave area—310 km^2 (120 mi^2). Nevertheless, the spring at 45 m, which is shown as No. 5 in Figure 2.27, discharges from the middle of this distributary and is hydrologically independent of it. We have named this the Waterworks Spring. Its recharge area is the Waterworks groundwater basin (Plate 1, basin 20).

As discussed by Shuster and White (1971) and then by others (White and White, 1974; Gaither, 1977; Quinlan and Ewers, 1985), there are two major end-members in the continuum of carbonate spring types: diffuse flow and conduit flow. Diffuse-flow springs respond slowly to precipitation, have small fluctuations in discharge, and exhibit little annual or storm-related change in either temperature or water chemistry. In contrast, conduit-flow springs respond rapidly to precipitation, have great fluctuations in discharge, and exhibit large variations in both temperature and water chemistry.

During the past 15 years the senior author and his field assistants have become familiar with more than 200 springs in the Mammoth Cave area. Only two of these springs, very minor in discharge, are known to be diffuse-flow, and only one of them is perennial. Because of the maturity of the principal aquifer, conduit-flow overwhelmingly prevails. The only known perennial diffuse-flow spring in the 2480 km^2 (970 mi^2) of the Mammoth Cave area is Waterworks Spring.

In 1964 the Northside Water District planned to build a treatment plant 160 m (0.1 mi) north of the tributary channel in order to utilize water from the Barren River. During foundation excavation, however, the Waterworks Spring was discovered beneath the site, and the decision was made to treat and use this water instead (Joseph Liles, oral communication, 1981). The plant was renovated in 1981. Since then almost one-half million gallons of spring water have been treated each day (20 l/sec). This plant serves about 2000 customers, mostly domestic households. The water is treated by chlorination and the addition of alum and lime to remove turbidity. The turbidity of the untreated water, which is usually quite low—from 2 to 3 nephelometric turbidity units (NTU)—can go up to about 10 NTU after a heavy rain. In comparison untreated river water during base flow would have a turbidity of about 40 NTU, and Graham Springs could have a turbidity of more than 1000 ntu after a heavy rain.

We do not yet understand the hydrogeology of the Waterworks groundwater basin, but we plan to study it. The temperature and conductivity of

Waterworks Spring and three of the four perennial springs in the Graham Springs distributary have been monitored. An elementary statistical analysis of these data indicates that the mean conductivity of Waterworks Spring is 488 micromhos, and its coefficient of variation (standard deviation/mean) is 2%. The mean conductivity of Graham Springs is 326 micromhos, and its coefficient of variation is 14%. The recharge has been traced from as far as 4.8 km away, as shown on Plate 1, and flow velocities to this spring are, as one might expect, significantly lower (by a factor of about 0.5–0.2) than in conduit-flow springs. It seems likely that some of the properties of Waterworks Spring could be a consequence of fracture permeability of the overlying Lost River chert and the fact that it is rarely breached by sinkholes. Less soil is flushed into the subsurface through this chert than is flushed into the subsurface through other stratigraphic units, so the turbidity is less than that of other springs.

SPECULATIONS ON THE EVOLUTION OF GROUNDWATER BASINS

The Mammoth Cave area south of Green River contains 27 recognized, major groundwater basins, plus additional sub-basins, all of which discharge at springs on the Green River, Barren River, and Little Barren River. The size of these basins ranges from less than 2.5 km^2 (1 mi^2), with an estimated base flow of 0.0028 cms (0.1 cfs) to 490 km^2 (190 mi^2) with an estimated base flow of 1.13 cms (40 cfs). Furthermore, these basins, as delineated by dye-tracing, exhibit a wide diversity of shapes, especially when considered separately from their areas drained by surface streams. For example, both the Turnhole Spring groundwater basin and the Bear Wallow groundwater basin have the appearance of an "inverted asymmetric mushroom" rooted in Green River. In contrast, the other basins are narrower and conform better to the pear-shaped proportions of a normal surface-drainage basin. It is interesting to speculate about the relationships between the distribution of these basin characteristics and the structure, surface-erosion history, stratigraphy, and probable hydraulic potential fields within the aquifer. This section complements the analyses by Deike (1967), Miotke and Palmer (1972), and Palmer (Chap. 12) of the age and development of Mammoth Cave.

Structural and Stratigraphic Setting

Figures 2.2, 2.4, and 3.2 depict the fundamental structural and stratigraphic relationships of the region. Of special interest are the capping sandstones and shales of the Big Clifty and Fraileys Members of the Golconda Formation, the Lost River Chert Bed of the uppermost St. Louis Formation, and the impermeable silty and shaly units in the lower third of the St.

Louis Limestone (Fig. 2.2). All of these units can be shown to function as aquicludes or, at some locations, as aquitards. These units, plus the intervening soluble limestones were deformed into a series of minor domes and en echelon east–west-trending monoclines, as shown in Figure 2.4. Overall, the dip is toward the north and northwest at about 0.011 to 0.019 (60–100 ft/mi).

Erosion History

The surface of the Chester Cuesta between its escarpment and the Green River is dissected by a series of dry valleys, which were formerly tributaries to the river. Some of the dry valleys are abruptly truncated along the escarpment, but their morphology suggests that they formerly extended into and perhaps across the present Sinkhole Plain. Subsequent to the establishment of these valley courses, a period of entrenchment occurred, which, based on paleomagnetic evidence, began at least 900,000 years B.P. (Schmidt, 1982). There is no geomorphic evidence of significant changes in the course of Green River or its two tributaries, the Barren River and Little Barren River, since that entrenchment. Finally, valley infilling has occurred to as much as 18 m (60 ft) at Mammoth Cave.

A Speculative Synthesis

In order to create a karst aquifer, meteoric water or allogenic surface runoff must have access to the soluble rocks. This must occur in a topographic setting where significant fluid potential can be achieved relative to active discharge points. The conditions of geologic structure and stratigraphy and the boundary conditions imposed by the Green River and the Barren River suggest a sequential opening of the limestones to karstification. If we assume that all of the valley bottoms have been at almost the same elevation at any given time during the entrenchment of Green River, Figure 3.10 depicts the sequence of exposure of the recharge and discharge areas within the region. This assumption is not unreasonable in view of the fact that present streams possess gradients less steep than the dip. In this figure no attempt has been made to imply the location of former valleys above the Sinkhole Plain. The sandstone–limestone boundaries on this figure separate the area of possible limestone exposure along any such valleys from areas completely covered by the Big Clifty for a given stage of valley deepening. The exposure of the limestone is presumed to have taken place by normal fluvial processes, not solutional mass-wasting.

At the arbitrarily selected valley-deepening stage of 275 m (900 ft), a large belt of limestone is exposed, but groundwater discharge from its rocks can take place only at the eastern margin, at Green River, and the western margin, at Barren River. At both discharge areas the subsurface flow from

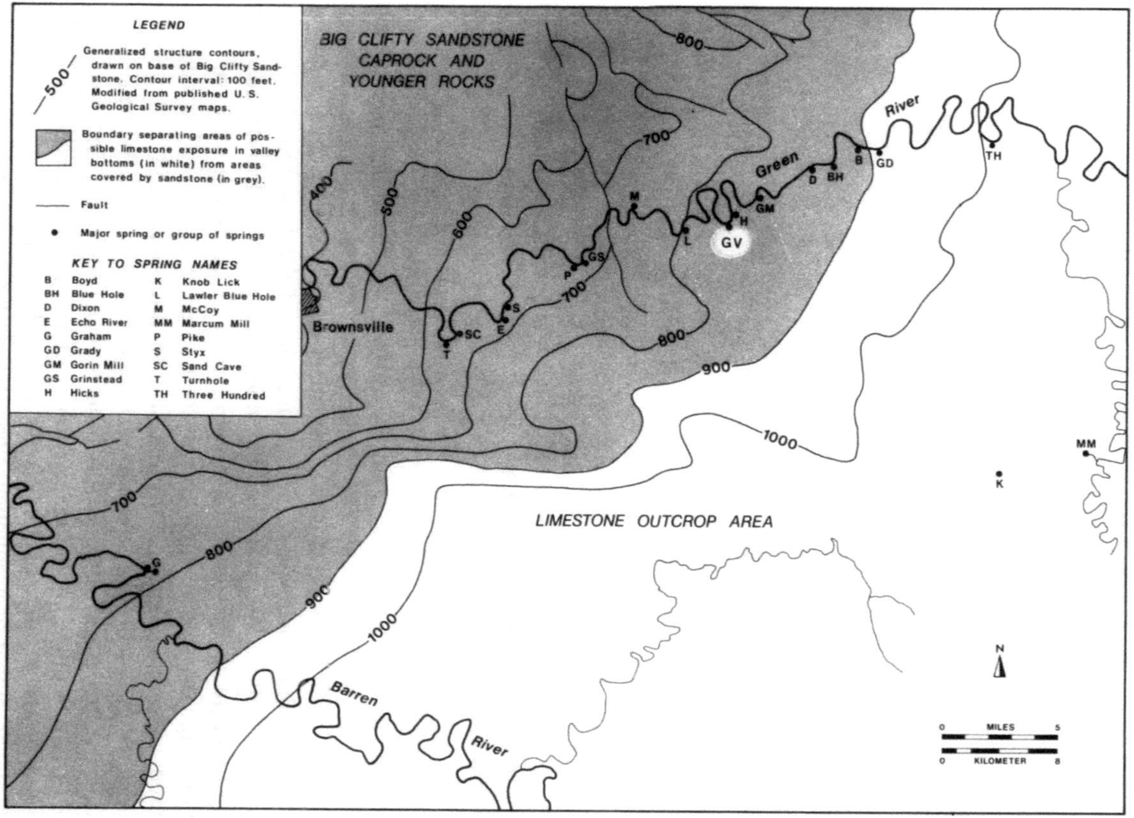

Figure 3.10 Map showing boundaries separating areas of possible limestone exposure in valley bottoms from areas completely covered by Big Clifty Sandstone during successive stages of valley deepening in the Mammoth Cave region. Sandstone-capped interstream ridges are not shown but can be inferred from Figure 3.1. The actual contact was serrate, with a probable amplitude of several miles. No attempt is made to show how the southeast margin of the Sinkhole Plain also migrated northwestward (downdip) as the terrain was lowered by erosion. GV = Garvin Spring; other abbreviations are explained in the legend. *(From Quinlan and Ewers, 1981)*

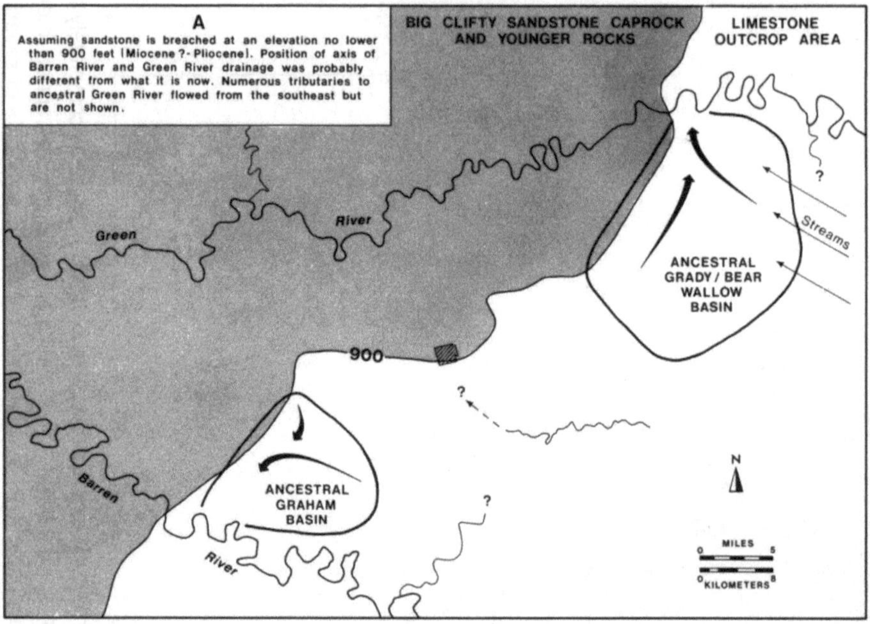

Figure 3.11 Evolution of groundwater basins in the Mammoth Cave area. The future site of Park City is shown as a square in the center of A, B, and C. *(From Quinlan and Ewers, 1981)*

local inputs should readily generate solution porosity. In the center of this exposed limestone belt, both the flow and the development of solution porosity will be comparatively sluggish. As the porosity, which develops at the discharge points, begins to drain the margins of the limestone belt, solution porosity will be integrated and extended gradually headward, thus establishing two principal drainage basins (Fig. 3.11*A*). Although we have drawn the drainage divide midway between the discharge points, reflecting the presumed water movement, we suggest that solutional modification of the limestone in the divide area had progressed only a short distance into the rock. At this stage, integrated solution voids should develop near the discharge points only. Continued evolution of the groundwater basins is shown in Figures 3.11*B* and 3.11*C*. More detailed expositions of the progressive development of groundwater conduits in the area can be found in Ewers (1982) and Ewers and Quinlan (1981).

Discharge points can be established along the north-central margin of the limestone belt at progressively lower stages of valley deepening. This sequence of development is depicted in greater detail in Figure 3.12. The ancestral spring sites at present-day Pike Spring and Echo River Spring, together with their respective distributaries, Grinstead Spring and Styx

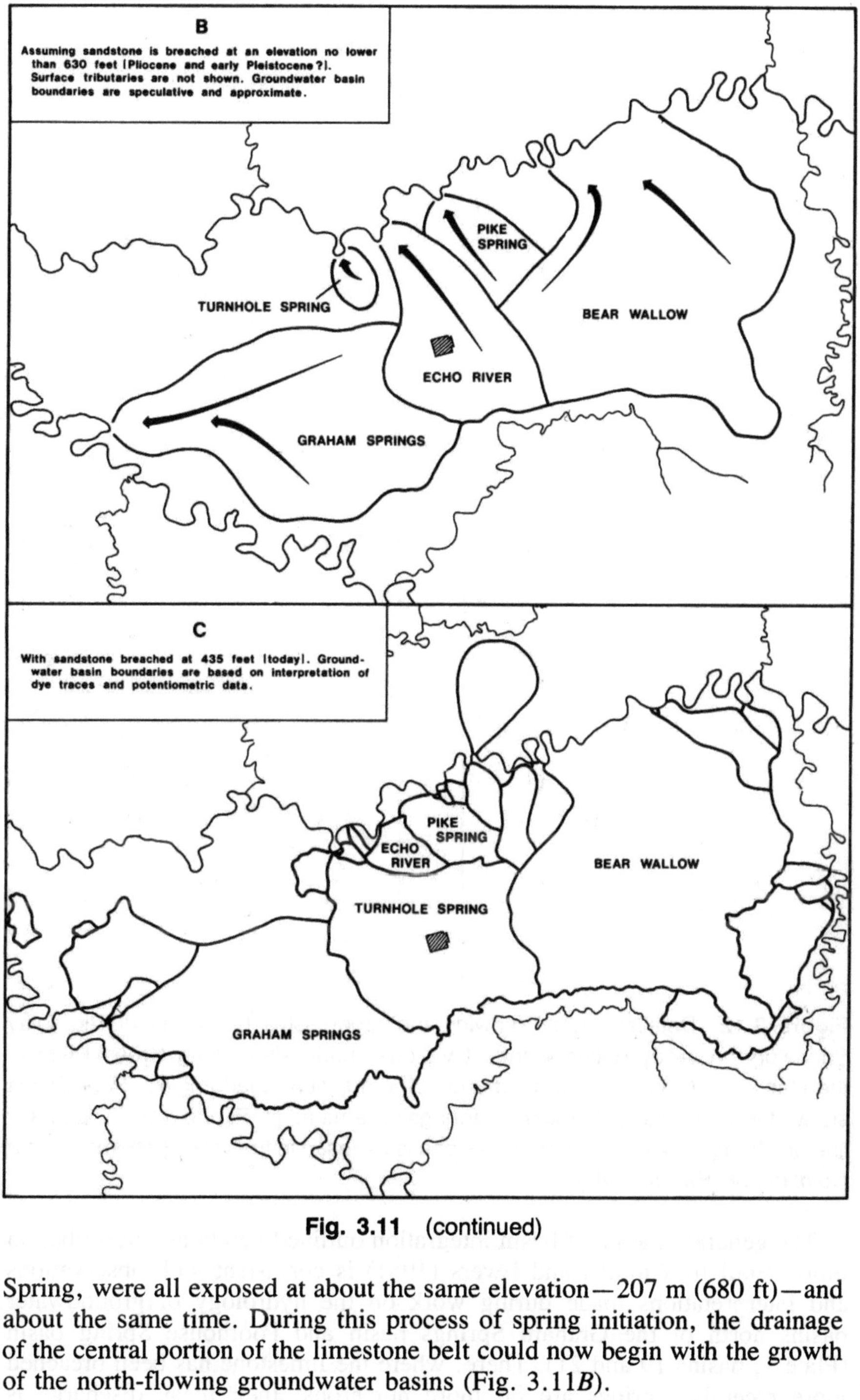

Fig. 3.11 (continued)

Spring, were all exposed at about the same elevation—207 m (680 ft)—and about the same time. During this process of spring initiation, the drainage of the central portion of the limestone belt could now begin with the growth of the north-flowing groundwater basins (Fig. 3.11*B*).

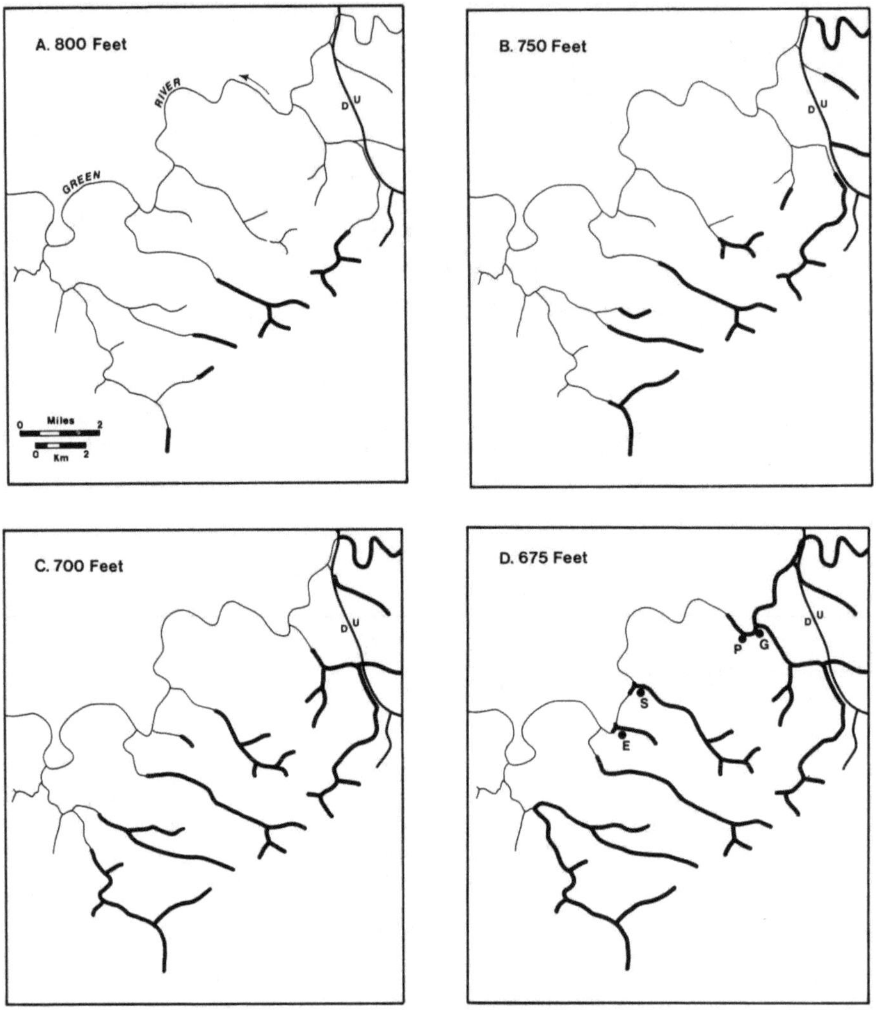

Figure 3.12 Detailed maps showing the progressive breaching of Big Clifty Sandstone in valley bottoms and at various spring sites along Green River in the Mammoth Cave 7.5-min quadrangle and adjacent quadrangles. Park City is shown for reference. Letter key for springs is same as for Figure 3.10. The heavy line on streams indicates where limestone is exposed in the valley bottoms. *(From Quinlan and Ewers, 1981)*

The general scheme of basin integration outlined herein and described in more detail by Quinlan and Ewers (1981) is consistent with observations and interpretations made during work on the hydrology of groundwater basins north of the Graham Springs basin and Poorhouse Spring basin (Plate 1, basins 19 and 21). There, where the limestone has been breached more recently, springs are far more numerous, their mean discharge is

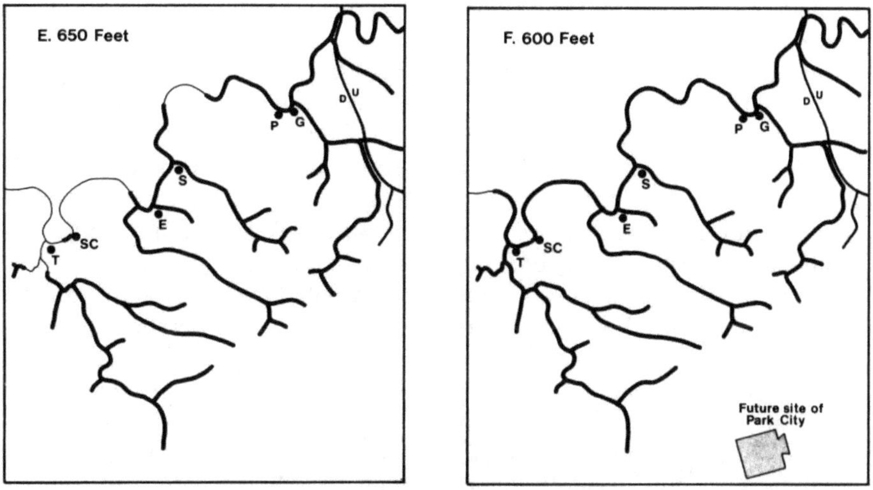

Figure 3.12 (continued)

much less, and their recharge area is much smaller than in terrains to the east. The youthful character of these smaller basins is enhanced by perching that occurs within the Ste. Genevieve Limestone.

For the purpose of our analysis we have straightened the 700-ft and 600-ft structural contour on the peculiar anticlinal "bump" at Pike Spring, which is shown in Figure 10.11 (Haynes, 1964), even though by doing so we have eliminated the conceptually attractive probability that Pike Spring was the first to form. The "bump" is not present in the subjacent limestones (Arthur N. Palmer, oral communication, 1978).

We have assumed that any springs that formed at the bottom of the tributary streams, at spillover points formed by the sandstone–limestone contact, were unimportant. Such springs are generally fed by shallow, weakly integrated circulation, whereas all of the basins under consideration herein are characterized by deeper, more integrated circulation.

Discussion

The present-day groundwater basins (Fig. 3.11*C*) appear to be consistent with this interpretation, with one exception (the Turnhole Spring basin, which will be discussed later). The Graham Springs basin and Bear Wallow basin are the largest; their ancestral spring-points were almost certainly the first to be initiated. The present springs in both basins, however, are lower than the points of earliest possible breaching. The Bear Wallow basin contains springs farther upstream, which are closer to the point of first possible breaching; Boyd Spring (Fig. 3.10) is a good example. Progressive movement of a basin spring-point downstream is not unexpected and has long been recognized. No comparable higher-level ancestral spring has yet been identified in the Graham Springs basin, except at Graham Springs itself (see Fig. 2.27).

It is also interesting to note that the Bear Wallow basin extends headward of the Garvin–Beaver basin and Lawler basin, giving it, in plan view, a distinctive inverted-asymmetric-mushroom shape. We interpret this to be a consequence of its time-related competitive advantage in integrating the groundwater flow of this area, an advantage conferred by the structure in concert with the erosion processes. If the Garvin–Beaver basin and Lawler basin ancestral springs had been initiated contemporaneously with the Bear Wallow springs, we suggest that the former would have evolved to a greater extent southeastward and the Bear Wallow basin would be smaller than it is. Although this explanation seems satisfactory, it should be noted that the headward boundary of the Garvin–Beaver basin and Lawler basin with the Bear Wallow basin is adjacent to the remarkably straight northeast–southwest-trending portion of the Chester Escarpment, which we refer to as the Horse Cave lineament. The lineament also nearly coincides with a major trough in the potentiometric surface (Plate 1). The cause of the lineament is unknown, but it is possible that it localized the development of solution permeability in the aquifer, a permeability that, because of its efficiency, has not yet been pirated by the smaller, younger, higher-gradient basins.

The large size of the Turnhole Spring basin is somewhat perplexing. Like the Bear Wallow basin, it extends headward of the adjacent basins, but it does so in spite of its apparently late initiation, after Green River was cut down to about 200 m (640 ft). It could be that some structural feature such as a joint or fault compromised the integrity of the caprock or increased the transmissivity of the limestone south of the spring-point. Either condition could have given the basin a competitive advantage, but no evidence for such features exists. The proximity of Turnhole Spring to the monocline west of Park City (see Fig. 2.4) could have provided a shorter, higher-gradient path for groundwater flow. Also, the paleo-valley crossing the Turnhole basin, Cedar Spring Valley, is larger than the paleo-valleys associated with the Pike basin and Echo Spring basin. The larger and presumably deeper valley would have increased the probability of inputs occurring close to the spring and at an earlier time than in the adjacent valleys. These conditions could have given the basin a competitive advantage.

The position of Turnhole Spring, downstream from the Echo River basin and Pike Spring basin, might have conferred a slight head advantage, thus allowing it to extend headward at a faster rate or to pirate the headwater conduits of the adjacent basins. This hypothesis is supported by the fact that present-day Echo River Spring has captured Styx Spring, and Pike Spring is interpreted to have captured Grinstead Spring. In each case the capturing spring is downstream about 1.5 km. Also, large paleo-drainage trunk passages, such as Kentucky Avenue in Mammoth Cave (see Fig. 10.11), suggest a larger basin than now exists upstream from these springs.

Dye-tracing has shown that the southernmost river in Roppel Cave (Fig. 3.1) flows to Turnhole Spring. A storm-related rise of this stream by less than 1 m diverts some of the otherwise Turnhole-bound water to Pike Spring. This suggests that piracy of the Pike Spring basin by Turnhole drainage is occurring now. Indeed, the Pike Spring, Echo River, and Sand Cave groundwater basins are all linked to the Turnhole Spring basin by high-level overflow routes, as shown on Plate 1 and in Figure 3.1.

Evidence for piracy of the headwaters of the Graham Springs groundwater basin by Turnhole drainage is given by the nature and location of Crump Cave, slightly less than 1 km northeast of Smiths Grove (Plate 1). This 1.5 km-long cave is a large, dry high-level trunk passage that has been abandoned by the underground river that formed it. The passage is 12 m (40 ft) wide and 9 m (30 ft) high, but the depth of its fill, locally at least 6 m (20 ft) where it has been flushed to inaccessible lower levels by vertically percolating water, is unknown. Wells (1973) showed that the paleoflow of this huge passage was to the west and southwest as indicated by scallops on its walls and by the 0.0034 (18-ft/mi) gradient of its floor, which is at an elevation of 165–168 m (540–550 ft), as discussed in Chapter 2. The potentiometric surface beneath the cave (Plate 1) is at 137 m (457 ft), about 27 m (90 ft) lower than its floor. By using best-fit exponential curves to extrapolate the present profile of Little Sinking Creek, Wells showed that it is graded to the floor of Crump Cave, and that Sinking Creek is graded to present-day Graham Springs as discussed in Chapter 2. The swallet of each of those surface streams is shown in Figure 3.9. These correlations are interesting and consistent with our interpretation of surface piracy of the south fork of Little Sinking Creek by its north fork, but we believe they are coincidental. We believe this because extrapolation of the profile of other sinking streams in the Graham Springs basin, using Wells' data and procedures, yields spring elevations that are as much as 23 m (75 ft) above or below the present-day outlet of Graham Springs. Also, Wells' procedures ignore complications related to alluviation of the Barren River Valley by about 10 m. In brief, Wells' mathematical analyses of stream gradients is clever and potentially a very significant contribution, but it was not taken as far as it could and should have gone; it lacks the rigor needed for uncritical acceptance. In addition, it should be noted that Sinking Branch (Fig. 3.9) flows to Graham Springs, not to Turnhole Spring as indicated in Table 2.2. Nevertheless, the proximity of this 12 m × 15 m cave passage to the boundary of the Turnhole Spring groundwater basin 2.4–3.2 km to the east, certainly requires that the former headwaters of the Graham Springs groundwater basin—the recharge area for the Crump Cave passage when a river flowed through it—have been captured by Turnhole Spring.

Although the caprock was breached at Pike Spring, Echo River Spring, and Styx Spring at about the same time (207 m; 680 ft), it is tempting to conclude that Pike Spring was the first of those springs related to Mammoth Cave to flow. Collins Avenue (No. 1 in Fig. 12.1)—elevation 210 m (690 ft)—is the highest passage in the Crystal Cave part of the Flint Ridge section of Mammoth Cave, and trends toward Pike Spring. Gothic Avenue (see Fig. 10.11)—at the much lower elevation of 183 m (600 ft)—is the highest passage in Mammoth Cave and is closest to Echo River and Styx Springs. This conclusion, however, would be inappropriate. It is based on the debatable assumption that the passages had to be graded to the level of Green River. They could well have been 6 m or even 18 m or more (20–60 ft) below the river when they formed.

As if matters were not already sufficiently complex, there remains the interpretation of the influence of the Lost River chert. This unit, which is frequently associated with the perching of small springs, reduces the infiltration of water to caves that it overlies. By preventing the infiltration of soil, it markedly impedes the formation of sinks (Woodson, 1981). Figure 3.13, which is a summary diagram, shows the breaching of the sandstone caprock relative to some present-day spring-points. Springs of the Bear Wallow basin emerge from below the chert, yet the basin must have been established above this horizon. This leads us to conclude that the influence of an established set of subsurface drainage trunks is so strong that a major lithologic discontinuity is not sufficient to cause major disruption of the established system. We further support this conclusion with the fact that water sinking in the Lawler basin descends through the Lost River chert before rising at Lawler Blue Hole Spring ("L" in Fig. 3.10).

Although we have made no attempt to discuss the absolute age of the development of the various groundwater basins, it should be mentioned that $^{230}Th/^{234}U$ age determinations by Hess and Harmon (1981) indicate that some intermediate-level passages within Flint Ridge, at an elevation of about 150 m (500 ft), are more than 350,000 years old. The lowest passage dated, at about 140 m (460 ft), is at least 140,000 years old and is related to the Illinoisan glaciation. If we naïvely use these dates and assume a constant rate of downcutting by Green River, and if we simplistically assume that the 12 m of downcutting took 210,000 years, then 67 m of downcutting would have taken 5.5 times longer, and the highest cave passages in the Pike Spring groundwater basin must have become a host for stalagmite deposition about 1.2 million years ago. This number is consistent with Schmidt's (1982) estimate that the highest cave sediments are at least 900,000 years to perhaps 2,000,000 years old, as suggested by paleomagnetic data. Geomorphic, isotopic, and paleomagnetic lines of evidence independently imply, therefore, that initiation of conduit development in the highest levels of the northwest-flowing Mammoth Cave system began during the late Pliocene.

The sequence of stripping of the caprock, in conjunction with structural evidence, suggests the order of initiation of the principal groundwater basins in the Mammoth Cave area; this sequence is summarized in Figure 3.13.

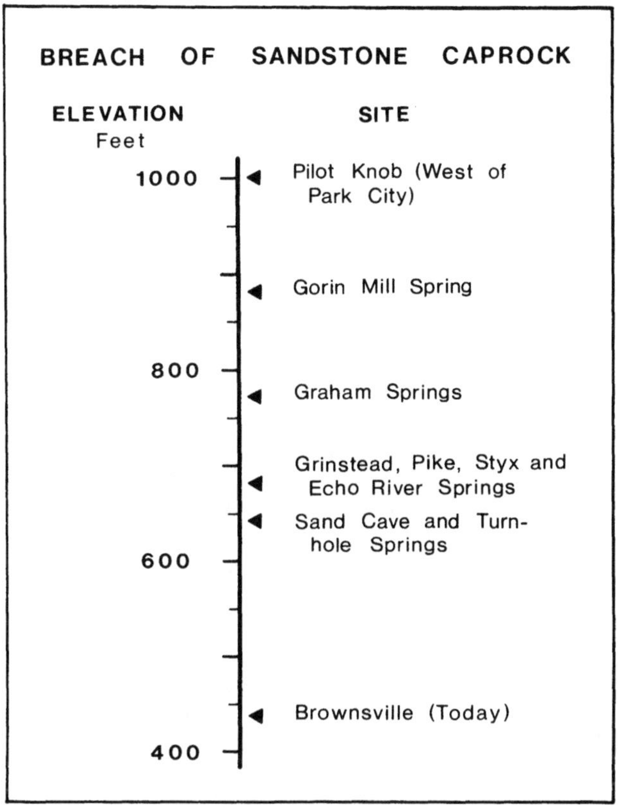

Figure 3.13 Summary of elevations (and therefore relative times) at which Big Clifty Sandstone caprock was breached at spring sites along Green River and Barren River, and which were, therefore, available to compete for regional discharge. *(From Quinlan and Ewers, 1981)*

RECENT CHANGES IN SUBSURFACE DRAINAGE

Until May 1984, all drainage that went to Cedar Sink was discharged only at Turnhole Spring (as shown in Fig. 3.1 and Plate 1). Similarly, all drainage from the adjacent Double Sink groundwater basin, a small 11 km^2 (4.3 mi^2) basin immediately to the west, was discharged at Sand House Cave Spring and at Notch Spring (as shown in Fig. 3.14 and Plate 1); none of the discharge from the Double Sink basin included water from the Turnhole Spring basin. This observation was confirmed repeatedly during more than 25 dye-traces at river stages up to 6 m above base flow. In May 1984, however, there was a major storm that might have been a 20-year event (see Figs. 5.4 and 5.5). Water levels in some cave streams rose more than 30 m. The following is a summary of the major effects of this exceptional storm on the hydrology of the Turnhole Spring groundwater basin:

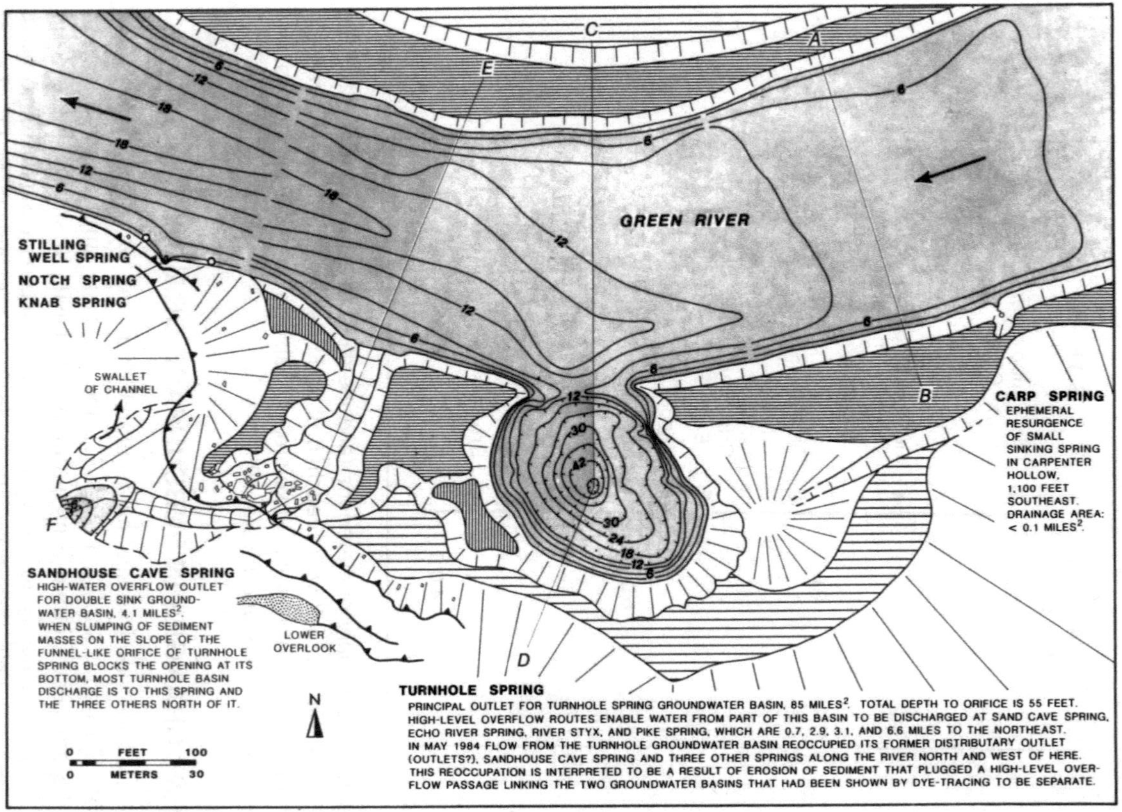

Figure 3.14 Sketch map of the landforms and bathymetry of the Turnhole Spring area, showing the effects of the May 1984 storm. *(From Quinlan and Ewers, 1981)*

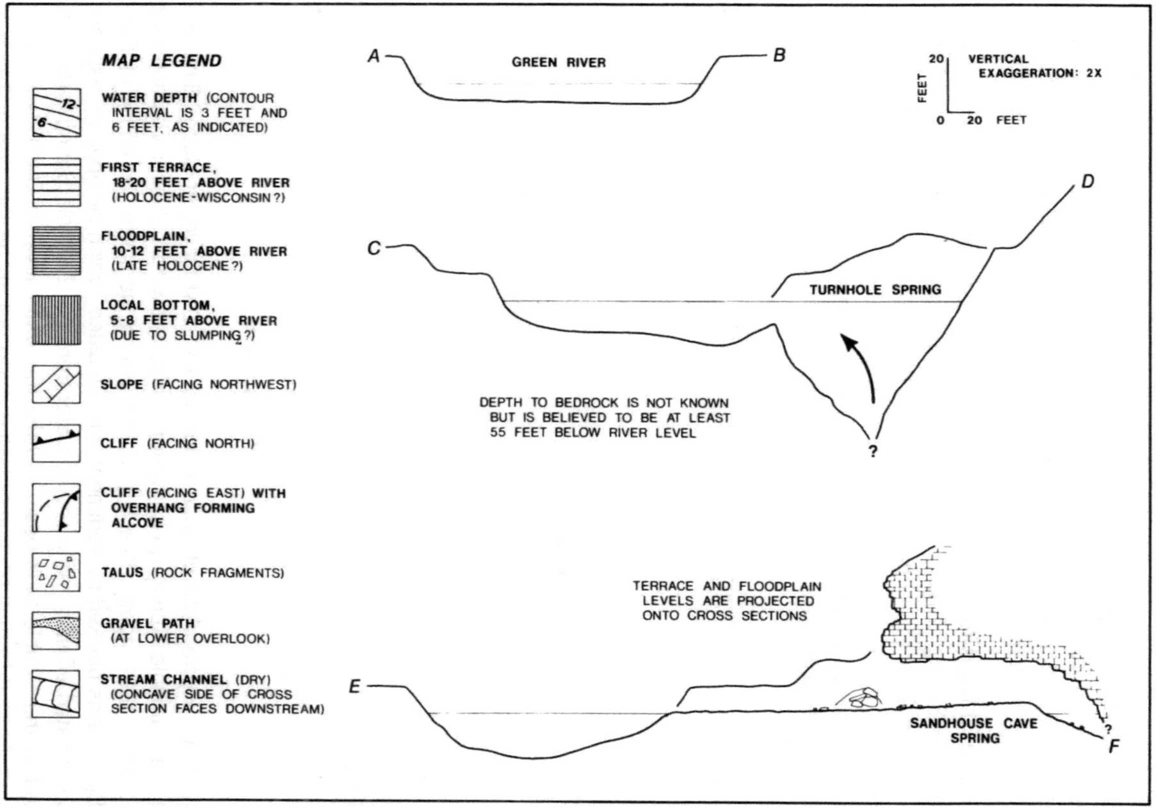

Figure 3.15 Legend and cross sections for Figure 3.14. *(From Quinlan and Ewers, 1986b)*

1. Fill in a sediment-blocked cave passage that once linked the conduit draining to Turnhole Spring with the conduit draining to Sandhouse Cave and its three associated springs was eroded and removed. The blockage in this distributary passage was at least partially removed. Now, but not within known time before May 1984, some or all of the water from the Turnhole Spring groundwater basin is discharged at Sandhouse Cave Spring, Stilling Well Spring, Notch Spring, and Knab Spring (as shown in Fig. 3.14).

2. Several landslides were triggered on the slope of the funnel-like orifice of Turnhole Spring (Figs. 3.14 and 3.15). The banks slumped, and trees were carried into the funnel. Sediment blocked the spring orifice, thus diverting all of what would have been the discharge of Turnhole Spring to the four springs in the Double Sink basin (as shown schematically in Fig. 3.14). This blockage, however, is intermittent. Some of the sediment plug is

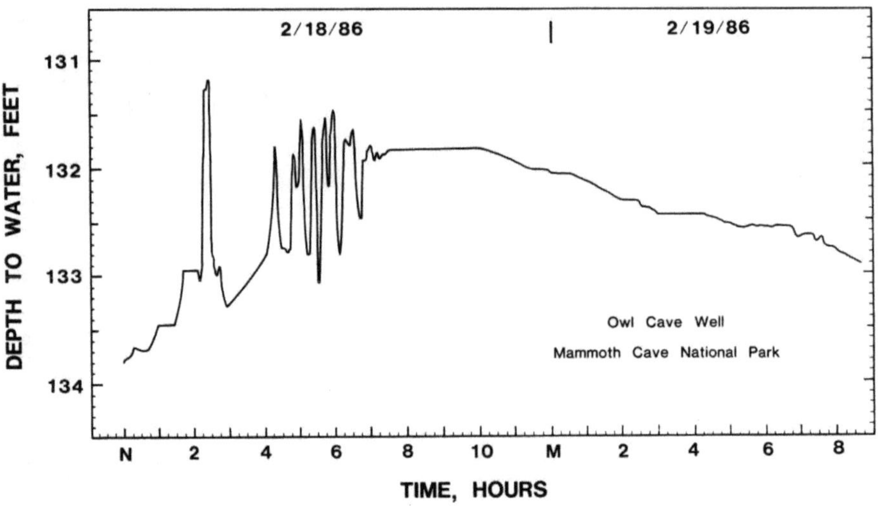

Figure 3.16 Hydrograph of observation well at Owl Cave, in Cedar Sink and within the Turnhole Spring groundwater basin, showing behavior characteristic since a May, 1981 storm. (See Fig. 3.14.) During the storm, the orifice of Turnhole Spring was blocked by sediment from its sides, which slumped in response to storm-related erosion. As a result, water from the Turnhole Spring basin was diverted to the Double Sink basin. When the water level in Owl Cave rose in response to storm runoff, thus increasing the pressure on the plug at Turnhole Spring, the sediment blockage was repeatedly blown out, sometimes at 15- to 30-min intervals. The abrupt falls in the hydrograph correspond to the release of water when the plug was blown out. The rises correspond to the buildup of head when the orifice was repeatedly blocked. *(From Quinlan and Ewers, 1986b)*

repeatedly blown out of the orifice when water levels upstream from it rise in response to storms. The now-inactive Turnhole Spring repeatedly blows its sediment plug and temporarily discharges surges of water. Sediment slumps back into the orifice (or orifices?) and blocks it (them) again. The result of this subaqueous flatulent phenomenon is depicted in Figure 3.16, which shows nine rises and falls of up to 60 cm (2 ft) over an 8-hr period. Other storms have triggered as many as 13 rises of 30–90 cm (1–3 ft) during a 48-hr period and eight rises of 3–30 cm (0.1–1 ft) during the same period. A rise of up to 1.8 m (6 ft) has been observed.

An understanding of the spiky behavior of the hydrographs at Owl Cave is very practical. Assume that a team of cave mappers is in Whigpistle Cave (Fig. 3.1) in one of the base-level streams, which has numerous segments on the order of hundreds of meters long in which there are only a few centimeters of air space. (Several such passages had to be traversed in order to run the dye-test necessary to identify the source of water in the newly discovered stream segment about 3 km north–northeast of Mill Hole.) Assume also that the cave mappers are in such a passage when the Turnhole Spring orifice is open. If the plug were to slump shut, there would be a consequent rise of 30–90 cm (1–3 ft) in water levels in these streams, and the mappers would be trapped or drowned.

Elsewhere in the Turnhole Spring basin during the May 1984 storm, at the Doyel Valley entrance drilled into Procter Cave, the sudden rise of water levels caused an increase in air pressure that blew a 160-kg (350-lb), 1-m diameter culvert out of the ground 5.5 m away.

APPLICATIONS OF RECENT HYDROLOGIC RESEARCH

The research described in this chapter has been applied to the following:

- Design of a 13-million-dollar regional sewage-treatment system for the towns of Horse Cave, Cave City, Park City, and Mammoth Cave National Park.
- Response to spills of hazardous materials along Interstate Highway 65 (an average of 1.5 per year; see Quinlan, 1986).
- Developing a strategy for reliable monitoring of pollutants in karst terrains (Quinlan and Ewers, 1985).
- Interpretation of the geomorphic history of the Mammoth Cave area.
- Environmental protection of Mammoth Cave National Park.
- Regional planning for solid-waste disposal and concomitant protection of groundwater supplies and the blind Kentucky Cave shrimp, *Palaemonias ganteri* (Hay), which is on the federal endangered species list.

NEW CONCEPTS OF GROUNDWATER MOVEMENT

There are four concepts of groundwater movement that have been recognized previously in other karst terrains, but are now described in the Mammoth Cave area better and in more detail than anywhere else. They are

1. Distributary flow (discussed in the Bear Wallow Groundwater Basin section).
2. Shunting of water by high-level overflow routes (discussed in the Turnhole Spring Groundwater Basin section).
3. Shared headwaters; sub-basin 14B is the headwaters of both basin 14A and 14C on Plate 1. In a sense, sub-basin 14B is similar to the overlap area in a Venn diagram.
4. Location of all major stream caves in troughs on the potentiometric surface and, likewise, association of all major troughs with axes of trunk drainage in the subsurface. This is clearly shown in:
   a. The area up-gradient and down-gradient from Mill Hole in Figure 3.1 and Plate 1 (basin 3A).
   b. The area north of Park City in Figure 3.1 and Plate 1 (basin 3B1).
   c. The area between Smiths Grove and Graham Springs on Plate 1 (basin 21) where three major caves are known and shown.
   d. The area between Cave City, Horse Cave, and the Green River in Figure 3.6 and Plate 1 (basin 14A). The major cave is Hidden River Cave, beneath the town of Horse Cave.

Although a new hydrologic concept is only remotely involved, the Hidden River groundwater sub-basin was the site of the first attempt, in 1977, to use the presence of optical brighteners and heavy metals in spring water as a prospecting tool in the search for effluent from a sewage-treatment plant (Quinlan et al., 1977; Quinlan and Rowe, 1977). Since then several investigators, independent of one another, have used the presence of optical brighteners in well and spring waters as indicators of pollution from septic tanks (e.g., Ashworth, 1984; Thrailkill et al., 1985; Scanlon, 1985; Aley, 1985).

Other significant "firsts," which are results of research sponsored by the National Park Service and described herein, include the first

1. Deliberate injection (in North America) of optical brightener as a tracer.
2. Application of CI Direct Yellow 96 as a tracer (Quinlan, 1977).
3. Published map showing the relations between surface drainage, numerous dye-traces, the potentiometric surface, springs, mapped caves, and groundwater basins in a karst terrain (Quinlan and Ray, 1981; included as Plate 1 of this volume).

4. Use of low-frequency electromagnetic induction equipment to locate sites for successful drilling of wells to monitor cave streams at depths ranging from 40 m to 143 m (130 ft to 470 ft).
5. Continuous monitoring for stage, conductivity, velocity, temperature, rainfall, and soil moisture at a genetically related series of sites. (Pioneering work on continuous monitoring of conductivity and temperature of a spring and a cave stream, plus stage of Green River, is described in Chap. 4.)

The occurrence of what is best described as the bedrock equivalent of bank storage in alluvium was a serendipitously discovered hydrologic concept. This discovery was made by interpretation of wells drilled into and near Mill Cave (shown as MC in Fig. 3.9). The Mill Cave well was drilled to intercept the cave. The Sunnyside well is an unused domestic well that is approximately 180 m north of the estimated position of the inaccessible upstream portion of the cave. The approximate difference in the elevation of base flow between the two streams is only 60 cm (2 ft). The hydrographs in Figure 3.17 show that, after floods, water in the Sunnyside well rises only to about 68% of the height of water in Mill Cave. The abrupt rises at Sunnyside lag 3 hours behind the rises at Mill Cave, and the crests at Sunnyside lag 17 hours behind Mill Cave. The flow of floodwater in the bedrock adjacent to the cave is analogous to bank storage in an alluvial aquifer. The spikes labeled A, B, C, and D on Figure 3.17 are probably a result of surface runoff or quick-flow within the soil flowing down the well casing from the depression in which the well is located; runoff from the drain of an adjacent sinkhole; or some combination of these sources (Quinlan et al., 1983).

A more significant new hydrologic concept, one that is based not only on the results of research in the Mammoth Cave area, but also experience in numerous other karst areas, is the strategy of monitoring for pollutants in karst terrains at springs and cave streams (Quinlan and Ewers, 1985; 1986*a*). The following is a summary of the rationale for this concept.

It is wrong to apply standard techniques of well placement and sampling to most karst terrains. The practice of placing three monitoring wells down-gradient from a site and one well up-gradient from it, and sampling them quarterly, is usually a waste of time and money—unless one seeks spurious data. These procedures generally give reliable results in sediments, but 99% of such wells drilled in most karst terrains are irrelevant because groundwater flow in most karst terrains is convergent to a dendritic conduit system that flows to a spring that might be many miles away. The probability of a randomly located well intersecting a cave stream is similar to that of a dart thrown at a wall map of the United

States hitting the Mississippi River. Reliable monitoring of water quality can only be attained by sampling springs and wells drilled to surveyed cave streams, all of which must be shown by dye-tracing to drain from the site.

Characterization of water quality at monitoring points must be based on many samples taken before, during, and after the hydrograph peak

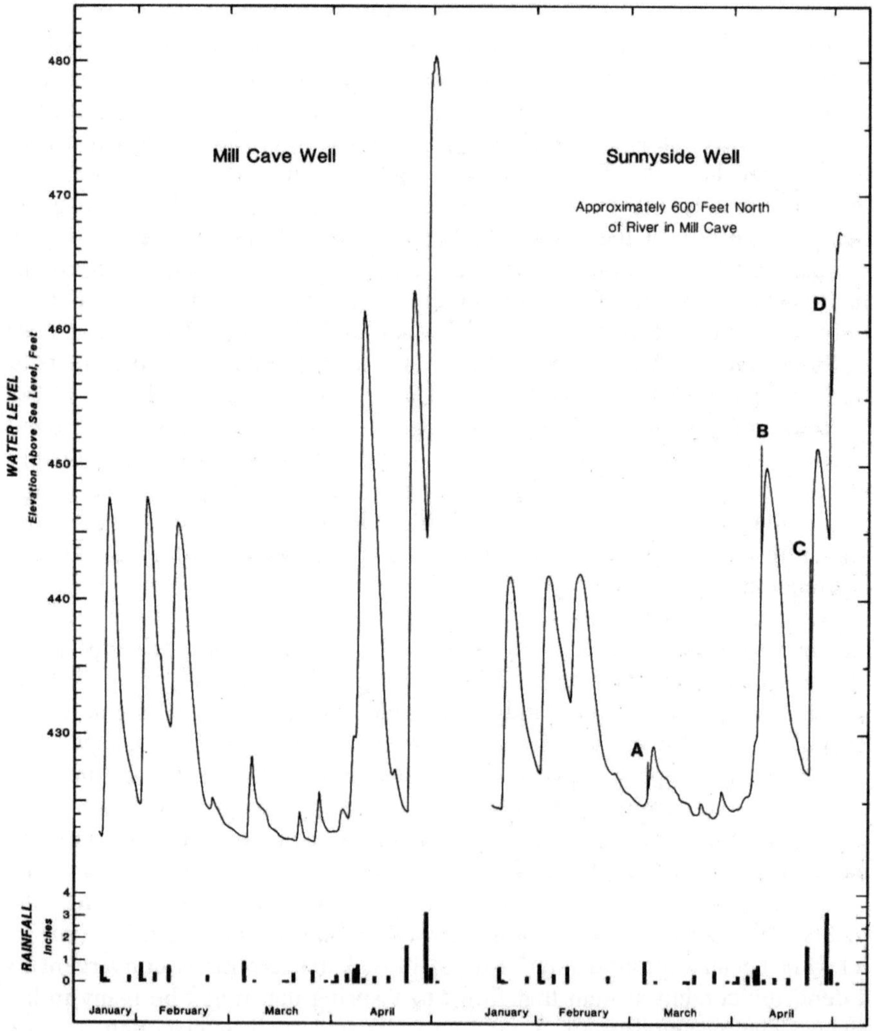

Figure 3.17 Hydrographs of observation wells at Mill Cave (shown as MC in Fig. 3.9) and Sunnyside, January–May 1983. *(From Quinlan et al., 1983)*

of significant storms. This is because the storm-related concentration of some pollutants in conduit-flow aquifers might temporarily increase to 1000 to 10,000 times their pre-storm level. When the source of pollutants is agriculturally related, the spiky maxima of some (such as pesticides) tend to coincide with the hydrograph maxima, suspended sediment maxima, and conductivity minima. In contrast, concentrations of other pollutants (such as nitrates) are minimal during the hydrograph peak and reach slowly decaying maxima that occur significantly after the hydrograph peak (Quinlan and Alexander, 1987). All of these responses are considerably attenuated in diffuse-flow aquifers.

RESEARCH IN PROGRESS

Our highest-priority current research is maintenance and interpretation of the continuous records of stage, conductivity, velocity, temperature, soil moisture, and rainfall at Hawkins Creek (a sinking stream at B of Section AB in Fig. 3.1), Brown River in Parker Cave, Mill Hole, Cedar Sink, and Notch Spring (discussed by Ewers et al., 1986; Meiman et al., 1988), and the installation of additional similar instrumentation in Parker Cave (at Parker River; Fig. 3.3) and in each of the two confluent rivers in Procter Cave. As shown in Figure 3.1, one of these rivers in Procter Cave drains from Park City; the other drains from the Interstate Highway 65 interchange area at Cave City.

ACKNOWLEDGMENTS

Like Newton, we have stood on the shoulders of giants and built on the results and observations of those who worked before us. We acknowledge our debts to them. Most of the research on the Mammoth Cave area was funded by the National Park Service, but some essential early work was funded by the Kentucky Water Resources Research Institute, the Barren River Area Development District, and four local banks that were "patrons of science."

The work described in this chapter could not have been done as efficiently and reliably if it were not for the dedicated help of Joe Ray and more than 20 field assistants and volunteers, many of whom came from the ranks of the Cave Research Foundation. The staff of the U.S. Geological Survey office at Louisville helped many times and in many ways. Frank Reid has graciously used a cave radio for locating drill holes and cave entrances. Several graduate students at Eastern Kentucky University, Joe Meiman, Scott Rucker, and Gary Andres have started monitoring studies that will, we believe, increase the sophistication of our ignorance by an order of magnitude.

REFERENCES

Aley, T. J., 1985, Optical brightener sampling: A reconnaissance tool for detecting sewage in karst groundwater, *Hydrol. Sci. Tech., Short Papers* **1**:45–48.

Aley, T., J. F. Quinlan, and J. E. Vandike, 1989, *The Joy of Dyeing: A Compendium of Practical Techniques for Tracing Groundwater, Especially in Karst Terrains*, Dublin, Ohio: National Water Well Association (in prep.).

Ashworth, R. A., 1984, Methodology delineating groundwater contamination by household wastewater, University of Arkansas, Ph.D. dissertation, 209p.

Crawford, N. C., 1985, *Groundwater Flow Routes, Lost River Groundwater Basin, Warren County, Kentucky*, Bowling Green, Ky.: Western Kentucky University, Center for Cave and Karst Studies, Map.

Crawford, N. C., 1986, Karst hydrologic problems associated with urban development: Groundwater contamination, hazardous fumes, sinkhole flooding, and sinkhole collapse in the Bowling Green area, Kentucky, in *Environmental Problems in Karst Terrains and Their Solutions: Field Trip B*, J. F. Quinlan and R. O. Ewers, eds., Dublin, Ohio: National Water Well Association.

Davies, W. E., and E. C. T. Chao, 1959, Report on sediments in Mammoth Cave, Kentucky, *U.S. Geol. Survey Admin. Rept.*, 117p. (Unpublished; copy held by Natl. Park Service at Mammoth Cave and by Geology Library at University of Texas at Austin.)

Deike, G. H., III, 1967, The development of caverns of the Mammoth Cave region, The Pennsylvania State University, Ph.D. dissertation, 235p.

Ewers, R. O., 1982, Cavern development in the dimensions of length and breadth, McMaster University, Ph.D. dissertation, 398p.

Ewers, R. O., and J. F. Quinlan, 1981, Cavern porosity development in limestone: A low dip model from Mammoth Cave, Kentucky, in *Proc. 8th Internat. Congress Speleol. (Bowling Green, Ky.)*, vol. 2, pp. 727–731.

Ewers, R. O., J. F. Quinlan, and D. C. Ford, 1978, The origin of distributary and tributary flow within karst aquifers (abstract), *Geol. Soc. America Abs. with Programs* **10**:398–399.

Ewers, R. O., J. J. Meiman, and J. F. Quinlan, 1986, Karst aquifer instrumentation system in the Mammoth Cave region, in *Environmental Problems in Karst Terrains and Their Solutions: Field Trip A*, J. F. Quinlan and R. O. Ewers, eds., Dublin, Ohio: National Water Well Association, pp. 90–93.

Gaither, B. E., 1977, The relation of spring discharge behavior to the hydrologic properties of carbonate aquifers, The Pennsylvania State University, M.S. thesis, 210p.

Gale, S. J., 1984, The hydraulics of conduit flow in carbonate aquifers, *Jour. Hydrology* **70**:309–327.

Haynes, D. D., 1964, Geology of the Mammoth Cave quadrangle, Kentucky, *U.S. Geol. Survey Geol. Quad. Map GQ–351*.

Hess, J. W., and R. S. Harmon, 1981, Geochronology of speleothems from the Flint Ridge Mammoth Cave System, Kentucky, U.S.A., in *Proc. 8th Internat. Congress Speleol. (Bowling Green, Ky.)*, vol. 2, pp. 433–436.

Lambert, T. W., 1976, Water in a limestone terrane in the Bowling Green area, Warren County, Kentucky, *Kentucky Geol. Survey, Rept. Inv. No. 17*, 43p.

Meiman, J., R. O. Ewers, and J. F. Quinlan, 1988, Investigation of flood pulse movement through a maturely karsted aquifer at Mammoth Cave National Park: A new approach, in *Environmental Problems in Karst Terrains and Their Solutions Conference, 2nd, Proceedings*, Dublin, Ohio: National Water Well Association (in press).

Miotke, F.-D., and A. N. Palmer, 1972, *Genetic Relationship Between Caves and Landforms in the Mammoth Cave National Park Area*, Würzburg: Böhler Verlag, 69p.

Miotke, F.-D., and H. Papenberg, 1972, Geomorphology and hydrology of the Sinkhole Plain and Glasgow Upland, central Kentucky karst: Preliminary report, *Caves and Karst* **14**:25–32.

Palmer, A. N., 1981, *A Geological Guide to Mammoth Cave National Park*, Teaneck, N.J.: Zephyrus Press, 210p.

Pickle, J. D., 1985, Dynamics of clastic sedimentation and watershed evolution within a low-relief karst drainage basin, Mammoth Cave region, Kentucky, University of New Mexico, M.S. thesis, 147p.

Plebuch, R. O., R. J. Faust, and M. A. Townsend, 1985, Potentiometric surface and water quality in the principal aquifer, Mississippian Plateaus region, Kentucky, *U.S. Geol. Survey Water Resources Inv. Rept. 84–4102*, 45p.

Quinlan, J. F., 1977, New fluorescent direct dye suitable for tracing groundwater and detection with cotton, *3rd Internat. Symp. Underground Water Tracing, Ljubljana-Bled, Yugoslavia, 1976, Papers* **2**:257–262.

Quinlan, J. F., 1981, Hydrologic research techniques and instrumentation used in the Mammoth Cave region, Kentucky, in *GSA Cincinnati '81 Field Trip Guidebook 3*, T. G. Roberts, ed., Falls Church, Va.: American Geological Institute, pp. 502–504.

Quinlan, J. F., 1985, Qualitative water-tracing with dyes in karst terrains, in *Practical Karst Hydrology, with Emphasis on Groundwater Monitoring*, vol. 6, J. F. Quinlan, ed., Dublin, Ohio: National Water Well Association, pp. E1–E26.

Quinlan, J. F., 1986, Recommended procedure for responding to spills of hazardous materials in karst terrains, in *Environmental Problems in Karst Terrains and Their Solutions Conference, Proceedings*, Dublin, Ohio: National Water Well Association, pp. 183–196.

Quinlan, J. F., and E. C. Alexander, Jr., 1987, How often should samples be taken at relevant locations for reliable monitoring of pollutants from an agricultural, waste disposal, or spill site in a karst terrane? A first approximation, in *Karst Hydrogeology: Engineering and Environmental Applications*, B. F. Beck and W. L. Wilson, eds., Rotterdam: A. A. Balkema, pp. 277–286.

Quinlan, J. F., and R. O. Ewers, 1981, Hydrogeology of the Mammoth Cave region, Kentucky, in *GSA Cincinnati '81 Field Trip Guidebook 3*, T. G. Roberts, ed., Falls Church, Va.: American Geological Institute, pp. 457–506.

Quinlan, J. F., and R. O. Ewers, 1985, Ground water flow in limestone terrains: Rationale for a reliable strategy for efficient monitoring of ground water quality in karst areas, in *Proceedings of the 5th National Symposium on Aquifer Restoration and Ground Water Monitoring*, Dublin, Ohio: National Water Well Association, pp. 197–234.

Quinlan, J. F., and R. O. Ewers, 1986*a*, Reliable monitoring in karst terrains: It can be done, but not by EPA-approved methods, *Ground Water Monitoring Rev.* **6**(1):4–6. (Reply to discussion **6**(4):42.)

Quinlan, J. F., and R. O. Ewers, 1986*b*, Ground water flow in the Mammoth Cave area, Kentucky, with emphasis on principles, contaminant dispersal, instrumentation for monitoring water quality, and other methods for study, in *Environmental Problems in Karst Terrains and Their Solutions: Field Trip A*, J. F. Quinlan, ed., Dublin, Ohio: National Water Well Association, pp. 197–234. (This work is an updated and partially revised version of that given in the Quinlan et al., 1983 reference.)

Quinlan, J. F., and R. A. Ray, 1981, Groundwater basins in the Mammoth Cave region, Kentucky, *Friends of the Karst Occasional Pub. No. 1*, Map. (Plate 1 of this book.)

Quinlan, J. F., and D. R. Rowe, 1977, Hydrology and water quality in the central Kentucky karst: Phase I, *Univ. Kentucky Water Resources Research Inst., Research Rept. No. 101*, 93p. (Reprinted by the National Park Service, with corrections, as *Uplands Field Research Laboratory Management Rept. No. 12*, 1977.)

Quinlan, J. F., M. R. McCann, W. M. Andrews, and J. A. Branstetter, 1977, Heavy metals and optical brighteners as ground water tracers in the central Kentucky Karst (abstract), *Internat. Assoc. Hydrog. Mem.* **12:**535–536.

Quinlan, J. F., R. O. Ewers, and J. W. Saunders, 1978, Distributary flow within karst aquifers: Description, occurrence, and function of groundwater dispersal conduits and later stages of their development (abstract), *Geol. Soc. America Abs. with Programs* **10:**475.

Quinlan, J. F., R. O. Ewers, J. A. Ray, R. L. Powell, and N. C. Krothe, 1983, Groundwater hydrology and geomorphology of the Mammoth Cave region, Kentucky, and of the Mitchell Plain, Indiana, in *Field Trips in Midwestern Geology*, vol. 2, R. H. Shaver and J. A. Sunderman, eds., Geological Society of America and Indiana Geological Survey, pp. 1–85.

Quinlan, J. F., R. O. Ewers, and A. N. Palmer, 1986, Hydrogeology of Turnhole Spring groundwater basin, Kentucky, in *Decade of North American Geology, Centennial Field Guide,* vol. 6, *Southeastern Section*, Boulder, Colo.: Geological Society of America, pp. 7–12.

Reid, F. S., 1984, Caveman radio, *73, Amateur Radio's Technical Jour., No. 281*, pp. 42–49.

Scanlon, B. R., 1985, Chemical, physical, and microbiological characteristics of groundwater in wells and springs of the inner Bluegrass karst region, University of Kentucky, Ph.D. dissertation, 208p.

Schmidt, V. A., 1982, Magnetostratigraphy of sediments in Mammoth Cave, Kentucky, *Science* **217:**827–829.

Shuster, E. T., and W. B. White, 1971, Seasonal fluctuations in the chemistry of limestone springs: A possible means for characterizing carbonate aquifers, *Jour. Hydrology* **14:**93–128.

Thrailkill, J., 1985, Flow in a limestone aquifer as determined from water tracing and water levels in wells, *Jour. Hydrology* **78:**123–136.

Thrailkill, J., R. F. Wiseman, and B. R. Scanlon, 1985, Investigation of pollution in a karst aquifer utilizing optical brightener, *Kentucky Water Resources Research Inst., Research Rept. No. 158*, 39p.

Wells, S. G., 1973, Geomorphology of the Sinkhole Plain in the Pennyroyal Plateau of the central Kentucky karst, University of Cincinnati, M.S. thesis, 122p.

White, E. L., and W. B. White, 1974, Analysis of spring hydrographs as a characterization tool for karst aquifers, in *Proceedings of the 4th Conference on Karst Geology and Hydrology*, H. W. Rauch and E. Werner, eds., Morgantown, W. Va.: West Virginia Geological Survey, pp. 103–106.

Williams, P. W., 1983, The role of the subcutaneous zone in karst hydrology, *Jour. Hydrology* **61**:45–67.

Woodson, F. J., 1981, Lithology and structural controls on karst landforms of the Mitchell Plain, Indiana, and Pennyroyal Plateau, Kentucky, Indiana State University, M.A. thesis, 132p.

4

WATER BUDGET AND PHYSICAL HYDROLOGY

John W. Hess and William B. White

Some information on the behavior of karst aquifers can be obtained from quantitative consideration of water inputs and outputs during median and base-flow regimes. The textbook version of the hydrologic cycle distributes precipitation as an input term into three components: evapotranspiration, surface runoff, and infiltration. Infiltration water remains in storage and then replenishes surface flow during periods of low precipitation. In a mature karst aquifer there is no surface runoff, and infiltration is more complicated. Internal runoff from sinking streams, internal runoff through sinkholes, and diffuse infiltration through the soils are partitioned between an internal drainage system that has some of the water residing in smaller fractures and joints and some in larger conduits with an unknown distribution of storage volume between the various types of "porosity."

The objective of this chapter is to use available data on river-flow and spring discharge to deduce something about the internal organization of the aquifer and in particular to assess the role of conduits in the overall water balance. Long-term records were taken from public sources, and the primary results were obtained from 1971–1973 as part of a Ph.D. research project (Hess, 1974).

CLIMATOLOGICAL DATA

Long-Term Precipitation and Temperature Data

Central Kentucky has a humid, temperate climate with a mean annual precipitation of 1264 mm and a mean annual temperature of 14.0° C. The mean number of freeze-free days ranges from 180 to 210. Storms cross the area from southwest to northeast. Precipitation is generally associated with low-pressure systems containing moisture transported from the west Gulf Coast. May through September is the thunder-shower period (Faller, 1969).

Table 4.1 Summary of Long-Term Precipitation Data for Central Kentucky (in mm)

Month	Bowling Green FAA Airport[a]	Mammoth Cave National Park[b] Minimum	Average	Maximum
January	144.8	23.1	131.6	508.8
February	104.6	2.8	115.4	262.4
March	133.6	35.6	147.3	290.8
April	103.1	28.7	112.3	174.3
May	95.8	33.8	101.8	268.0
June	105.7	16.0	120.5	230.9
July	107.4	20.8	114.4	258.6
August	91.4	20.1	102.2	236.5
September	78.0	1.3	84.8	180.3
October	60.5	—	64.8	155.4
November	91.7	28.7	105.2	333.0
December	110.2	18.8	102.2	271.3
Annual	1225.8		1302.5	

[a]Based on the period 1931–1960 (NOAA, 1972)
[b]Based on the period 1935–1968 (Faller, 1969)

Table 4.2 Summary of Long-Term Temperature Data for Central Kentucky (in °C)

Month	Bowling Green FAA Airport[a]	Mammoth Cave National Park[b] Minimum	Average	Maximum
January	3.4	−7.8	1.5	8.0
February	4.5	−3.0	3.2	8.3
March	8.6	0.5	8.2	13.5
April	14.5	10.7	14.0	17.1
May	19.6	15.3	18.4	22.4
June	24.4	19.4	22.4	26.8
July	26.2	21.2	24.2	26.9
August	25.5	20.3	23.8	26.9
September	21.9	17.8	20.4	24.0
October	15.6	10.7	14.5	18.1
November	8.3	4.5	7.6	10.2
December	3.9	−3.2	3.1	8.8
Annual	14.7		13.4	

[a]Based on the period 1931–1960 (NOAA, 1972)
[b]Based on the period 1935–1968 (Faller, 1969)

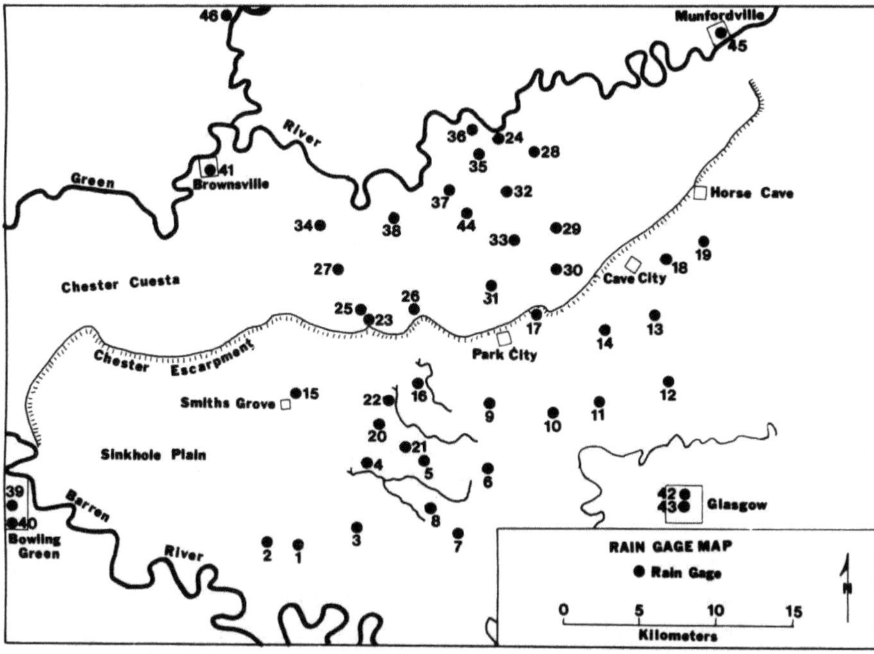

Figure 4.1 Locations of rain gages in the south-central Kentucky karst.

Long-term precipitation data for central Kentucky are available from two stations, Bowling Green FAA Airport (NOAA, 1972) and Mammoth Cave National Park (Faller, 1969; Table 4.1). The average values used for calculations of water budget for central Kentucky are averages of the data from the Bowling Green and Mammoth Cave stations. Precipitation is somewhat unevenly distributed throughout the year; the wettest months are January and March and the driest months are September and October. Table 4.2 gives the long-term temperature data. The years of record and the sources of data are the same as for the precipitation.

Precipitation Data in 1971–1973

Precipitation and rainfall distribution were monitored within the various drainage basins by a rain gage network installed for the purpose of this study. The network consisted of up to 27 installed gages for any given month for the period of operation of September, 1971 through October, 1973 along with the eight stations operated for the National Oceanic and Atmospheric Administration (NOAA, 1972–1973). The installed network consisted of small wedge-type gages that were read once a day by cooperating landowners (Fig. 4.1). The mean monthly values were calculated as arithmetic

Table 4.3 Precipitation Data for Central Kentucky from 1971–1973

Month	*1971*	*1972*	*1973*
January		209.8	64.8
February		141.9	61.0
March		134.3	108.9
April		133.5	142.8
May		81.8	182.3
June		59.0	155.8
July		175.3	158.0
August		43.6	31.1
September	51.3	99.3	68.4
October	23.5	104.5	63.7
November	42.2	159.6	
December	102.5	223.4	
Annual		1556.0	

Note: Monthly averages (in mm) are of rain gage network plus eight NOAA stations.

means based on data from 8 to 35 stations (Table 4.3). The monthly values from each individual station can be found in appendix B of Hess (1974). The deviations from normal rainfall of the monthly data are presented in Figure 4.2. The precipitation period as a whole was 5.6% higher than normal and the year 1972 was 23.9% higher.

Distribution of precipitation among the gages was fairly uniform on a yearly basis. The coefficient of variation (= standard deviation/mean) for the eight NOAA stations for the period September, 1971 through October, 1973 was 5%. Looking at individual months for the entire rain gage network, the coefficients of variation ranged from about 10% for the winter to about 40% for the summer thunder-shower period. The variability of individual storms can be quite high; for example, the storm of July 22, 1973, had a coefficient of variation of 48% as contrasted to the storm of February 13–15, 1973, which had a coefficient of variation of 21%.

Evapotranspiration

Potential evapotranspiration for the central Kentucky karst was calculated by the Thornthwaite (1948) method. The potential evapotranspiration was first derived from monthly temperature and latitude and then modified to an estimate of actual evapotranspiration through consideration of measured monthly precipitation (Table 4.1) and estimated soil moisture (Fig. 4.2). Evapotranspiration rates can also be estimated from lake evaporation, from Class A pan evaporation, and from water balance calculations for the entire basin.

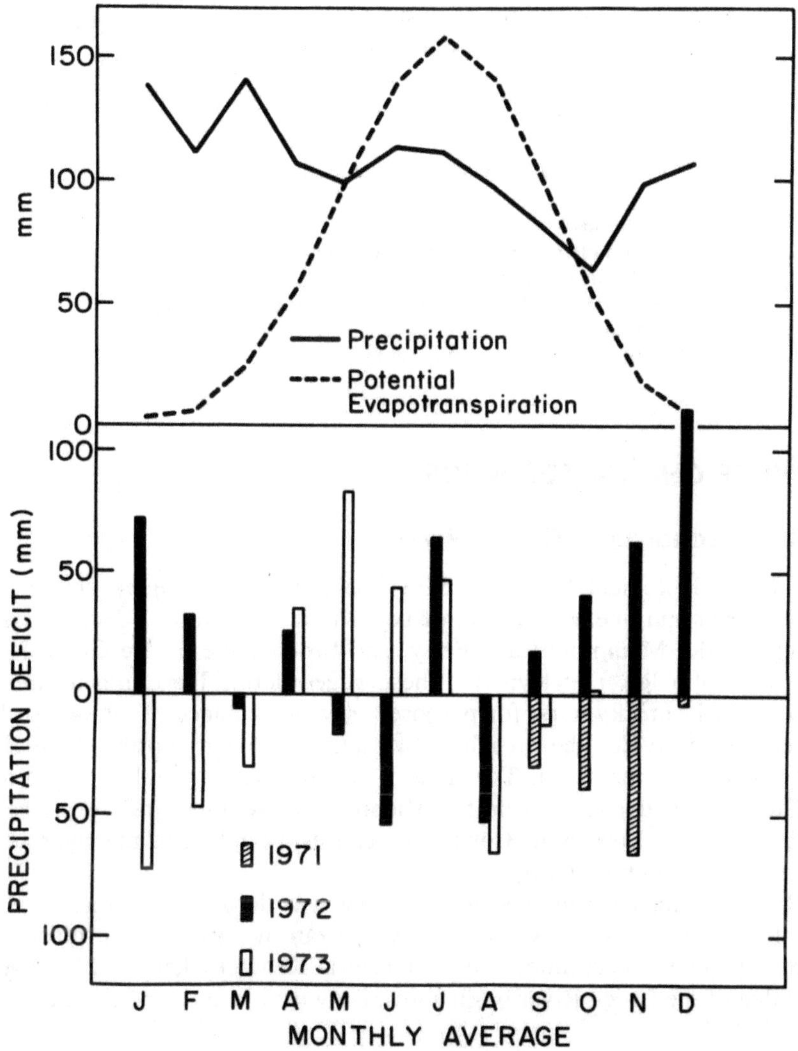

Figure 4.2 Precipitation deficit and potential evapotranspiration loss for the south-central Kentucky karst. The plotted precipitation curve is for 1972.

All of the various methods for estimating the evapotranspiration (E-T) loss for the central Kentucky karst are compared in Table 4.4 along with the actual evaporation loss from Nolin Reservoir, which was averaged for 1970–1972. The Class A pan and lake E-T losses would be expected to be higher than those estimated by other methods. The values obtained by the Thornthwaite method and from the runoff values disagree by 23%.

Table 4.4 Comparisons of Evapotranspiration (E-T) Loss for Central Kentucky

Method	*E-T Loss (mm)*
Class A evaporation pan	1010
Lake	940
Nolin Reservoir, mean 1970–1972	1050
Thornthwaite (1948) potential evapotranspiration, Bowling Green	830
Thornthwaite (1948) potential evapotranspiration, Mammoth Cave National Park	770
Precipitation minus runoff mean for central Kentucky	649

RUNOFF CHARACTERISTICS

Mean Discharge of Green River

The U.S. Geological Survey (USGS) maintains three gaging stations on Green River and one on the Nolin River within the central Kentucky karst: Munfordville, Mammoth Cave Ferry, and Brownsville on the Green River and the Nolin River at Kyrock. These gages define the drainage area of the central Kentucky karst for purposes of water balance (Table 4.5). The Green River divides the area into two parts, 246 km^2 north of the river and 744 km^2 to the south. The flow of Green River is regulated by a dam at Greensburg, upstream from the Mammoth Cave area, and by the Nolin River reservoir. Lock 6 at Brownsville causes a pooling of the river as far up river as Mammoth Cave.

Runoff from the central Kentucky karst was derived by subtracting the Munfordville and Nolin River discharges from the Brownsville discharge. The normalized mean runoff from the central Kentucky karst is 21% higher than that of the Green River basin above Brownsville as a whole. The runoff

Table 4.5 Green River Basin Discharge Data

	Area		*Mean Annual Discharge*		*Normalized Mean Runoff*		*Years*
	km^2	*mi^2*	*l/sec*	*cfs*	*l/sec/km^2*	*csm*	*of Record*
Green River basin above Brownsville	7154	2762	115,149	4056	16.1	1.47	40
Green River basin above Munfordville	4333	1673	72,640	2565	16.8	1.53	45
Nolin River basin	1831	707	23,222	820	12.7	1.16	23
Central Kentucky karst	989	382	19,286	681	19.5	1.78	—

for the Nolin River basin is 21% lower than that for the Green River as a whole. These differences are more than can be explained by differences in precipitation. The mean-annual precipitation upriver at Greensburg is 1243 mm and the runoff at that point is 503 mm as compared to 1264 mm precipitation for the karst area and 508 mm runoff at Brownsville. There is little difference in the two stations. Therefore, the runoff differences in the karst must be due to the hydrogeologic setting. The karst area must have a high infiltration rate; precipitation is rapidly conducted downward and away from the surface where evapotranspiration can take place. There is also a lack of surface water to contribute to the evaporation losses. On the other hand, the Nolin River basin is within the clastic rocks that tend to restrict groundwater infiltration and thus enhance runoff. The Green River basin as a unit is an average of various hydrogeologic settings and, as expected, exhibits an intermediate runoff value.

Base Flow

The low-flow runoff was obtained by a similar technique. The monthly average runoff at each of the three gaging stations was tabulated for the low-flow months, September and October, over the 6 years of record, 1966–1971. The average monthly flows for September and October at the Nolin and Munfordville gages were deducted from the monthly flow at Brownsville, and the residual was assigned to the low-flow runoff from the central Kentucky karst (Table 4.6). These residuals were averaged over the 12 records to yield a value of 1440 l/sec (50.9 cfs) over this period. However, Green River is wide and shallow for much of the 64 km through the central Kentucky karst, and evaporation losses are significant. A calculation using the lake formula and the effective area of the river surface yields an additional 140 l/sec (5 cfs) loss. The base-flow runoff contributed to Green River from the catchment area of the karst is therefore 1580 l/sec (55.9 cfs) or 1.64 l/sec/km^2.

Table 4.6 Green River Basin Low-Flow Summary

Station	*Drainage Area km^2*	*Low Flow l/sec*	*Normalized Low Flow l/sec/km^2*	*Normalized Mean Flow l/sec/km^2*
Nolin River at White Mills	924	2688	2.91	13.1
Green River between Munfordville and Greensburg	2427	9799	4.04	17.4
Central Kentucky karst	989	1442	1.46	19.5
Central Kentucky karst with correction for evaporation losses	989	1583	1.64	

The differences in low flow can be explained by the differences in the storage and yield of the groundwater aquifers. The karst has low storage and drains rapidly, which explains the low base-flow runoff. The value of 1.64 l/sec/km^2 compares favorably with a value of 1.12 l/sec/km^2 for the southern Indiana karst (Palmer, 1969) and with White's (1977) measurements on base flow in other karstic drainage basins.

Base-Level Backflooding

Spot observations of the springs revealed intervals of reversed flow. The estimated peak reversed flows were 150, 900, and 150 l/sec for Pike, Styx,

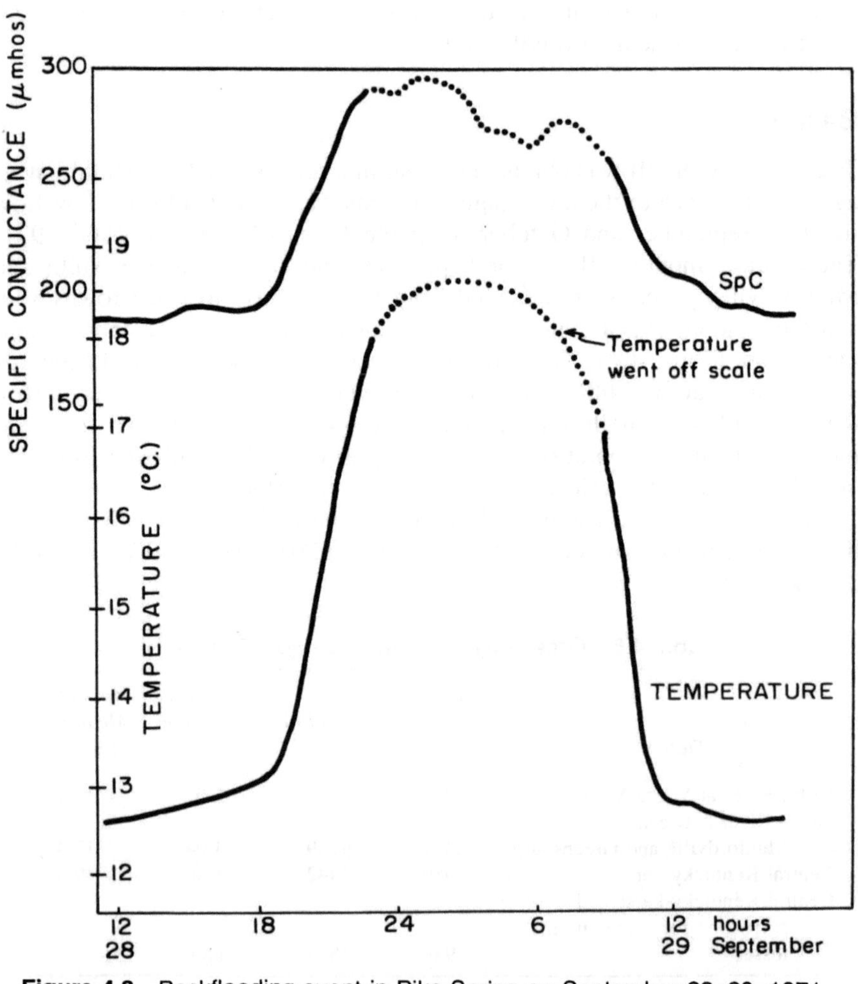

Figure 4.3 Backflooding event in Pike Spring on September 28–29, 1971.

and Echo springs, respectively. Styx Spring in general had reversed flow for the period of mid-December, 1972 through the end of March, 1973, during the winter when the Green River stage was high.

Base-level backflooding provides a fifth source of recharge to the carbonate aquifer in addition to the four catchments discussed in Chapter 2. How far flow reversal reaches into the aquifer is open to question. Owl Cave, 1.7 km back from Turnhole Spring, was never observed to have a reversal of flow. Eyeless Fish Trail in the Flint Ridge section of the Mammoth Cave system, which extends some 500 m back from Pike Spring was observed to be backflooding by Palmer (personal communication, 1973). The water levels rose 0.6 m in 5 hours with a reversed flow velocity of 0.6 m/sec. When Palmer first observed the stream earlier in the day, it was flowing out swiftly. Therefore, backflooding extends somewhere between 0.5 km and 1.7 km back from Green River during annual stages of river rise. More extreme floods should have proportionally larger effects.

The backflooding phenomenon can be observed visually. One visit to Pike Spring showed murky green water flowing out of the spring at a velocity of a few cm/sec, forming a green streamer in the rising muddy water of Green River. It was autumn, and colored leaves floating on the water made good flow markers. As we watched, the outward current slowed, stopped, and then reversed. First, the leaves floating on the green water of the outlet channel were drawn into the spring, followed by tongues of muddy water from the river. The entire event took about 15 minutes, at the end of which the spring orifice was entirely filled with muddy water flowing inward at a velocity of a few cm/sec.

The instrument package, which was installed in Pike Spring during the period 1971–1973, also caught flow-reversal events (Fig. 4.3). The instrument package was installed some 10 m inside the bedrock conduit feeding the rise pool at a depth of 5 m below river level.

WATER BALANCE

Discharge–Evapotranspiration Balance

By making use of the effective drainage area between the Green River gage at Brownsville and the gage at Munfordville, it is possible to use long-term USGS records to determine water budget. The water balance was carried out for the entire area to avoid the difficulty of proportioning the parts north and south of the river. Long-term precipitation data were obtained from 34 years of records taken at Mammoth Cave National Park and 9 years of records taken at Bowling Green. The mean average precipitation is 1264 mm (49.77 in.; Table 4.1). The Thornthwaite (1948) estimate of evapotranspiration loss (Table 4.4) was taken as most representative of the karst area.

Table 4.7 Runoff and Evapotranspiration (E-T) Loss

	Runoff		*E-T Loss*	
Basin	*in.*	*mm*	*in.*	*mm*
Green River at Brownsville	19.99	508	29.76	756
Green River at Munfordville	20.82	529	28.94	735
Nolin River basin	15.75	400	34.02	864
Central Kentucky karst	24.20	615	25.55	649

The total annual runoff was obtained from USGS records at each of the gaging stations and averaged over the 40 years of record (Table 4.7). Evapotranspiration loss can then be obtained by balancing the runoff against the precipitation, and these numbers are also tabulated. The evapotranspiration loss for the entire Green River basin, calculated at both gaging stations, agrees with the estimates from the Thornthwaite potential. E-T losses from the Nolin River basin are higher and approach those of a lake estimate, presumably because of the influence of the Nolin Reservoir. Of greatest interest is the observation that the E-T loss from the karst is lower by some 13% than the basin average. It can be argued that rapid infiltration of water into sinkholes is responsible. Normalized runoff is correspondingly higher, although it is, of course, an underground runoff.

Conduit-Flow and Diffuse-Flow Balance

It is now possible to compare the discharges of the springs with runoff from the entire area to evaluate the contribution of the big conduit springs to the total water budget. Of 81 springs along Green River (Hess et al., 1974), 37 had a finite flow during the summer months of 1971–1972. The total estimated flow of the 37 springs was 2750 l/sec (97 cfs), although this number is somewhat uncertain because the springs were observed at different times. The nine largest springs account for approximately 95% of the spring discharge. The two largest springs draining from the south, Gorin Mill Spring and Turnhole Spring, together account for approximately 56% of the total spring discharge and an even larger fraction of the drainage from the area south of Green River.

The six large springs that enter Green River from the south were measured at various times during the summer of 1973 in a search for low-flow conditions, a difficult task in an exceptionally wet year. The set of measurements finally accepted was obtained in September, already established as one of the low-flow months in this region (Table 4.8). The total discharge from the group of six springs during low flow was 950 l/sec (33.6 cfs) compared with 1390 l/sec for this same group of springs during the complete spring

Table 4.8 Base-Flow and Catchment Areas for Major Springs

Spring	*Base Flow* l/sec	*Base Flow* cfs	*Calc. Area* km^2	*Calc. Area* mi^2	*Measured Area*[a] km^2	*Measured Area*[a] mi^2	*Percent*
Gorin Mill Spring	731	25.8	445	172	473	182	94%
Garvin Spring	42.5	1.5	26	10	28	10.8	93%
Lawler Blue Hole	59.5	2.1	36	14	37	14.3	97%
Pike Spring	68.0	2.4	41	16	58	22.4	71%
Echo River Spring	51.5	1.8	31	12	33	12.7	94%
Turnhole Spring[b]	396	14	241	93	340	131	71%

[a]From groundwater drainage basins outlined by Quinlan and Ray (1981).
[b]Discharge estimated at stream in Owl Cave, now known to be only part of the flow because the underground stream is diverted around both sides of the collapse doline known as Cedar Sink.

inventory (see Chap. 2). If the total discharge of all springs is adjusted by the ratio 950/1390 = 0.68, an estimated low-flow discharge of 1870 l/sec (66 cfs) is obtained equivalent to 1.86 l/sec/km^2 (0.17 csm). Considering the rough nature of the spot measurements, this number is in good agreement with the 1.64 l/sec/km^2 (0.15 csm) obtained from a longer term averaging of river records.

If we accept the values in Table 4.8 as true base-flow discharges, and use the base-flow runoff calculated from the river records, an estimate can be made of the catchment areas that feed the big springs. The sum of the drainage areas is calculated as 820 km^2 (316 mi^2), which is 110% of the measured catchment of 744 km^2 (287 mi^2) south of Green River. There is an additional error of about 5% that arises from having ignored the contributions of the small springs. The most obvious source of this overestimate of drainage area is that the flows listed in Table 4.8 are not true base flows. Because records gathered over a long period of time are not available on individual springs, this point cannot be checked.

The main conclusion of this set of measurements and calculations, however, seems unequivocal. To within experimental error, all runoff from the karst is discharged through the springs, and most of that is discharged through a few large springs. The two regional springs actually account for more than 80% of the discharge. Therefore, it can be concluded that the bulk of the groundwater in this mature karst is being delivered to the conduit system. The diffuse component reaching the river, which one can argue must be present, is lost in the experimental uncertainties. Diffuse flow within the aquifer is toward the conduits that act as low-gradient drains as Ewers and Quinlan (1981) have postulated.

STORM RESPONSE

Because the response time for chemical or thermal equilibrium between groundwater and the limestone wall rock is comparable to the transit time of injected groundwater through the open conduit system, individual storm events can be used to probe the internal arrangement of the drains from infeeders and the master trunk drains leading to the springs. The following section is based on a study by the authors on the Turnhole Spring drainage system (Hess, 1974; Hess and White, 1988). The experiment was to record electrical conductance (a measure of total dissolved solids) and temperature so that a continuous record would be available through various storm events.

Turnhole Spring is a regional drain whose catchment was shown in Figure 3.1. Two of the larger surface streams, Gardner Creek and Little Sinking Creek, discharge through the Turnhole drainage line. The main underground conduit also can be sampled at Mill Hole. Water-balance data indicate that runoff from the entire Sinkhole Plain segment of the Turnhole catchment appears in Mill Hole. A portion of the regional drain is again observed as a stream that crosses the bottom of Owl Cave, a short fragment of trunk passage in the north wall of Cedar Sink (see Fig. 3.4). The regional drain empties into Green River at Turnhole Spring, a large rise pool in the riverbank. Because diving operations in Turnhole Spring revealed no well-defined conduit opening, the instrumentation was installed at Owl Cave, some 1.7 km to the south.

Instrumentation and Data Collection

Rainfall distribution and storm magnitude were monitored by a rain gage network (Fig. 4.1). Monthly totals are summarized in Table 4.3; daily precipitation is indicated by bar graphs on the time series records.

The submerged conduit leaving Owl Cave on its way to Turnhole Spring was monitored with an automatic recording device, which was attached to sensors placed in the conduit. Temperature was measured with a thermistor, and specific conductance (Spc) was measured with a standard Beckmann electrode. A complete wiring diagram with the electrical characteristics of the system is given in Hess (1974).

The monthly charts were first examined by recording the chart values for the specific conductance and temperature at noon each day and noting when a pulse had occurred. Data for the pulses were recorded at whatever time interval was necessary to define the shape and detail of the specific conductance and temperature hydrographs. On the plots of these data (in Figs. 4.4 and 4.5), the solid lines indicate continuous data, the dashed lines indicate either no data or that the temperature and/or specific conductance were based on spot measurements. The spot points are indicated by x's.

Temperature data reduction was a straightforward chart reading followed by scale conversion to ° C. The error in the temperature measurements was 0.2° C. Specific conductance was first read from the charts and then converted to a raw Spc with a calibration chart. The raw Spc was then normalized to 25° C by a second conversion chart. The accuracy was considered to be within 5%.

Aquifer Response to Storm Inputs

Figures 4.4 and 4.5 show some of the more interesting records. Intense storms appear in the downstream conduit drainage as depressions in specific conductance, peaks in the water temperature, and peaks in the stage record at Green River. In characteristic hydrograph form, the depression in conductance is abrupt followed by a much slower recovery that spans several weeks. The most useful storm events, therefore, are those in which an abrupt, intense storm is followed by several weeks of dry weather so that the entire recovery curve can be observed without interruption. Such events are quite rare and throughout the entire 1.5-year period of observation, only eight storms occurred that produced useful data. These are labeled events A–H in the records (the records containing events D and E are not shown).

A high-resolution plot of events A and H (Fig. 4.6) reveals a considerable amount of detail including a surprising amount of fine structure on both specific conductance and temperature curves. If the water levels in the aquifer have stabilized, the specific conductance remains at a constant value for a time past the input pulse (which defines time zero). This lag time is a measure of the length of time required for the first undersaturated water to reach the recording station. Once the fresh runoff reaches the recorder, a series of complicated dips and peaks appear in the record. These are interpreted as arrival times of specific conduit inputs upstream along the trunk. The conductance continues to decrease until a broad minimum is reached, which defines a response time. The minimum corresponds to the maximum dilution of the groundwater by fresh surface water input, but does not necessarily correspond to the maximum in the discharge when the bulk of the Sinkhole Plain runoff has arrived at the recording station. A long period of slow recovery follows, during which the conductivity gradually rises to its pre-storm value and the discharge gradually falls. We can define, rather imprecisely, a recovery time to pre-storm conditions.

It has been shown by Chambers (1973) that the discharge peak leads the hardness minimum (Spc and hardness are directly related) in all spring types from conduit flow to diffuse flow. However, the hardness minimum should be a more accurate estimate of the arrival of the fresh water. The discharge will increase because of water driven out of storage by the increased hydraulic head in the catchments before the actual storm water appears at the spring.

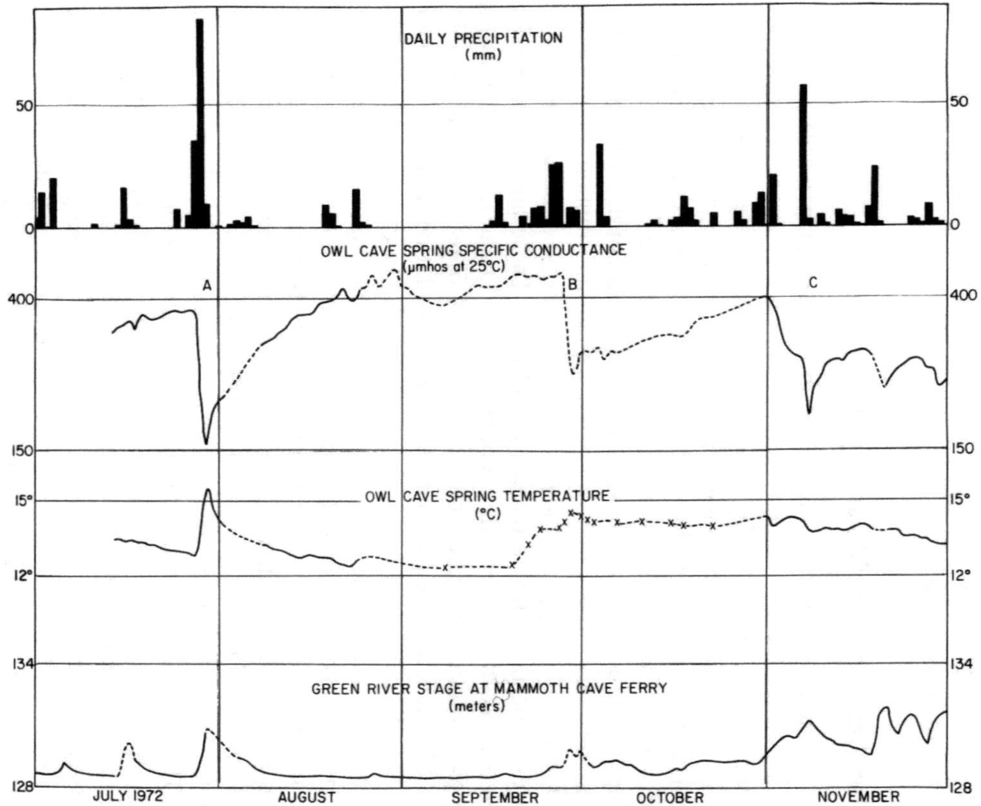

Figure 4.4 Annual continuous records for Owl Cave (Cedar Sink), July–November, 1972.

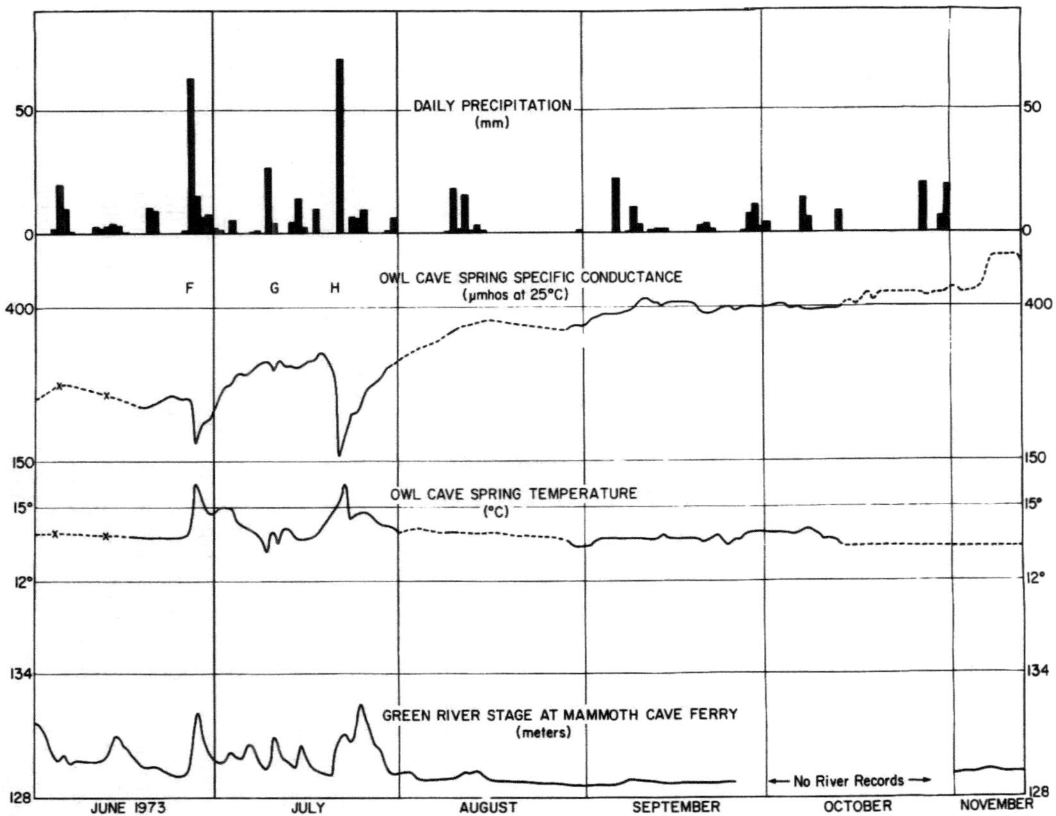

Figure 4.5 Annual continuous records for Owl Cave (Cedar Sink), June–November, 1973.

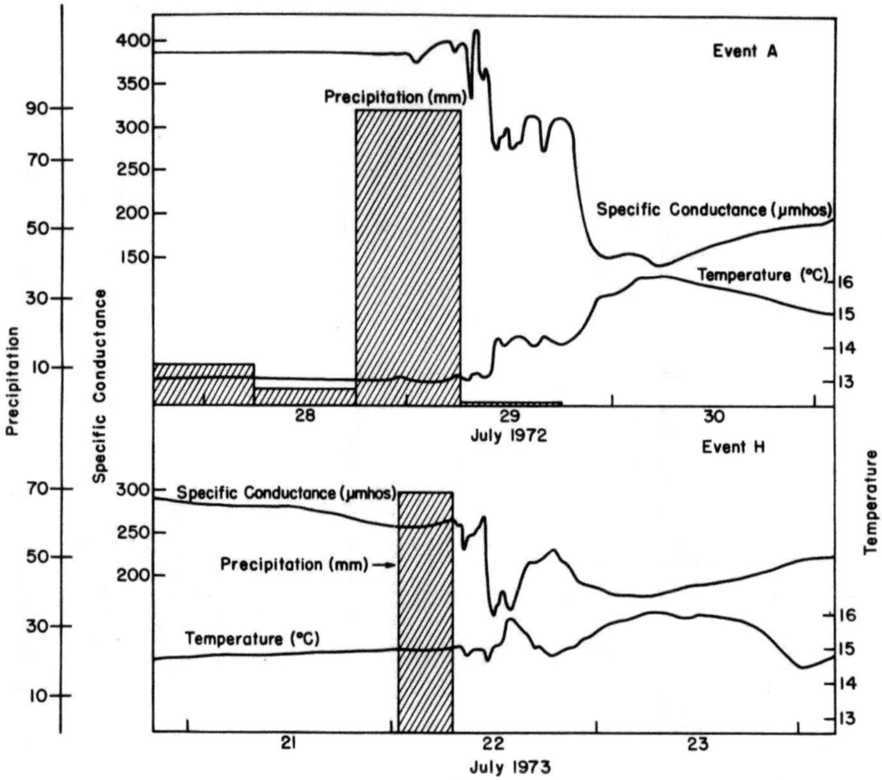

Figure 4.6 Detailed record for events A and H showing precipitation as bar graphs (with a resolution of only 1 day) and continuous recordings of specific conductance and temperature.

A similar relationship is seen in the temperature hydrograph that appears to be in phase with the Spc hydrograph (Figs. 4.4 and 4.5). The temperature changes can be either positive or negative, depending on the season. There appears to be some loss of small detail in the temperature record as compared to the Spc hydrograph, which can be explained by the relative changes of magnitude of the two measurements.

The Owl Cave Spc hydrographs (Fig. 4.6) can be broken down into three parts. There are two early stages of fine structure, which are thought to represent inputs from the karst valleys on the plateau and from well-defined trunk-conduit tributaries farther upstream along the drainage line. The third smooth part of the curve represents water from the sinking streams. Even if the fine structure is lost, the two early events can be distinguished. The shape, duration, and intensity of the precipitation pulse, as well as antecedent aquifer and soil moisture conditions, affect the detail and shape of the Spc and temperature hydrographs and explain why the curves do not reproduce exactly from storm to storm. The best detail in the fine structure of the hydrographs is achieved during short, intense precipitation events onto a land surface at field capacity so that there is rapid runoff and recharge to the system.

These discrete sources of water must be points of concentrated recharge to the aquifer, such as streams flowing off the edge of the clastic caprock and sinking streams in the karst valleys. It is interesting to see if a reasonable location for these input points can be deduced. The straight line distance from Little Sinking Creek to Mill Hole and then to Cedar Sink is 11 km. The time for water to arrive at Owl Cave as represented by the broad minimum in the Spc pulse is an average of 35 hours. This yields an average flow velocity of 0.3 km/hr (0.01 m/sec), which is in the range of high-flow transit times calculated from tracer experiments (see Turnhole Spring Groundwater Basin in Chap. 3).

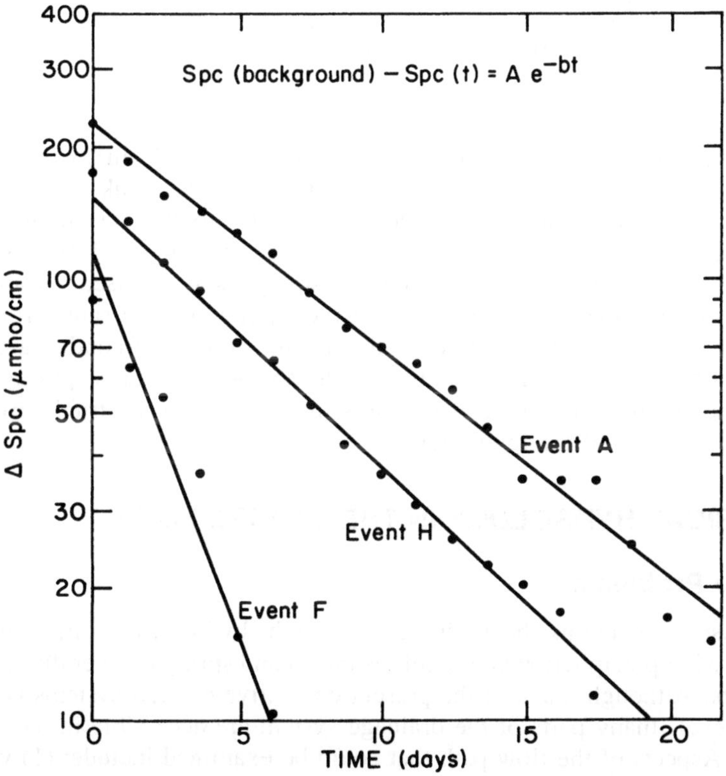

Figure 4.7 Exponential fitting of specific-conductance recovery curve. The response times (1/b in units of days) are: event A, 8.3; event H, 7.0; and event F, 2.5.

Aquifer Relaxation and Storage

The recovery time of the Spc and temperature hydrographs is on the order of 2–3 weeks for the Turnhole drainage. This is an indication of a very open system with conduit flow and very little long-term storage above the elevation determined by the stage of Green River. Figure 4.7 was obtained by subtracting the instantaneous Spc from the background Spc and plotting the logarithm of these values versus time for three of the storm events. It demonstrates that the shape of the recovery response is the same for different events, but that each has a characteristic response time.

An estimate of the recharge to the aquifer during a storm can be made by correcting the precipitation value for evapotranspiration and interception losses. This estimate has been done for two storms:

	Event A *July 28–30, 1972*	*Event H* *July 22, 1973*
Precipitation	113.9 mm	70.5 mm
Evapotranspiration	−5.4	−5.4
Interception	−17.4	−9.2
Recharge	91.1 mm	55.9 mm

The drainage area for Cedar Sink is approximately 336 km^2, which allows a small correction for the catchment between Cedar Sink and Turnhole Spring. The total volume of water recharged into the system was 30.6 million and 18.8 million m^3, respectively, for the 1972 and 1973 storms. This is the volume held in temporary storage and released through the springs during the relaxation period of the aquifer. By the same argument, it must also represent a minimum value for the volume of conduit available to hold the water. Such conduits include not only the large passages but also the small tributaries and fractures that drain into the trunks over the recovery period of the hydrograph.

PHYSICAL HYDROLOGY IN THE VADOSE ZONE

Some Problems

The discussion in this chapter has dealt with the hydrology of one entire sub-basin using precipitation and runoff as inputs and spring or river discharge as outputs. Although much of the present-day active conduit systems is below base level, many parts of the drainage system are accessible in the vadose zone. Aspects of the flow path that could be examined include: (1) volume and time of retention of diffuse infiltration in the soil and weathered-rock zone near the surface, which was Williams' (1983, 1985) "subcutaneous zone"; (2) hydrologic characteristics of vertical shafts and other open fast-travel time routes through the vadose zone; (3) travel times of diffuse infiltration and an overall water balance between vadose seepage and flow in the open shafts and solution crevices; and (4) hydrologic characteristics of vadose streams—the shaft and valley drains under the plateau and the free-surface cave streams on the Sinkhole Plain. Base-level cave streams tend to flood with appalling quickness as the explorers of such low-level conduits as Whigpistle Cave and the Hidden River (Hicks Spring) complex can testify. Does this imply that storm waters are simply filling the groundwater troughs (shown on Plate 1) rather than causing a region-wide rise in the water table?

The measurements required to address these problems cannot be done with casual observation. Recently, however, Quinlan and Ewers (Ewers et al., 1986) began a systematic program of instrumentation and continuous recording of the active underground drainage system.

Flow in Vertical Shafts

This section which follows from Brucker et al. (1972) is about the active vertical shafts. These shafts contain films of water that slide down the walls, adhere to the wall by surface tension, but which, in the aggregate, are of a sufficient flow volume to account for the small free-surface streams in the shaft drains. These water films, responsible for the active dissolution of the walls of shafts, are quite thin. Rough measurements of depths range from 0.1 mm to 1 mm. Measurements of flow velocity by timing dye streamers injected into the water films range from 0.3 m/sec to 3 m/sec.

The standard engineering formula for the velocity of a free-sliding fluid film in a vertical pipe is

$$\bar{v} = \frac{\rho g \delta^2}{3\eta}$$

where ρ = fluid density, η = fluid viscosity, g = acceleration due to gravity, and δ = film thickness, all in dimensionally consistent units. The data on film thickness and flow velocity, which were collected from several sets of measurements in eight shafts in the Mammoth Cave system and in northern Alabama generally fit this equation. When the flow volume and thus the film thickness increase beyond some critical value, the water breaks free at the top of the shaft and there is a spray or waterfall. Waterfalls are rare in the Mammoth Cave area; most active shafts drain only through the films on the walls except during periods of high precipitation.

Most shafts have very large diameters compared to the thickness of the water film. If shafts are regarded as wide channels standing on end, the channel width is the circumference of the shaft and the flow depth is the thickness of the water film. In a wide channel the hydraulic radius and the hydraulic depth are directly related, and values can be calculated for both Reynolds and Froude numbers. These numbers, in turn, allow the calculation of a fence diagram (Fig. 4.8), which is divided into four regions by the transitions from subcritical to supercritical channel flow and by

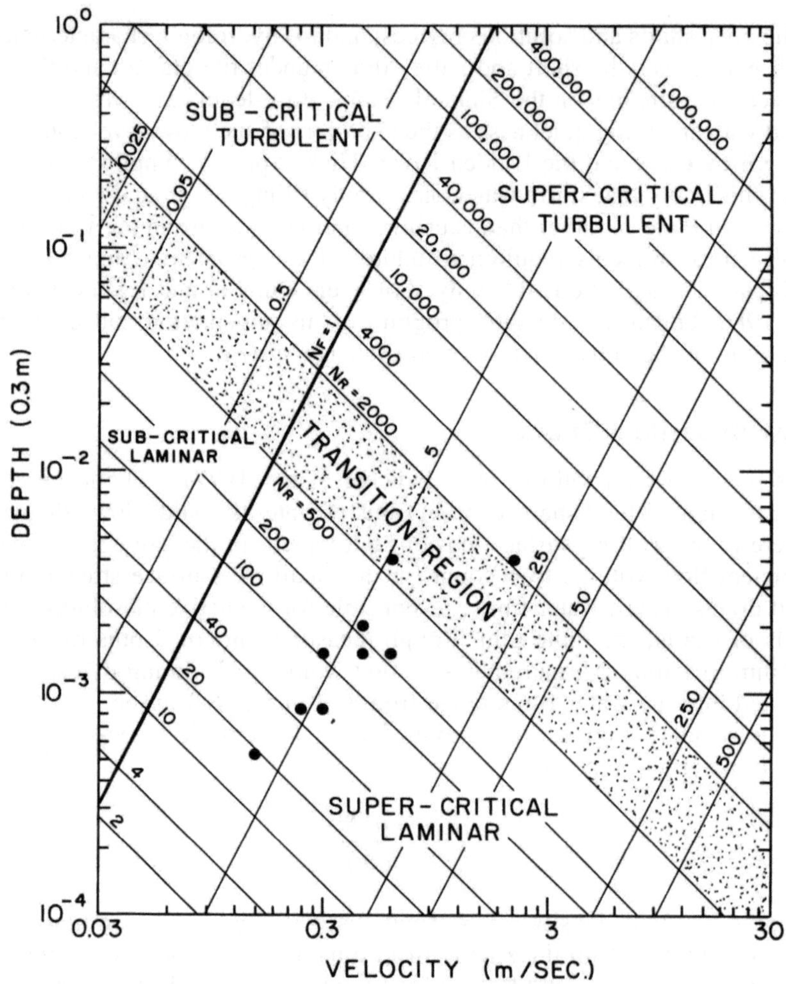

Figure 4.8 Fence diagram calculated with the assumption of a wide channel. The grid lines are contoured on Reynolds number N_R and Froude number N_F. Solid points show measured depth and velocity for various water films moving down vertical shafts. The metric scale has been added to a diagram originally calculated in units of ft (depth) and ft/sec (velocity).

transition from laminar to turbulent flow. The plotted data show that most water films in shafts fall into the unusual supercritical laminar regime. Probing the water films with small disturbances confirms that they are in a regime of supercritical flow.

The flow regime does much for accounting for the vertical walls characteristic of vertical shafts. Projections of any sort would break up the thin high-velocity sheet flow, increasing the effective channel depth, and causing a transition to subcritical flow. The hydraulic jump at the transition with its large energy release would increase the corrosive power at that point with the result that projections are preferentially dissolved away, leaving vertical walls.

CONCLUSIONS

Existing river-flow and precipitation records, combined with additional precipitation data from a rain gage network and additional observations on spring discharges and spring specific conductance and temperature, allow a few general conclusions to be drawn about the hydraulic response of the shallow karstic-carbonate aquifer in south-central Kentucky.

Evapotranspiration is lower, and runoff is higher for the karst area than for other parts of the Green River basin by 10–15%. This effect results from rapid movement of water underground through sinkholes and sinking streams with consequent reduced opportunity for evaporative and transpirational losses. The runoff is an underground runoff.

Normalized base flow from the karst springs is almost an order of magnitude lower than for other drainage basins. The low-gradient open conduits drain quickly and retain less water in storage.

To within the error of measurement, all groundwater discharge from the karstic conduit aquifer can be accounted for by the large springs. Diffuse flow within the aquifer must be directed toward the conduits that provide a groundwater trough rather than leading to the river.

High-resolution plots of specific conductance and temperature for the Turnhole Spring drainage system, in response to high intensity storms, shows that the overall response time of the aquifer to a storm event is about 2 weeks. The existence of well-defined tributaries on the drainage line is revealed by sharp peaks and dips on the conductivity curve. Data at present are insufficient to translate these into specific details of conduit geometry.

The physical hydrology of the active underground drainage system is poorly known at the present time, and details of water motion along various parts of the flow path would be an interesting area for research.

ACKNOWLEDGMENTS

This work was supported by the U.S. Office of Water Resources Technology through the Environmental Resources Research Institute of The Pennsylvania State University. Thanks are extended to Richard Rigg and David Jagnow for their assistance in the scuba diving operations needed to install the instrumentation. The Cave Research Foundation allowed use of their field station on Flint Ridge and gave much additional logistical support. We thank the superintendent and staff of Mammoth Cave National Park for their cooperation and for permission to carry out this investigation within the park.

REFERENCES

Brucker, R. W., J. W. Hess, and W. B. White, 1972, Role of vertical shafts in the movement of ground water in carbonate aquifers, *Ground Water* **10**(6):5–13.

Chambers, W. J., 1973, Limestone springs and individual flood events. *Cave Research Group Great Britain Trans.* **15:**91–97.

Ewers, R. O., and J. F. Quinlan, 1981, Cavern porosity development in limestone: A low dip model from Mammoth Cave, Kentucky, *Proc. 8th Internat. Congress Speleol. (Bowling Green, Ky.)* **2:**727–731.

Ewers, R. O., J. J. Meiman, and J. F. Quinlan, 1986, Karst aquifer instrumentation system in the Mammoth Cave region, in *Environmental Problems in Karst Terranes and Their Solutions: Field Trip A*. J. F. Quinlan, and R. O. Ewers, eds., Dublin, Ohio: National Water Well Association, pp. 90–93.

Faller, A., 1969, An introduction to the climate of Mammoth Cave National Park, Mammoth Cave National Park, unpublished report, 8p.

Hess, J. W., Jr., 1974, Hydrochemical investigations of the central Kentucky karst aquifer system, The Pennsylvania State University, Ph.D. dissertation, 218p.

Hess, J. W., and W. B. White, 1988, Storm response of the karstic carbonate aquifer of south-central Kentucky, *Jour. Hydrology* **99:**235–252.

Hess, J. W., S. G. Wells, and T. A. Brucker, 1974, A survey of springs along the Green and Barren Rivers, central Kentucky karst, *Natl. Speleol. Soc. Bull.* **36**(3):1–7.

NOAA climatological data (1970–1973) for Kentucky, Washington, D.C.: National Oceanic and Atmospheric Administration, pp. 65–67.

Palmer, A. N., 1969, A hydrologic study of the Indiana karst, Indiana University, Ph.D. dissertation, 181p.

Quinlan, J. F., and R. A. Ray, 1981, Groundwater basins in the Mammoth Cave region, Kentucky, *Friends of the Karst Occasional Pub. No. 1.*, Map.

Thornthwaite, C. W., 1948, An approach toward a rational classification of climate, *Geog. Rev.* **38:**55–94.

White, E. L., 1977, Sustained flow in small Appalachian watersheds underlain by carbonate rocks, *Jour. Hydrology* **32:**71–86.

Williams, P. W., 1983, The role of the subcutaneous zone in karst hydrology, *Jour Hydrology* **61:**45–67.

Williams, P. W., 1985, Subcutaneous hydrology and the development of doline and cockpit karst, *Zeitschr. Geomorphologie* **29:**463–482.

5

FLOOD HYDROLOGY

Elizabeth L. White

Many cave conduits, which are below the base level and subject to solution, are within the Flint–Mammoth Cave system. Above the base level of both surface and groundwater systems is a floodwater zone. The floodwater zone is defined as the elevation between low stage of the regional water table and the highest flood level per year. The role of the seasonal rise and fall of base level allows the cave passages to drain, breakdown to fall, and sediment to be transported (both erosion and deposition). The passages subject to flooding are still enlarging with the rise and fall of base level.

It is of interest to know the recurrence intervals for floods at a particular elevation within the cave passages as extrapolated from the height of the surface stream at Green River. Another point of interest is to identify the distribution function of the annual maximum floods to show the importance of the extreme catastrophic floods as well as the repeated short-return period flooding on the formation of cave passages. The flood behavior of the Green River basin in relation to the karst drainage system is discussed in this chapter.

A SHORT TUTORIAL ON EXTREME VALUE STATISTICS

Hydrologic phenomena, such as floods, are uncertain events in nature; therefore, one must use statistical methods such as probability theory or frequency analysis to extract meaningful generalizations from a mass of observational data. Flood peaks can be predicted by flood-frequency analyses if stream gage data are available. Empirical or semi-empirical methods can be used for ungaged watersheds. A special model for the Green River basin was proposed by Katz and Toebes (1980). Empirical methods have been proposed by the U.S. Geological Survey, the Federal Highway Administration, the Pennsylvania Department of Environmental Resources, and the Pennsylvania Department of Highways. These empirical methods utilize regression equations for estimating flood peaks for watershed outlet stations. Runoff modeling methods by the Hydraulic Engineering Center (in Davis, California) and the Soil Conservation Service or other models such as PSU–IV (Aron et al., 1981), used by the Pennsylvania Department of Highways; SWMM (Huber et al., 1987), a storm water management model; and the Penn State Runoff Model (PSRM) (Aron, 1987), all simulate physical runoff processes but utilize empirical coefficients and thus might be termed semi-empirical.

Within flood-frequency analyses one must determine the frequency with which a particular flow will be exceeded using the annual maximum floods as the input data. Underlying any flood-frequency analysis are random or stochastic processes. Stochastic processes are made up of two basic types: one is descriptive, the other is inferential. The descriptive portion is a direct application of statistical methods; the inferential portion requires an understanding of the statistical methods as well as the dangers (risk) involved in predicting and estimating. In order for these stochastic processes to better predict future flood events, the longest record data set should be used. A large data set of flood events would ensure the stochastic processes to be a better predictor of the peaks, would show a much smoother distribution, and also, more importantly, would define the shape of the distribution curve more precisely.

Possible flood-frequency distributions used to define hydrologic data include the normal (Gaussian), log normal, extremal value (Gumbel, 1941), and gamma (Log Pearson Type III) distributions. If flood events followed a normal distribution, then approximately 68% of the floods would fall within plus or minus one standard deviation of the mean annual flood (the arithmetic mean of the series of maximum floods per year). However, a normal distribution has zero skew and no provision for upper and lower boundaries of the sampled event magnitudes. Because the distribution of maximum annual floods are definitely bounded by values well above zero and are also usually positively skewed with a long tail for the frequency distribution, flood-frequency analyses must be analyzed differently. The annual maximum flood-frequency distributions are reasonably represented by a log normal, extremal, or gamma distribution, which are extremely flexible and all three are positively skewed. When the gamma distribution uses the transform $y = \log x$ to reduce the skewness, then the distribution is called Log Pearson III. The Gumbel (1941) distribution is essentially a log normal distribution with a constant skewness. This distribution has been used with success to describe the populations of many hydrologic events. A full discussion of the fundamental theorem can be found in standard textbooks such as Viessman et al. (1977), Linsley et al. (1958), or Linsley and Franzini (1979).

Recurrence Interval (Return Period) and Recurrence Probability

The recurrence interval, T_r, or return period are terms given to the reciprocal of the probability of a flood of some specified magnitude. In an annual series, recurrence interval is the average number of years in which an event of a given size or greater recurs as an annual maximum. For example, a flood that appears to have been exceeded on the average once in 20 years has the probability of recurring in any year of 1/20 or 0.05. In other words, an annual peak flood that is exceeded on the average of 1 year in 20 years has a percent probability of 5% (0.05×100) of being exceeded in any year. However, it does not mean that every 20 years a flood of that magnitude will occur.

In the analysis of an annual series (a set of annual maximum floods per year) one would intuitively assume the recurrence interval to be number of years of record/flood rank (n/m). For example, when $n = 100$ years of record, the maximum recorded flood has a rank, $m = 1$, and a recurrence interval, $T_r = 100/1 = 100$ years; the tenth largest flood ($m = 10$) would have $T_r = 10$ years and the lowest flood of record would have a $T_r = 1$ year. In many hydrologic investigations, other of the more frequently used plotting-position formulas, as shown in Table 5.1, compare the number of recurrence intervals with their corresponding probability of occurrence ($1/T_r$).

Using the example in Table 5.1 with $m = 1$ and $n = 10$, note that the Hazen plotting position assumes the probability of occurrence to occur exactly in the center (e.g., probability = 0.05 between 0.00 and 0.10) within each segment of a series, thus generally accounting for the best estimation. The probabilities of occurrence are stacked at the nodal points on a probability line centered at about 0.65 for the Blom, Tukey, and Chegadayev methods and centered between 0.09 to 0.10 for the Weibull and California methods.

Mean Annual Flood

Consider the annual maximum discharges of any river at a particular point on the river. These data can be displayed graphically as a histogram using

Table 5.1 Probability of Occurrence or Exceedence and Return Period Formulas in Common Use

Method	*Plotting position formulas*	*Probability* $m = 1$	*Return period* $n = 10$	*Probability* $m = 10$	*Return period* $n = 10$	*References*
Hazen	$\frac{2n}{(2m-1)}$	0.05	20.0	0.95	1.059	Hazen, 1930
Gringorten	$\frac{(n+0.12)}{(m-0.44)}$	0.055	18.1	0.94	1.059	Gringorten, 1963
Blom	$\frac{(n+1/4)}{(m-3/4)}$	0.061	16.4	0.94	1.065	Blom, 1958
Tukey	$\frac{(3n+1)}{(3m-1)}$	0.065	15.5	0.935	1.069	Tukey, 1962
Chegȧdayev	$\frac{(n+0.4)}{(m-0.3)}$	0.067	14.9	0.93	1.072	Alekeyev, 1955; Leivikov, 1955; Benard and Bos-Levenbach, 1953
Weibull	$\frac{(n+1)}{m}$	0.091	11.0	0.909	1.1	Weibull, 1939*a*, 1939*b*
California	m^n	0.10	10.0	1.00	1.00	California, 1923

Note: n = total number of items; m = order number of items arranged in descending magnitude; therefore, $m = 1$ for the largest item.

discharge intervals or they can be displayed graphically as discharges in cumulative terms. The data used can be considered a sample of all the annual maximum flows that have occurred or will occur at that location. If a much longer record were available, smoother curves could be drawn. If the exact form of the curve were known, the curve could be discussed instead of merely the mass of numbers collected from the records.

Curves in statistics are usually established by the mean, standard deviation, and skew. The mean locates the centroid of the area under the curve. The average of annual maximum discharges is called the mean annual flood. Using extreme value statistics, the mean annual flood corresponds to a flood with a return period of approximately 2.33 years, which approximates bank-full stage.

FLOODS IN CARBONATE DRAINAGE BASINS

Flood Damping

Flood flow is strongly influenced by the presence of conduit-type underground drainage systems. In a previous study of Pennsylvania drainage basins underlain by noncarbonate rocks, the mean annual flood per unit area of catchment (usually given in $ft^3/sec/mi^2$, csm) ranged from 60–122 csm, whereas those basins underlain by carbonate rocks ranged from 9–20 csm (White, 1969; White and Reich, 1970).

Watersheds underlain by carbonate rocks are among the most peculiar. Rainfall that infiltrates into the absorbing soils is temporarily stored at the corroded bedrock surface that Williams (1983, 1985) called the subcutaneous zone. Because the open-conduit system has a storage potential for a higher proportion of the precipitation from a storm, initial flood pulses from sinking streams and sinkhole internal runoff are absorbed into the conduit system, thereby reducing the total volume of flow at any position of the surface stream (flood damping). The flood damping is proportional to the extent of karstic development of the drainage basin. It is reflected in the flood hydrograph, which tends to have a lower peak value and a somewhat more drawn out recession limb than hydrographs for an equivalent drainage basin without the karstic component as first observed by White and Reich (1970). This, of course, assumes that the return of the subsurface water to the surface drainage through springs occurs upstream from the gage point where the flood is being observed. The quantity of subsurface water that is held by the underground system depends on both the extent of karst development and the geohydrologic setting. The Green River basin has an additional complication in that excess runoff (flood flows) is channeled into the cave system by backflooding as shown in Figure 5.1.

When individual hydrographs of comparable-size flood discharges were plotted from the noncarbonate Susquehanna River basins (Pennsylvania) and the Green River basin, they were approximately the same size and shape. This reinforces an earlier finding (White, 1975) that when the quantity of limestone within the basin is less than about 30% of the total basin area then the flood damping effect of the limestone on individual storm hydrographs is lost in the background noise.

On the other hand, a high-magnitude, low-recurrence interval rainfall can fill the conduit system and force sinking streams to spill over onto the surface. Some surface routes are either not well developed or do not exist at all in areas underlain by karstic carbonate rocks. In a well-developed karst, surface streams are diverted into the underground. During normal floods, the old channel downstream from the swallow hole sometimes completely degrades and grows over; all trace of the channel might be lost. In a developed or urbanized area the presence of the abandoned water courses might not be recognized and various structures—houses, buildings, or streets—might be constructed directly in the route of the old drainage channel. When the rare event of an extreme flood happens, the spillover water has no available channel and must carve a new route directly through whatever has been constructed in its path. Thus, there is the peculiar situation that a moderate flood might be less severe in a karstic basin, while a large flood might do more damage than in a nonkarstic basin.

Figure 5.1 Infrared photograph showing backflooding of Green River into Echo River rise on March 24, 1968. *(Photo by Gordon Smith)*

Flooding of Closed Depressions

The capture of the internal runoff from an extreme value flood via sinkhole input into the subsurface drainage has another effect. The groundwater trough along the conduit system fills rapidly when subjected to high intensity precipitation. If the rapidly rising water table reaches the surface, the drain channels of sinkholes can reverse their flow, filling the closed depressions and transforming them into small lakes. Phantom Lake (local nomenclature), located about 6 km to the northeast of Pleasant Gap in central Pennsylvania, is one example that has been investigated by the author (White and White, 1984).

In central Kentucky the city of Bowling Green is built on the Sinkhole Plain with major portions of the city within large, shallow sinkholes with large catchment areas (Crawford, 1984). Crawford stated that (p. 283),

> during periods of short duration, intense rainfall flooding occurs 1) when the quantity of storm water runoff flowing into sinkholes exceeds their outlet capacities and they cannot drain into underlying caves fast enough to prevent ponding, 2) when the capacity of the çave system to transmit storm water is exceeded, and water must be stored temporarily in sinkholes since it cannot spread out onto flood plains like surface streams, and 3) when there is a backwater effect upon groundwater flow from sinkholes with bottoms lower than the level of surface streams at flood stage.

The landscape resembles large funnels (sinkholes); storm water accumulates at the bottom of some sinkholes then drains slowly through the soil into the cave streams below. Increased runoff into the underground system is due primarily to urbanization; sinkhole flood plains have been filled, thereby reducing their storm-water storage capacities by often clogging their drains, filling the sinkholes, or increasing the impervious areas to increase the storm water runoff. Filled sinkholes are not generally recognized by an untrained eye; therefore, when flooding occurs, unsuspecting property owners sustain major damages primarily because structures have been built within the natural sinkhole floodplains. The size of the flood pulse that will saturate the system and either run down the dry valley or saturate the underground system and cause water to back up onto the surface depends on the development of the underground drainage paths.

For flood insurance purposes, the Department of Housing and Urban Development (HUD) has defined the 100-year expected flood elevation along surface streams as the flood-plain elevation. For Bowling Green, HUD has defined the sinkhole floodplain as the 3-hour, 100-year flood elevation assuming no drainage from the sinkhole (Crawford, 1984).

GREEN RIVER FLOODS

Flood Statistics for the Green River Basin

The Green River flows in a narrow valley through much of its route through the central Kentucky karst (Fig. 5.2). The rating curves for the Green River at Munfordville (area = 4333 km^2) and at Brownsville (area = 7154 km^2) were compared with two Pennsylvania basins that drain the dissected Appalachian Plateau (Fig. 5.3). The areas of the Pennsylvania basins are comparable: 3787 km^2 and 7705 km^2. The stage height rises per

Figure 5.2 Mammoth Cave ferry on March 24, 1968. River flow at this point is approximately 990 cms with a return period of about 3–4 years. *(Photo by Gordon Smith)*

unit discharge for the Susquehanna River basins are approximately 5.6 m/cms/km^2, whereas the Brownsville Basin is twice that at 11.3 m/cms/km^2 and the Munfordville basin is three times that at 18.5 m/cms/km^2.

The discharge vs. return periods (probability of occurrence or exceedence) for the Green River basins are given in Figures 5.4 and 5.5 for their entire period of record. Upstream of the Munfordville gage, the Green River Lake was constructed in 1969. The effect of the Green River Lake on the stream flows at both gages has generally only influenced the flows with return periods of less than 5 years; therefore, the regulated flows are included in these analyses. If the post-1969 flows are omitted from the analyses, the mean annual flood changes very insignificantly from 845 cms to 861 cms (from 29,850 cfs to 30,410 cfs) at Munfordville; at Brownsville it changes from 1080 cms to 1140 cms (from 38,149 cfs to 40,285 cfs). When the Hazen plotting positions were used for these plots the data follow both the Gumbel and the Log Pearson III distributions. At the Munfordville gage the Gumbel-predicted 10,000-year return period flood is 3930 cms (139,000 cfs); the Log Pearson III-predicted flood is 2380 cms (84,000 cfs). However, at the Brownsville gage the Gumbel and Log Pearson III 10,000-year predicted floods are both approximately 5700 cms (200,000 cfs).

An exceptional flood occurred in 1962 with a peak discharge of 2175

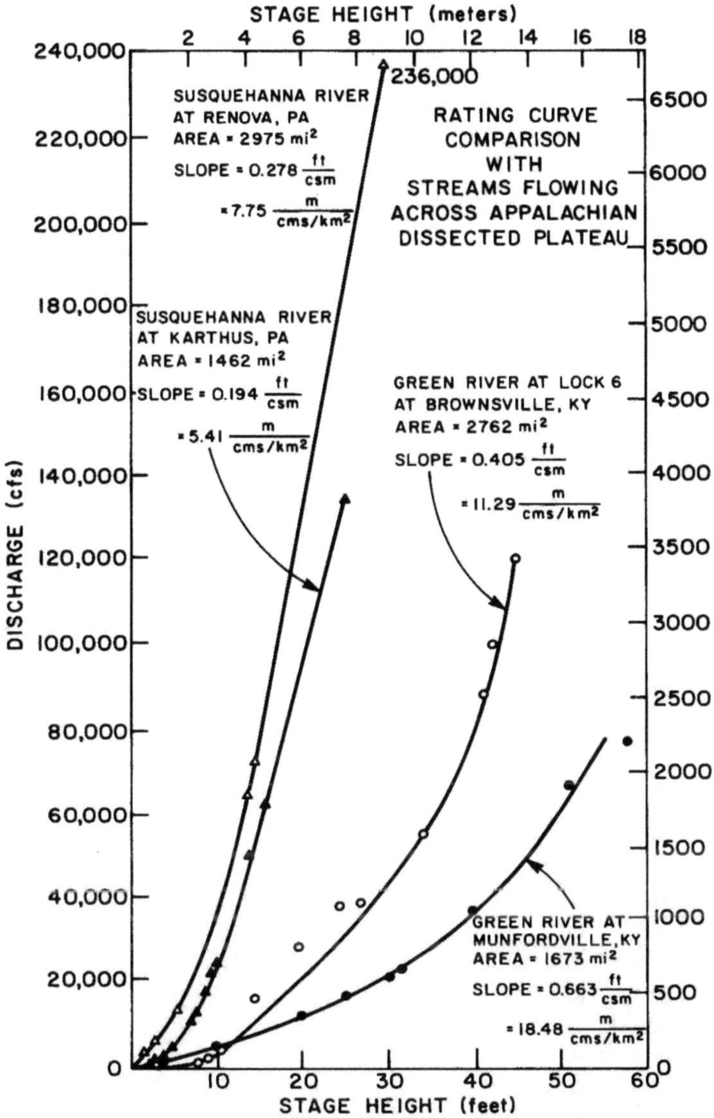

Figure 5.3 Comparison with rating curves from the Susquehanna River, which flows through the dissected Appalachian Plateau. Drainage areas of Green River at Munfordville and Susquehanna River at Karthus are 4333 and 3787 km^2, respectively; drainage area of Green River at Brownsville and Susquehanna River at Renova are 7154 and 7705 km^2, respectively.

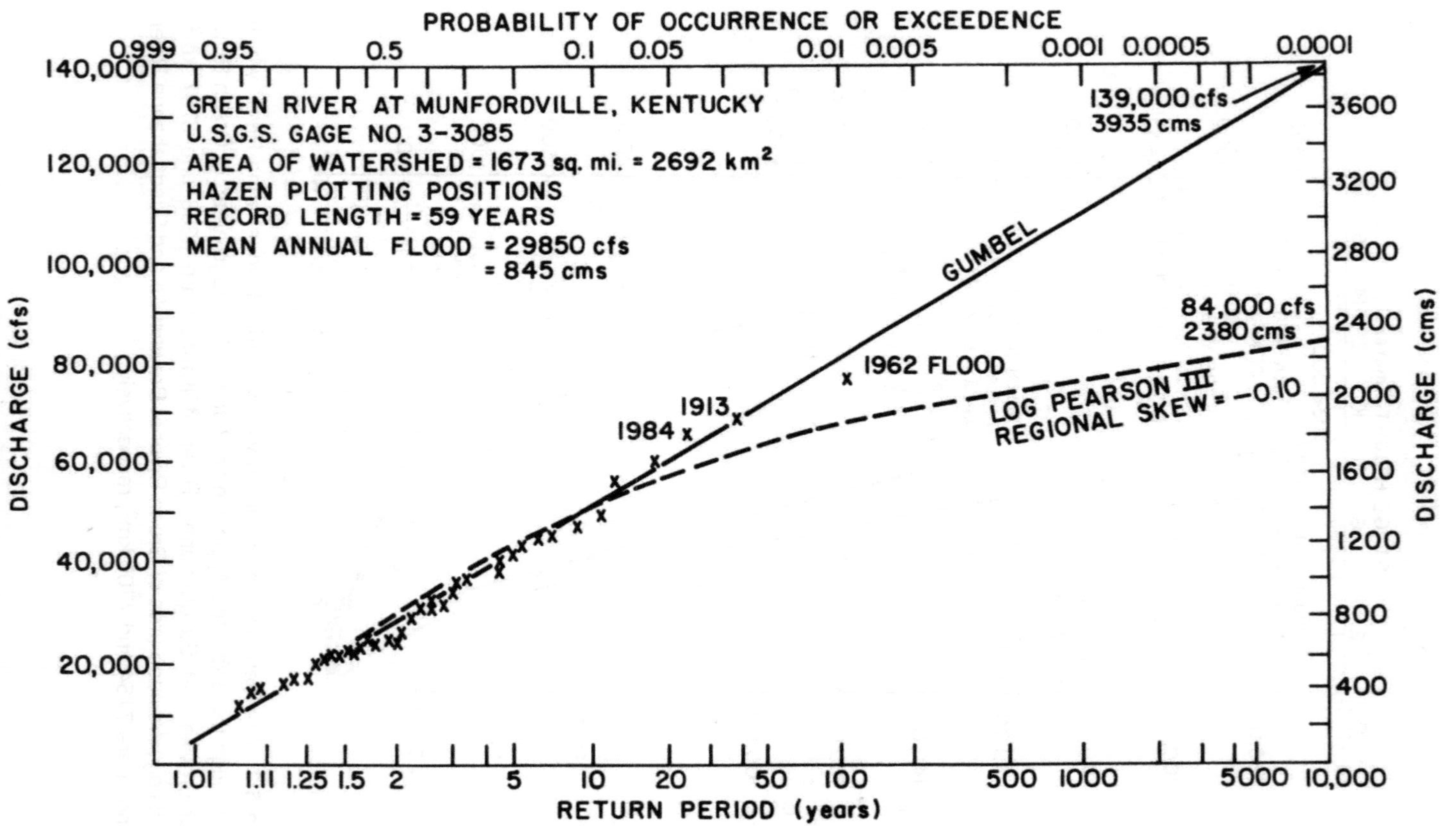

Figure 5.4 Probability of occurrence or exceedence for flood discharges at Munfordville, Kentucky gage. Extrapolated flood flows for Gumbel and Log Pearson distributions were calculated using ME3A extreme-value software package.

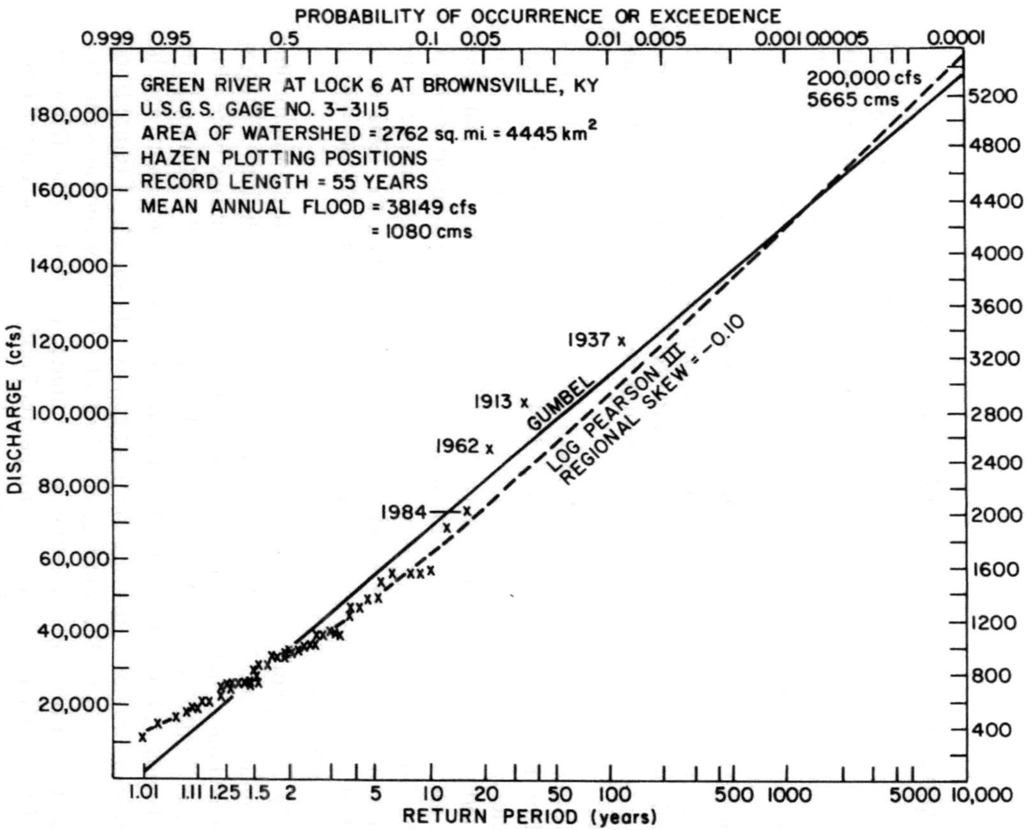

Figure 5.5 Probability of occurrence or exceedence for flood discharges at Brownsville, Kentucky gage. Extrapolated flood flows for Gumbel and Log Pearson distributions were calculated using ME3A extreme-value software package.

cms (76,800 cfs). This flood was equivalent to more than a 1000-year return period flood using the Log Pearson III distribution and equivalent to a 90-year return period flood using the Gumbel distribution. Even though the Log Pearson III distribution usually gives a better prediction for the extreme high value floods in nonlimestone areas, I believe the Gumbel estimate is more realistic.

Underground Flooding in the Green River Basin

Because the Green River flows in a narrow valley, floods are accompanied by large rises in river stage with accompanying flooding of many cave passages that lie within the flood range. Low-gradient trunk passages are arrayed in tiers from below pool stage at 125 m to 250 m (400–700 ft) elevation with major levels separated by only 20 m (White et al., 1970; Quinlan, 1970; see Chap. 12). Because of the low gradients, flooding extends a considerable distance back into the aquifer. Of particular interest is the relationship between, and the importance of, the long-return period catastrophic floods as well as the short-return period, more frequent floods to the cave-level elevations.

By using standardized federal (Benson, 1967) flood-frequency distributions, Log Pearson III with a regional skew, and the Gumbel (1941) analyses, the long-term 1000-year and the 10,000-year expected floods were calculated and related to cave elevations (Fig. 5.6). The equivalent cave elevation/stage height relationships were based on personal observations of the 1962 flood levels in the cave at Pohl Avenue and on the surface at the ferry crossing of the Green River. The water level during the 1962 flood filled Pohl Avenue completely to an elevation of 137–146 m (450–480 ft; Miotke and Palmer, 1972).

The 59 years' worth of discharge data were analyzed using a microcomputer program called ME3A, which was written by the Hydrology Group at The Pennsylvania State University. The Hazen plotting positions were plotted for the Gumbel and the Log Pearson III distributions at the Munfordville gage in Figure 5.4. Log Pearson III, with a regional skew coefficient, and the Gumbel analyses have provided the most consistent estimate of future flood frequencies. The Log Pearson regional skew of −0.10 for central Kentucky was taken from Interagency Advisory Committee on Water Data (Newton, 1982). The expected 1000-year return period floods were 2380 csm (84,000 cfs) for the Log Pearson III distribution and 3930 cms (139,000 cfs) for the Gumbel distribution. Using the rating curve derived in Figure 5.6, these are equivalent to 148 m and 151 m (485 ft and 497 ft) cave elevations for the Log Pearson and Gumbel distributions, respectively. These cave elevations correspond to the cave passages, which include Pohl Avenue, Columbian Avenue, and Mud Avenue levels in Flint Ridge (Fig. 5.7) and Echo River and River Hall in Mammoth Cave.

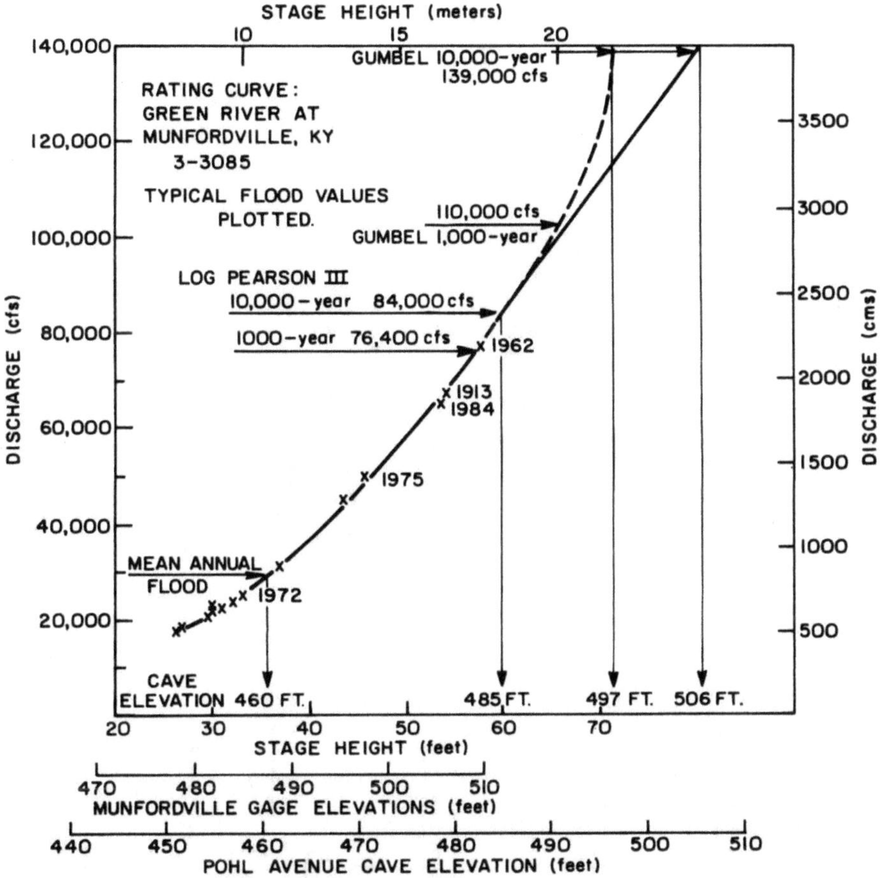

Figure 5.6 Stage-discharge relationship (rating curve) for the Green River at Munfordville, Kentucky (3-3085) showing the predicted 1000-year and 10,000-year return period floods and the equivalent cave elevations in Mammoth Cave.

Wolman and Miller (1960) showed that the end-member catastrophic events and the more frequent events of smaller magnitude are relatively important in controlling geomorphic processes such as erosion and sediment transport. Analyses of the transport of sediment indicated that a large portion of the "work" was performed by events of moderate magnitude, which recur relatively frequently rather than by rare events of unusual magnitude. However, in many valleys, rivers scour to bedrock only during the rare high-magnitude floods. The magnitude and frequency of events responsible for one feature might be very different from the magnitude and frequency of events responsible for another feature.

Sediments are currently being deposited within a zone extending from low water at 128 m (421 ft) to a flood zone of 150 m (485 ft), creating an

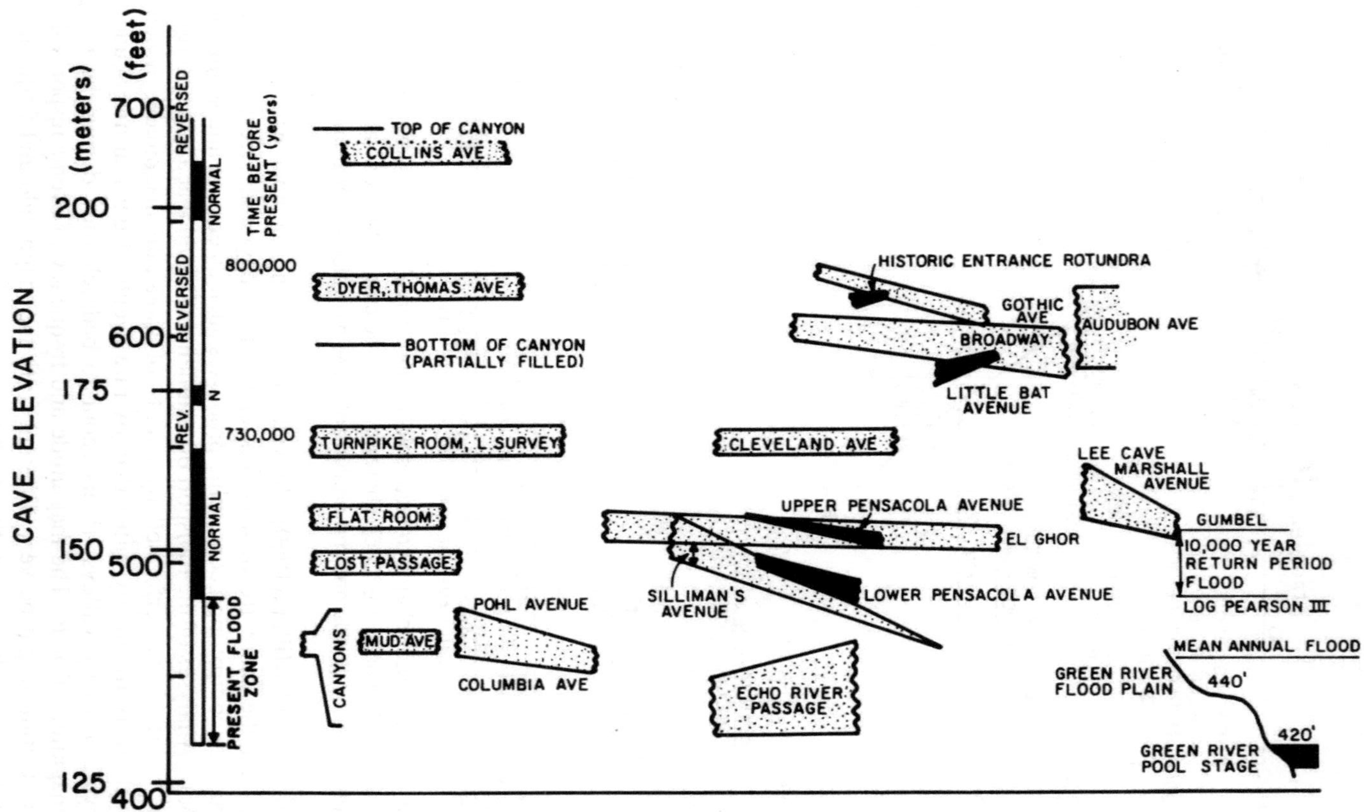

Figure 5.7 Schematic drawing of Flint–Mammoth Cave levels showing present and predicted flood zones with respect to existing cave levels.

interval of approximately 20 m. The sedimentary sequence in the base-level passages of the cave are intermixed by flood events so that precise time intervals on a very small scale are not to be expected.

Floods of moderate frequency have two effects on the sedimentation process. First they carry residual material from the Green River into the cave and might deposit some of it in the base-level passages as proposed long ago by Hendrickson (1961). Second, the flooded passages create a ponding situation that greatly reduces the load-carrying capacity for the conduit flows draining from the Sinkhole Plain. Thus, sediments derived from the upstream reaches of the drainage net are deposited during flood events. The sediment-carrying capacity of the conduits is actually decreased during flood events in contrast to the usual behavior of a surface stream.

Based on the Gumbel extreme-value analyses, the 1000-year flood at Munfordville corresponds to a discharge of 3100 cms (110,000 cfs). This corresponds to a maximum flood level at Mammoth Cave of about 150 m (493 ft). If the stage/discharge relationship is considered to be constant and one extrapolates to a 10,000-year return period flood, the highest elevation reached by this flood would be within the uncertainty range of 151–155 m (497–506 ft) elevation. The admittedly speculative 10,000-year flood is one that would have occurred only 2 or 3 times since the close of the Wisconsin Ice Age. Water could have reached passages as high as 152 m (500 ft). However, little influence on the sediments is expected because the events are so rare.

One curious bit of evidence is in agreement with these calculations. At the downstream end of Marshall Avenue, a major trunk conduit in Lee Cave, is a partially burnt log at an elevation of about 155 m. The charcoal gave a radiocarbon age of 6000 years B.P. (Freeman et al., 1973). Although the transport of the log into the cave by aboriginal peoples cannot be discounted, the age of the log preceded most active aboriginal use of the caves by several thousand years (Watson, 1974). A very large flood could have floated a log into Marshall Avenue in post-Wisconsin time.

The relationship between the sedimentation sequence and the flood events can be compared with recent paleomagnetic studies (Schmidt, 1982). All sediments within the 130–170 m (420–560 ft) elevation are normally magnetized, while a reversed magnetic zone extends from 170 m to 200 m (from 560 ft to 650 ft) as shown in Figure 5.7. Flooding during any interval of geologic time would have influenced only a narrow band of the sedimentary column within the contemporary floodwater zone. The overall magnetic stratigraphy is unlikely to be disturbed.

REFERENCES

Alekeyev, G. A., 1955, O formule dlya vychisleniya obespechenosti gidrologicheskikh velichin (Formulas for the calculation of the confidence of hydrological quantities), *Metrologiia i Gidrologiia (Leningrad) No. 6*, pp. 40–43.

Aron, G., 1987, *Penn State Runoff Model for IBM PC: Users Manual,* University Park, Pa.: Department of Civil Engineering, The Pennsylvania State University, 35p.

Aron, G., D. F. Kibler, E. L. White, A. C. Cavacas, R. P. Craig, D. C. Froehlich, C. I. Moore, M. R. Morton, and K. Yandrasitz, 1981, *Field Manual of Procedure PSU-IV for Estimating Design Flood Peaks on Ungaged Pennsylvania Watersheds,* University Park, Pa.: Department of Civil Engineering and Environmental Resources Research Institute, The Pennsylvania State University (for the Pennsylvania Department of Transportation and the Federal Highway Administration), 37p.

Benard, A., and E. C. Bos-Levenbach, 1953, The plotting of observations on probability paper (in Dutch), *Statistica (Rijkswijk)* **7:**163–173.

Benson, M. A., 1967, *A Uniform Technique for Determining Flood Flow Frequencies,* Washington, D.C.: Hydrology Committee, Water Resources Council, Bulletin 15, 15p.

Blom, G., 1958, *Statistical Estimates and Transformed Beta-Variables,* New York: John Wiley & Sons, 176p.

California, 1923, Flow in California streams. *California State Department Public Works Bull. 5.*

Crawford, N. C., 1984, Sinkhole flooding associated with urban development upon karst terrain: Bowling Green, Kentucky, in *Sinkholes: Their Geology, Engineering, and Environmental Impact,* B. F. Beck, ed., Rotterdam: A. A. Balkema, pp. 283–292.

Freeman, J. P., G. L. Smith, T. L. Poulson, P. J. Watson, and W. B. White, 1973, Lee Cave, Mammoth Cave National Park, Kentucky, *Natl. Speleol. Soc. Bull.* **35:**109–125.

Gringorten, I. I., 1963, A plotting rule for extreme probability paper, *Jour. Geophys. Research* **68:**813–814.

Gumbel, E. J., 1941, Probability interpretation of the observed return period floods, *Am. Geophys. Union Trans.* **21:**836–850.

Hazen, A., 1930, *Flood Flows, A Study of Frequencies and Magnitudes,* New York: John Wiley & Sons, 199p.

Hendrickson, G. E., 1961, Sources of water in Styx and Echo Rivers, Mammoth Cave, Kentucky, *U.S. Geol. Survey Prof. Paper 424-D,* pp. 41–44.

Huber, W. C., J. P. Heaney, S. J. Nix, R. E. Dickinson, and D. J. Polmann, 1987, *Storm Water Management Model, Version 3,* Cincinnati, Ohio: Report Project No. CR-805664 to Storm and Combined Sewer Section Systems and Engineering Evaluation Branch, Wastewater Research Division, Municipal Environmental Research Laboratory, 503p.

Katz, P. G., and G. H. Toebes, 1980, *Green River Basin Flow Forecasting Models,* West Lafayette, Indiana: Purdue University Water Resources Research Center, Tech. Rept. No. 136, 193p.

Leivikov, M. L., 1955, *Meterologiia, gidrologiia i gidrometriia* (Meteorology, Hydrology, and Hydrometry), 2nd edition, Moscow: Sel'khozgiz.

Linsley, R. K., and J. B. Franzini, 1979, *Water-Resources Engineering,* 3rd edition, New York: McGraw-Hill, 716p.

Linsley, R. K., M. A. Kohler, and J. L. H. Paulhus, 1958, *Hydrology for Engineers*, 2nd edition, New York: McGraw-Hill, 482p.

Miotke, F.-D., and A. N. Palmer, 1972, *Genetic Relationship Between Caves and Landforms in the Mammoth Cave National Park Area,* Würzburg: Böhler Verlag, 69p.

Newton, D. W., 1982, *Guidelines for Determining Flood Flow Frequency,* Reston, Virginia: U.S. Geological Survey Office of Water Data Coordination, Bulletin 17B, 28p.

Quinlan, J. F., 1970, Central Kentucky karst, *Études et Travaux de Méditerranée* **7**:235–253.

Schmidt, V. A., 1982, Magnetostratigraphy of sediments in Mammoth Cave, Kentucky, *Science* **217**:827–829.

Tukey, J. W., 1962, The future of data analysis, *Annals Math. Statist.* **33**:1–67.

Viessman, W., Jr., J. W. Knapp, G. L. Lewis, and T. E. Harbaugh, 1977, *Introduction to Hydrology,* 2nd edition, New York: Harper & Row, 704p.

Watson, P. J., 1974, *Archeology of the Mammoth Cave Area,* New York: Academic Press, 255p.

Weibull, W., 1939*a*, A statistical theory of the strength of materials, *Ingeniörsvetenskapsakad. (Stockholm) Handl.* **151**:15.

Weibull, W., 1939*b*, The phenomenon of rupture in solids, *Ingeniörsvetenskapsakad. (Stockholm) Handl.* **153**:17.

White, E. L., 1969, Regional trends in flood parameters for small watersheds in Pennsylvania, The Pennsylvania State University, M.S. thesis, 117p.

White, E. L., 1975, Role of carbonate rocks in modifying extreme flow behavior, The Pennsylvania State University, Ph.D. dissertation, 164p.

White, E. L., and B. M. Reich, 1970, Behavior of annual floods in limestone basins in Pennsylvania, *Jour. Hydrology* **10**:193–198.

White, E. L., and W. B. White, 1984, Flood hazards in karst terrain: lessons from the Hurricane Agnes storm, *Internat. Contributions to Hydrogeology* **1**:261–264.

White, W. B., R. A. Watson, E. R. Pohl, and R. Brucker, 1970, The central Kentucky karst, *Geog. Rev.* **60**:88–115.

Williams, P. W., 1983, The role of the subcutaneous zone in karst hydrology, *Jour. Hydrology* **61**:45–67.

Williams, P. W., 1985, Subcutaneous hydrology and the development of doline and cockpit karst, *Zeitschr. Geomorphologie* **29**:463–482.

Wolman, M. G., and J. P. Miller, 1960, Magnitude and frequency of forces in geomorphic processes, *Jour. Geology* **68**:54–74.

6

CHEMICAL HYDROLOGY

John W. Hess and William B. White

Karst aquifers, more than all others, are dynamic systems. There is a continuous chemical interaction between the moving groundwater and the carbonate wall rock. As a result of this, the internal porosity distribution and flow regimes of carbonate aquifers are continuously changing on a time scale of tens of thousands to millions of years. This phenomenon can be contrasted with silicate rock aquifers where there is indeed an influence of the aquifer wall rock on the chemistry of the moving groundwater, but there is relatively little change or at best an extremely slow change in permeability of the aquifer itself. Carbonate aquifers compensate for this problem by the accessibility with which the various types of water in the flow system can be sampled. Rather than simply depending on a well bore penetrating what one hopes is a representative location in the flow field, the direct exploration of the cave systems permits the analysis of many different inputs. It is therefore possible to understand the chemical evolution of the groundwater in carbonate aquifers in considerable detail.

The chemistry of the water in the karstic aquifer of the Mammoth Cave area has been investigated by the authors, Thrailkill (1968, 1972), Thrailkill and Robl (1981), Miotke (1975), and others. The objective of this chapter is to first set forth and review the chemical principles on which the chemistry is to be interpreted, then to review what has been done, and finally to attempt an overall interpretation of the evolution of groundwater chemistry of the karsted carbonate aquifer of the Mammoth Cave area. The chapter concludes with a short commentary on the environmental chemistry of the aquifer, which is becoming increasingly more important due to growth of the towns of Horse Cave, Cave City, Park City, and particularly Bowling Green on the Sinkhole Plain in the recharge area of most of the aquifer.

CHEMICAL PRINCIPLES

Carbonate rocks in south-central Kentucky are made up primarily of calcite, in the form of small calcite crystals (sparite), extremely fine-grained lime mud (micrite), and calcite oolites. There are isolated pods and lenses of dolomite within the carbonate section, which tend to be of limited horizontal and vertical extent. There is considerable silica in the stratigraphic column in the form of dispersed quartz grains, clay minerals, silicified fossil fragments, and in the lower part of the section, particularly in the St. Louis Limestone, extensive bedded chert. Although the chert plays a physical role in diverting groundwater flow on the Sinkhole Plain, the chemical influence of silica is minor and the chemistry of carbonate groundwater is primarily that of the dissolution of two minerals: calcite and dolomite.

The Dissolution of Calcite and Dolomite

Calcite and dolomite can be regarded as ionic salts, which in pure water dissociate according to the reactions

$$CaCO_3 \rightleftharpoons Ca^{++} + CO_3^{-2} \qquad (6.1)$$

$$a_{Ca}a_{CO_3} = K_c \qquad (6.2)$$

$$CaMg(CO_3)_2 \rightleftharpoons Ca^{++} + Mg^{++} + 2CO_3^{-2} \qquad (6.3)$$

$$a_{Ca}a_{Mg}a_{CO_3}^2 = K_d \qquad (6.4)$$

These dissolution reactions can be described by solubility product constants, K_c and K_d. The a's are thermodynamic activities of the ions in solution. For a review of the thermodynamics and appropriate solution chemistry see Stumm and Morgan (1981), Nordstrom and Munoz (1985), and for a more engineering-oriented detailed discussion of carbonate chemistry, see Loewenthal and Marais (1976).

If one substitutes numerical values for K_c and K_d and calculates the solubility of calcite and dolomite, it is found to be extremely small: 6–7 mg/l for calcite and approximately the same for dolomite. This solubility is actually lower than that for quartz, which is on the order of 10 mg/l at representative groundwater temperatures of 10–15° C. In completely pure water, the solubility of carbonate rocks would actually be less than the solubility of sandstones. This is clearly not the case, and the difference is due to the ready reaction of carbonate minerals with weak acids in contrast to quartz-bearing rocks where solubility is essentially independent of pH over very wide ranges.

Various acids have been proposed for the dissolution of carbonates in nature including organic acids and sulfuric acid derived from the oxidation of sulfide minerals or hydrogen sulfide. Carlsbad Caverns, New Mexico, for example, is a large cave system that was excavated primarily by reaction of a deep-seated hydrogen sulfide-bearing groundwater rising into an oxidizing mixing zone at the level of the cavern (Hill, 1981).

In spite of a possible role of pyrite oxidation in the formation of gypsum and other sulfate minerals in the cave system, the chemical evidence is that the cave systems of south-central Kentucky are the result of dissolution of carbonate bedrock by carbonic acid-bearing groundwater. The sources of the CO_2 are the atmosphere and the soil. The concentration of CO_2 in the atmosphere is 0.03 volume percent, often written as a carbon dioxide partial pressure. The carbon dioxide partial pressure of the atmosphere has been rising gradually since the industrial revolution (Clark, 1982), but remains nearly constant over wide areas of the earth except close to cities and other major CO_2 producers. Carbon dioxide forms in the soil from the plant roots, from oxidation of decaying organic matter, and from the activity of microorganisms.

CO_2 dissolves in water according to Henry's law. Aqueous CO_2 hydrates in water to form a dilute solution of carbonic acid. Experimentally, these two reactions are difficult to separate and so the custom has been to write them as a combined reaction described by a single equilibrium constant.

$$\mathrm{CO_2(gas)} \rightleftharpoons \mathrm{CO_2(aqueous)} \tag{6.5}$$

$$\frac{a_{\mathrm{CO_2}}(\mathrm{aq})}{P_{\mathrm{CO_2}}} = K_h \tag{6.6}$$

$$\mathrm{CO_2(aqueous)} + \mathrm{H_2O} \rightleftharpoons \mathrm{H_2CO_3} \tag{6.7}$$

$$\frac{a_{\mathrm{H_2CO_3}}}{a_{\mathrm{CO_2}}} = K \tag{6.8}$$

$$\mathrm{CO_2(gas)} + \mathrm{H_2O} \rightleftharpoons \mathrm{H_2CO_3} \tag{6.9}$$

$$\frac{a_{\mathrm{H_2CO_3}}}{P_{\mathrm{CO_2}}} = K_{\mathrm{CO_2}} \tag{6.10}$$

where K_h is the Henry's law constant, K is the hydration constant, and K_{CO_2} is the equilibrium constant for the combined reaction. Numerical values are known for the Henry's law constant and for the hydration constant. If concentrations of the two dissolved species are calculated, it is found that the bulk of dissolved carbon dioxide is present in solution simply as aqueous CO_2. The carbonic acid represents approximately 0.17% of the dissolved carbon and is a rather minor species. This relationship takes on some importance in the chemistry of carbonate aquifers because the hydration reaction is quite sluggish with a rate constant of 0.03 sec^{-1} (Kern, 1960). This means that the time scale for the hydration of carbon dioxide is approximately 30 seconds. If reactions rapidly deplete the available carbonic acid, the overall reaction rate might be controlled by the rate at which the much more abundant aqueous carbon dioxide can be hydrated to form a fresh supply of carbonic acid.

Carbonic acid dissociates in water according to the reactions

$$H_2CO_3 \rightleftharpoons H^+ + HCO_3^- \tag{6.11}$$

$$\frac{a_{CO_3^-} a_{H^+}}{a_{H_2CO_3}} = K_1 \tag{6.12}$$

$$HCO_3^- \rightleftharpoons H^+ + CO_3^{-2} \tag{6.13}$$

$$\frac{a_{CO_3^{-2}} a_{H^+}}{a_{HCO_3^-}} = K_2 \tag{6.14}$$

where K_1 and K_2 are the first and second ionization constants for carbonic acid. These reactions are the source of protons and form mildly acid solutions, which can then attack the carbonate minerals. When all of the above chemical reactions are considered together, the net reactions for the dissolution of calcite and dolomite are

$$CaCO_3 + CO_2 + H_2O \rightleftharpoons Ca^{++} + 2HCO_3^- \tag{6.15}$$

$$CaMg(CO_3)_2 + 2CO_2 + 2H_2O \rightleftharpoons Ca^{++} + Mg^{++} + 4HCO_3^- \tag{6.16}$$

Note that these reactions have been written with the bicarbonate ion as the product species. When numerical values are substituted for the equilibrium constants, it is possible to solve equations 6.10, 6.12, and 6.14 as functions of pH and calculate relative proportions of the various carbonate species in solution. Over the pH range typical of karst water, which extends from about pH 5 for CO_2-rich waters through pH 8–9 for carbonate-supersaturated waters, we are dealing primarily with the equilibrium between carbonic acid and the bicarbonate ion. The carbonate ion itself is a minor species in all carbonate groundwater. It becomes important only in highly alkaline waters, which are not usually found in carbonate aquifers.

If the carbonate system is at thermodynamic equilibrium, the above sets of equations are satisfied simultaneously. The equations describe a three-phase system: an aqueous solution that contains dissolved calcium, magnesium, bicarbonate, and carbon dioxide coexisting with a gas phase with a specified

Table 6.1 Equilibrium Constants for Carbonate Reactions

Temperature (°C)	pK_{CO_2}	pK_1	pK_2	pK_c	pK_d
5	1.19	6.52	10.55	8.39	16.63
10	1.27	6.46	10.49	8.41	16.71
15	1.34	6.42	10.43	8.43	16.79

Data used by Hess (1974); updated by results of Plummer and Busenberg (1982); $pK = -\log K$.

partial pressure of carbon dioxide and with a solid phase of calcite, dolomite, or both. Solutions of these equations have been investigated in considerable detail and numerical values for equilibrium constants are known with good precision (Table 6.1).

Chemical Measurements and Calculations

The chemical equilibria describing the interactions between groundwater and carbonate rock are written in terms of thermodynamic activities. Input from the field, however, are chemical analyses. The measurement and computational procedures necessary to connect field observations to theoretical chemistry follow.

Water samples are collected at surface streams, springs, wells, and underground streams, shafts, and drips. The samples should be filtered into clean bottles that can be tightly capped. Two samples are collected, one for the analysis of alkalinity and other anions and the other for the analysis of metals. The sample for analysis of metals is acidified with a few drops of nitric acid to prevent the precipitation of carbonates during storage and transport. At the sampling site, temperature is measured because all equilibria are temperature dependent. The specific conductance and the pH are also measured; pH measurements must be made precisely using a high-quality meter, ideally calibrated to 0.02 of a pH unit, so that the precision with which one knows the hydrogen ion activity is comparable to the precision of the other analyses. The main difficulties of pH measurements are the moisture conditions underground and temperature mismatches between samples, buffers, and electrodes.

The dissolved CO_2/carbonic acid/bicarbonate ion equilibria are the most unstable and for this reason one of the water samples should be titrated for alkalinity (essentially bicarbonate ion in most karst waters) as soon as possible after collection. Samples can be kept on ice after collection and then analyzed in the field. The acidified sample can be sent to a laboratory for analysis of calcium, magnesium, alkali metals, and any other metallic elements of interest using atomic absorption or atomic emission spectroscopy. The remainder of the alkalinity sample can also be sent to the laboratory for analysis of nitrate, chloride, sulfate, and other anions using ion chromatography.

The activities of dissolved species are directly related to the concentrations by

$$a_i = \gamma_i m_i \tag{6.17}$$

where γ_i is an activity coefficient. Most carbonate groundwaters are dilute, so it is possible to estimate the activity coefficients using the Debye–Hückel equation

$$-log\,\gamma_i = \frac{A z_i^2 \sqrt{I}}{1 + \mathring{a}_i B \sqrt{I}} \tag{6.18}$$

The coefficients were tabulated by Stumm and Morgan (1981) and by other works on aqueous chemistry. The ionic strength, I, is defined by

$$I = \frac{1}{2} \sum_i m_i z_i^2 \tag{6.19}$$

If a complete analysis of the water sample is available, then the ionic strength can be calculated directly from equation 6.19 and fed into the Debye–Hückel equation to produce activity coefficients, which allows the determination of the activities of the various species in solution. Many times, however, complete analyses are not available. It has been found (Jacobson and Langmuir, 1970) that there is an empirical relationship between the ionic strength and the specific conductance

$$I = 1.88 \times 10^{-5} \mathrm{Spc} \tag{6.20}$$

where Spc has been corrected to 25° C.

As a refinement on the calculations described above, account must be taken of ion pairs. The charged species in solution have a certain tendency to associate with each other, forming transient bonded species known as ion pairs. The most important ones in carbonate waters are $CaHCO_3^+$, $CaCO_3^0$, $CaSO_4^0$, $MgHCO_3^+$, $MgCO_3^0$, and $MgSO_4^0$. These do not change the overall chemistry of the solution, but they do change the ionic strength because they reduce the effective number of free charges. The reduction in ionic strength changes the activity coefficients, which in turn modify the activities. The dissociation constants for these and other ion pairs are known (Thrailkill, 1976), and it is possible to include them in the calculations.

The overall calculation of aqueous species by computer has evolved tremendously over the past 20 years. Detailed calculation of carbonate chemistry really only dates from the late 1960s. Before then nearly all interpretation was in terms of the raw measurements. One of the first computer programs was written by Roger Jacobson at The Pennsylvania State University. This program, although never formally published, was widely circulated in the geochemical community and forms the basis for the calculations used later in this chapter. Other programs written specifically for carbonate geochemistry include IONPAIR (Thrailkill, 1976) and WATSPEC (Wigley, 1977). During the same period, a much more elaborate program, WATEQ, was written by Blair Jones and Alfred Truesdell at the U.S. Geological Survey. WATEQ was intended for a much wider range of speciation chemistry and carbonate equilibria were only a special case. Its various updates and modifications, such as WATEQF (Plummer et al., 1978*b*), have also been used widely. In the past 10 years, many other aqueous speciation and reaction path codes have been written. Many of these programs have been critically compared by Nordstrom et al. (1979). Present microcomputers are very powerful machines, and the current practice is often to write programs on personal computers for the specific task at hand as an alternative to the more formal mainframe codes.

Geochemical Parameters

The raw chemical data extracted from water analyses give some insight into the behavior of carbonate groundwater, but do not directly answer some of the more important questions. It is more useful to reduce these data to parameters, which directly describe the state of saturation of the groundwater, its carbon dioxide partial pressure, and its load of dissolved carbonates.

The hardness of carbonate waters was originally defined in terms of a soap test. Hardness is a measure of dissolved carbonate and is now defined by

$$Hd = 100(m_{\mathrm{Ca}^{++}} + m_{\mathrm{Mg}^{++}}) \tag{6.21}$$

Hardness is a parameter of little theoretical significance in that one is expressing calcium and magnesium in terms of weight concentration as calcium carbonate. However, it is used widely to describe the total content of carbonate in solution.

The calcium/magnesium ratio is defined by

$$\mathrm{Ca/Mg} = m_{\mathrm{Ca}^{++}}/m_{\mathrm{Mg}^{++}} \tag{6.22}$$

and is simply the atomic ratio of these metals in solution. The ratio is important for indicating whether a groundwater has been predominantly in contact with calcite or with dolomite. All limestone contains some magnesium and typical Ca/Mg ratios for groundwater in limestone range from 5 to 8. Dolomite groundwater has a Ca/Mg ratio close to unity.

Groundwater in carbonate rocks is rarely in chemical equilibrium. It can be highly undersaturated and capable of dissolving more rock or it can be supersaturated, in which case it is capable of precipitating secondary calcite. Although chemical analyses for various groundwaters can be plotted on saturation curves, these do not allow for variations in temperature and ionic strength. The state of saturation can be reduced to a single number by use of the saturation index. The saturation index (SI) for any mineral is defined as the ratio of the ion activity product determined experimentally to the solubility product constant. Equations defining this quantity for calcite and dolomite are

$$SI_c = \frac{K_{iap}}{K_c} = \frac{a_{Ca^{++}} a_{CO_3^{-2}}}{K_c} = \frac{m_{Ca^{++}} \gamma_{Ca}\, m_{HCO_3^-} \gamma_{HCO_3^-}}{K_c K_2 10^{-pH}} \tag{6.23}$$

$$SI_d = \left[\frac{K_{iap}}{K_d}\right]^{\frac{1}{2}} = \left[\frac{a_{Ca^{++}} a_{Mg^{++}} a^2_{CO_3^{-2}}}{K_d}\right]^{\frac{1}{2}} \tag{6.24}$$

$$= \left[\frac{m_{Ca^{++}} \gamma_{Ca^{++}} m_{Mg^{++}} \gamma_{Mg^{++}} m^2_{HCO_3^-} \gamma^2_{HCO_3^-}}{K_d K_2^2 10^{-2pH}}\right]^{\frac{1}{2}}$$

The ion activity product is expanded into functions of the experimentally determined quantities. The carbonate ion is a minor species but its activity can be calculated from the measured bicarbonate concentration and the experimentally determined pH. This assumes that the kinetics of dissociation of carbonic acid are very fast. The time scales for reactions 6.11 and 6.13 are on the order of milliseconds, so for geological purposes it is safe to assume that carbonate, bicarbonate, and un-ionized carbonic acid are always in equilibrium. The saturation index for dolomite is written as a square root which allows SI_c and SI_d to be compared in terms of the same number of moles of dissolved carbonate.

The carbon dioxide partial pressure is also derived from the measured bicarbonate concentration and the pH. It is essentially a reverse calculation through reactions 6.11 and 6.9. The calculation gives the carbon dioxide partial pressure of a hypothetical gas phase with which the water sample would be in equilibrium. The calculated partial pressure is very unlikely to be the actual partial pressure of the cave atmosphere where the water sample was collected because of the relatively sluggish kinetics of CO_2 dissolution and de-gassing. The gas phase and the aqueous phase are often far out of equilibrium. On the other hand, the calculated CO_2 partial pressure is often representative of the carbon dioxide in the source area so this parameter tells something about the overall mass balance of CO_2 along the flow path and gives a minimum value for the CO_2 pressure at the source.

Chemical Kinetics

The rate at which carbonate reactions occur is very critical for the formation of underground drainage systems and also surface karst landforms. If the rates were extremely fast, equilibrium would be obtained almost instantaneously and there would be no karst. The available supply of carbonic acid would be consumed by reaction with the limestone bedrock at the base of the soil, only saturated water would enter the aquifer, and no further dissolution would take place. If the reaction kinetics were extremely sluggish as they are for most silicate-water reactions, there also would be no development of cave systems because water would require such a long time to achieve equilibrium that the localization of flow and the formation of conduits would not take place.

Karst landforms and well-integrated conduit drainage systems are uniquely found in carbonate rocks because of an optimum time scale for the kinetics of reaction. Figure 6.1 shows an experimental curve for the dissolution of calcite in carbon dioxide-saturated water under laboratory conditions. There is an initial rapid uptake of material. As the system approaches equilibrium, the rate, which is the slope of the curve, slows down. Times on the order of several days are required for the solution to approach equilibrium, but much longer times are required for equilibrium to be actually reached.

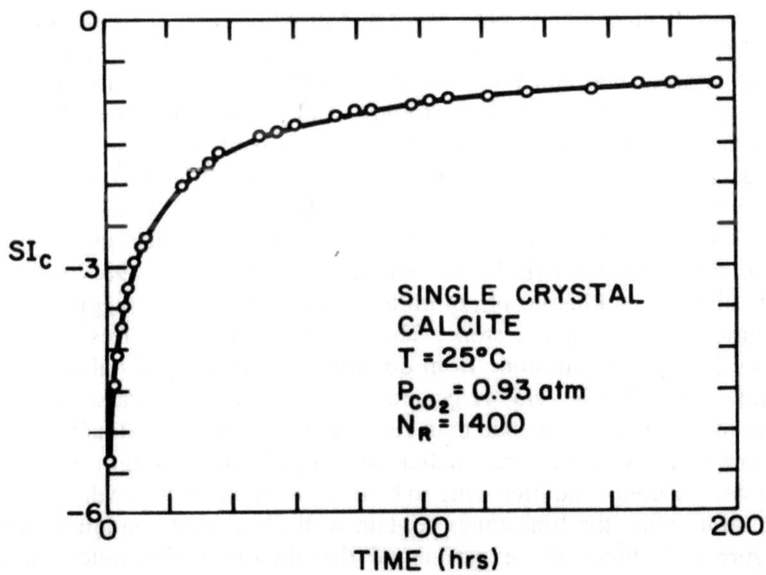

Figure 6.1 An experimental rate curve for calcite dissolving in CO_2-saturated water under known flow regime. *(After Herman, 1982)*

During the past 10 years there have been many studies of the dissolution kinetics of calcite. The analyses of the rate data are still somewhat controversial, and many different rate equations have been proposed. An early experimental study by Berner and Morse (1974) showed that there is a rather abrupt decrease in dissolution rate as reactions approached equilibrium. Over a wide range of undersaturations, the rate varied only slightly with undersaturation until the saturation index approached a value of −0.3 to −0.4. At values closer to saturation the rate plummeted rapidly, approaching zero exactly at saturation.

Of the various rate equations that have been proposed, that of Plummer et al. (1978*a*) is the one most commonly used for field interpretation of karst processes.

$$\text{Rate} = k_1\, a_{H^+} + k_2 a_{H_2CO_3} + k_3 a_{H_2O} - k_4 a_{Ca^{++}} a_{HCO_3^-} \qquad (6.25)$$

Of the three forward-reaction rate terms in the equation, one describes dissolution as a function of hydrogen ion activity and would be the dominant term when limestones are dissolved in strong mineral acids. This term is mass-transport controlled and the value for the rate constant is rather sensitively a function of fluid-flow velocity. The second term describes the dissolution of limestone in carbonic acid and is most relevant to groundwater in karst. This term appears to be mainly surface-reaction controlled and has only a small dependence on fluid-flow velocity. The third term represents the rate of dissolution of limestone in pure water and provides a background minimum rate when no other sources of protons are present. The back reaction term is more complicated and Plummer et al. (1978*a*) claim that this term also describes precipitation reactions (Reddy et al., 1981). Equation 6.25, like most other rate equations, does not work very well near equilibrium. The rapid decrease in rate near equilibrium found by Berner and Morse (1974) is not expressed very well in the equation. Wolfgang Dreybrodt, in a series of papers applying kinetic theory to many karst processes, has used equation 6.25 as a starting point (Buhmann and Dreybrodt, 1985*a*, 1985*b*).

The dissolution kinetics of dolomite are generally more sluggish than the dissolution kinetics of limestone. This is the basis for the freshman "acid test" to distinguish limestone from dolomite by the sample's ability to fizz in dilute hydrochloric acid. A general observation is that karst landscapes on dolomite are more subdued than karst landscapes on limestone: sinkholes are more shallow, caves are smaller, and the drainage system is less well integrated. Dolomite aquifers tend to be fracture aquifers somewhat modified by solution while the limestones contain well-developed conduit systems.

Figure 6.2 shows the experimental dissolution of dolomite. There is an initial rapid dissolution of the dolomite, which is nearly as fast as that of calcite, but as time goes on, when the dolomite has dissolved to about 1% of the saturation level, the rate curve flattens and the rate drops

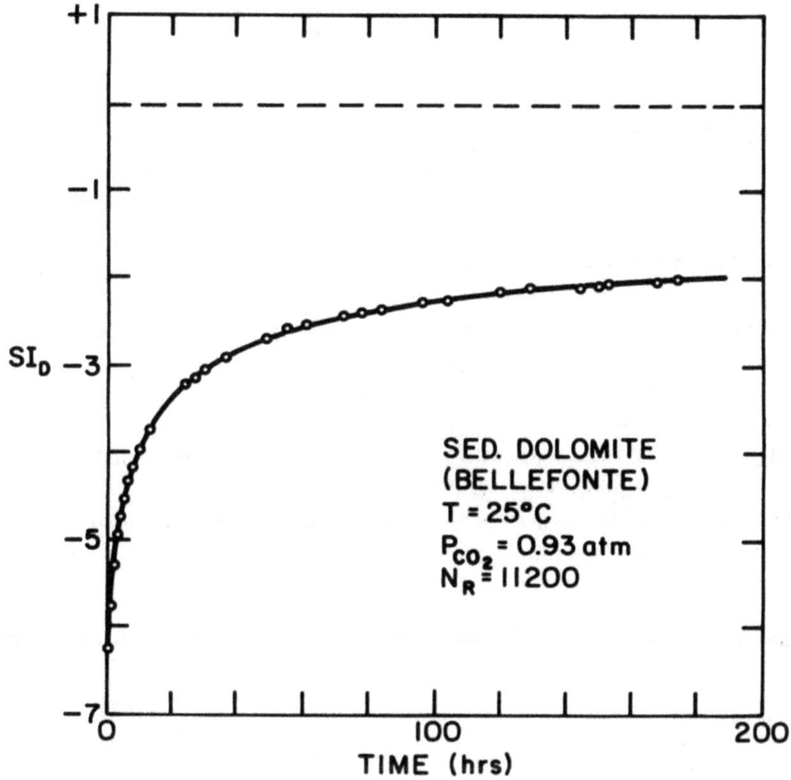

Figure 6.2 An experimental rate curve for dolomite rock (Ordovician Bellefonte Formation, central Pennsylvania) dissolving in CO_2-saturated water under known flow regime. *(After Herman, 1982)*

precipitously. The system will approach an $SI_d = -2$ in a few days, but then the rate becomes so sluggish that times on the order of weeks or years are required for the system to reach saturation with respect to dolomite. Analyses of dolomite groundwater indicate that the groundwater does eventually come into equilibrium with dolomite, but laboratory results show that very long times are required. The experimental dissolution kinetics of dolomite have also been reduced to a rate equation (Busenberg and Plummer, 1982; Herman and White, 1985).

CHEMISTRY OF CARBONATE GROUNDWATERS

Sources of Carbon Dioxide

Carbon dioxide, which is the primary chemical driver for all of the carbonate dissolution reactions, has its main origin in the soil. The soil types that overlie the central Kentucky karst are described in Chapter 2. The well-drained soils of the uplands of the Sinkhole Plain are relatively thick terra rosa type. The alluvial floodplains of the sinking streams are brown soils, which are quite different from the terra rosas above. The soils of the Chester Cuesta tend to be very thin and sandy, reflecting the source rocks and organic-poor forest soils. Karst valley soils are of unknown thickness and are a mixture of slope-washed sandstone soil and limestone residual soil. The carbon dioxide of the soil atmosphere can be measured by inserting gas lysometers and measuring the CO_2 content directly with a gas analysis device (Miotke, 1972). Most of the CO_2 analyses by karst researchers in the United States have been done with the Drager air pump device. A known volume of air is drawn through a glass tube containing the active agent, hydrazine hydrochloride for CO_2, and the volume percent is read directly from the advancement of a purple reaction front down the calibrated tube. Miotke (1974, 1975) measured CO_2 levels in different soil localities in the south-central Kentucky karst (Table 6.2).

The CO_2 partial pressures are not exceptionally high. Values occur up to 1 volume percent, but rarely more even during the summer months in the

Table 6.2 Carbon Dioxide Concentrations in Soil and Cave Atmospheres

Location	*Depth (Soil) (cm)*	*Volume Percent CO_2*
Flint Ridge, sandy residual soil	9	0.4
	15–20	0.35
	> 20	0.45
Cedar Sink, November data	20	0.48
	30	0.45
	50	0.6
	70–90	0.8
Echo River Spring, silt terrace	20	1.5
	28	1.1
	35	1.2
	46	2.8
	65	5.0
	70	3.9
	95	3.5
Sinkhole Plain, grassy soil	10	1.1
	15	1.0
	30	0.6
Mammoth Cave, historic section	30 data,	mean = 0.07, s.d. = 0.03
Flint Ridge section	31 data,	mean = 0.04, s.d. = 0.02

All data from Miotke, 1975.

peak growing season. Thrailkill and Robl (1981) found similar values in the Carter Caves area of eastern Kentucky in a similar geologic environment.

Groundwater in the Vadose Zone

Infiltration waters travel along three possible routes. Water that drains from the small surface catchments and the outlet waters from the Haney springs drain through air-filled vertical shafts, move rapidly to the base of the shafts, and then through small drains to underground streams in the main conduit system. The second vertical pathway is along the tight fractures and joints in which the water is not in contact with a gas phase. When such waters emerge from the roofs of cave passages, the seepage often de-gasses CO_2 with the formation of secondary calcite deposits. In the Mammoth Cave area, the secondary calcite is almost exclusively limited to the edge of the caprock. Where cave passages underlie the sides of the karst valleys, infiltration waters penetrate the limestone soils and react with the underlying limestone to limits dictated by the CO_2 in the soil atmosphere. These waters then move downward through vertical joints and fractures. When they emerge from the roof of the cave they are out of equilibrium with the cave atmosphere, so that calcite is deposited. The caprock, especially the Fraileys Shale, acts as an aquiclude that protects the cave passages beneath from active seepage, and therefore most of the cave passages are barren of calcite deposits.

Comparison of the vertical shaft waters with the seepage waters (what Thrailkill [1968] called "vertical flows" and "vertical seeps") shows that seepage waters are highly supersaturated with respect to calcite (Fig. 6.3), whereas the vertical shaft waters are highly undersaturated. However, the carbon dioxide partial pressures of both types of water are only about a factor of 10 above that of the surface atmosphere. Although higher CO_2 pressures are found in the soil, this amount of CO_2 apparently does not get into the infiltrating groundwater. The chemistry provides an immediate explanation for the coexistence of vertical shafts as sites of intense dissolution, and masses of stalactites and stalagmites in the same general region at the margin of the caprock (Fig. 6.4).

The third type of vertical seepage is that which moves beneath the caprock. Passages beneath the caprock are dry and relative humidity can fall as low as 80%. In these passages gypsum occurs as thick crusts, sometimes as curving masses of crystal fibers known as "gypsum flowers," and occasionally in the form of stalactites or columns. Although liquid water is rarely seen in these passages, the presence of the gypsum shows that there is some slow percolation of solutions. Where the Fraileys Shale is absent, percolating fluids can move down from the overlying caprock. One hypothesis for the origin of the sulfate minerals is that they are derived from the oxidation of pyrite that occurs near the top of the Big Clifty Sandstone (Pohl and White, 1965). Another is that gypsum is derived from evaporite beds in the St. Louis Limestone (George, 1974). The driest passages occur in the cores

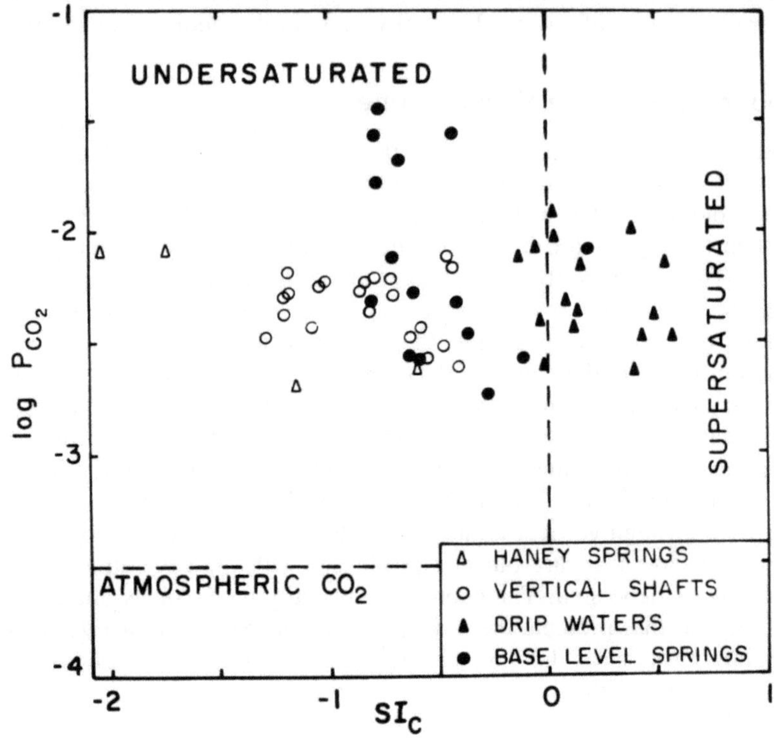

Figure 6.3 Relation of carbon dioxide pressure to saturation index for various waters from the Mammoth Cave area.

of the ridges where other highly soluble sulfate minerals are deposited (Table 6.3). Mirabilite occurs as spectacular water-clear stalactites and "flowers."

Undersaturation and Conduit Flow

The horizontal component of the conduit drainage system is surprisingly inaccessible in the south-central Kentucky karst. Free surface streams can be sampled in such places as Parker Cave under the Sinkhole Plain, where a series of underground rivers draining toward the plateau give access to some kilometers of stream passage (Quinlan and Rowe, 1978). Within the enlarged Mammoth Cave system, a major underground stream can be followed in the Hawkins–Logsden River complex. This would give access to a substantial length of active master trunk drainage system, as would a few other places such as streams in Roaring River and Mystic River in Mammoth Cave and the base-level streams in Whigpistle Cave and the Hidden River complex. Chemical data have not been collected from these

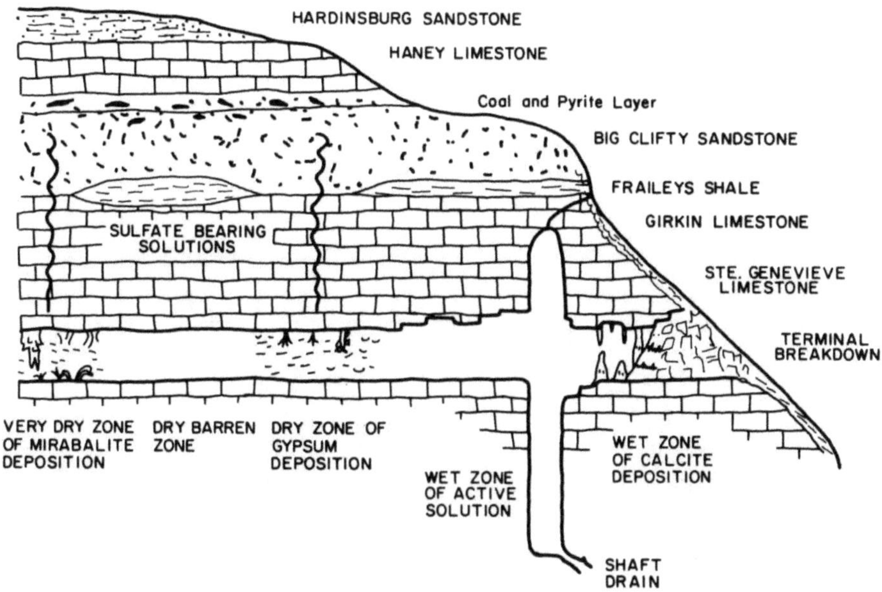

Figure 6.4 Sketch showing various types of vadose water migrating through the cavernous limestones.

Table 6.3 Sulfate Minerals in the Mammoth Cave System

Name	*Formula*	*Typical Occurrence*
Gypsum	$CaSO_4 \cdot 2H_2O$	Crusts, "flowers," dripstone
Epsomite	$MgSO_4 \cdot 7H_2O$	Crusts, "flowers," dispersed in soil
Hexahydrite	$MgSO_4 \cdot 6H_2O$	Crusts, loose crystals
Mirabilite	$Na_2SO_4 \cdot 10H_2O$	"Flowers," dripstone
Celestite	$SrSO_4$	Crusts
Blödite	$Na_2SO_4 \cdot MgSO_4 \cdot 4H_2O$	Crusts
"Labile salt"	$2Na_2SO_4 \cdot CaSO_4 \cdot 2H_2O$	Dispersed in mirabilite dripstone

sources and so the analysis of the chemistry of cave system depends only only on data from the vadose zone and from spring discharge points.

Data collected for the water year 1972–1973 (Hess, 1974) provides the basis for the following discussion. Water samples were collected, analyzed, and the analytical data processed according to the procedures described above. For a description of the streams and springs used as sampling sites, see Chapters 2 and 3. The most detailed investigation was along the Turnhole Spring drainage system, which is described in Chapter 3.

The saturation index for the water from the Haney Springs, the local springs, and the regional springs all fall well below the saturation line.

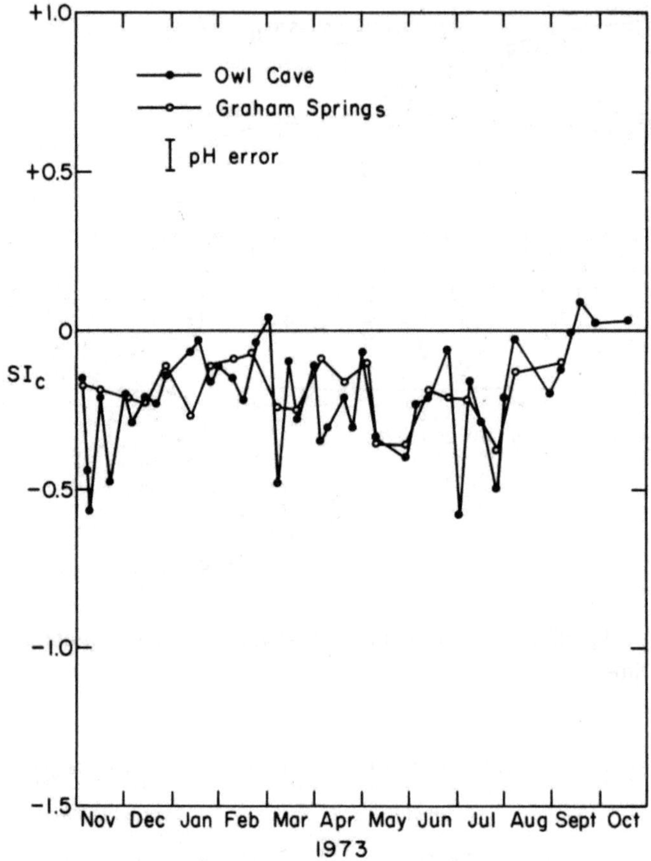

Figure 6.5 Saturation index for Owl Cave (Turnhole Spring drainage) and Graham Springs over the course of 1 year.

Figure 6.5 gives complete data for the undersaturation of the two large regional drainage systems: Graham Springs and the Turnhole Spring system (the latter sampled at Owl Cave in Cedar Sink). As might have been expected from the storm response analysis described in Chapter 4, there are substantial week-to-week fluctuations in the chemical data. The SI_c data for the four hydrogeologically grouped sets of waters are seen in Figure 6.6. The Haney Springs are far out of equilibrium. The saturation index is as low as –1 and sometimes gets as low as –2; it is never near saturation. Here we have waters seeping downward through the Hardinsburg Sandstone into solutionally widened fractures and joints in the Haney Limestone. There are short travel times and short travel distances from the dispersed and diffuse infiltration surface to the discharge around the margins of the caprock. Over this time scale, possibly on the order of hours, these waters do not come close to equilibrium.

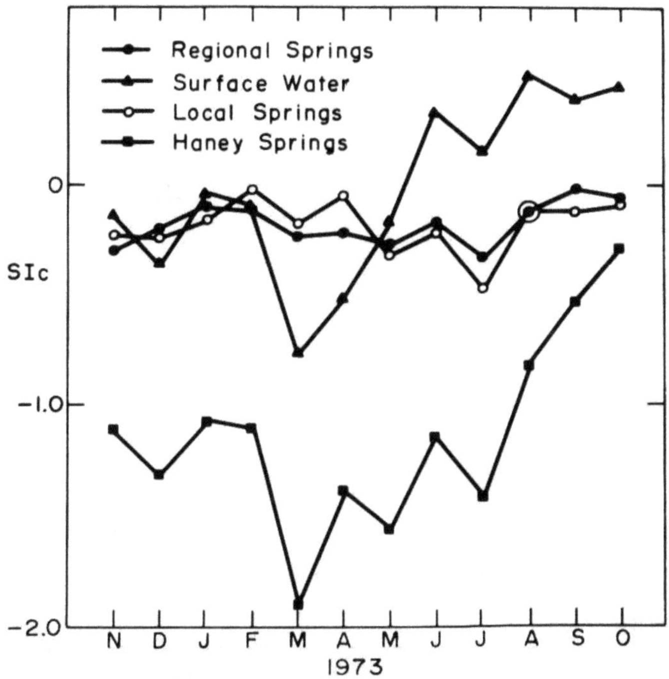

Figure 6.6 Saturation index for various waters in the Mammoth Cave area. Data have been averaged into 1-month time intervals and over several sampling points with the same hydrogeologic setting.

The regional springs' and local springs' (such as the Echo, Styx, and Pike) discharge points show a more or less constant undersaturation of about −0.2 to −0.3. This represents about the undersaturation at which the dissolution kinetics slow down substantially. The curve in Figure 6.1 begins to flatten out at residence times beyond 1 or 2 days, and the time to approach equilibrium stretches out to many days or weeks. This is a longer time than is required for flow from any of the recharge points down to the springs. Although the waters are integrated over long distances, they still reach the springs substantially undersaturated.

The approach to saturation of the local and regional springs is a little closer than might have been expected from the kinetics curves themselves. This could be due to the mixture of conduit waters from localized inputs with the diffuse input entering the conduits from the overlying Sinkhole Plain. Based on the chemistry of the seepage waters seen in the caves, this water is closer to saturation and therefore has the effect of pulling up the average of the overall saturation of the springs. The relationship between saturation index and mixing volume is not linear, and the increase in saturation will be less than expected from the chemistry of the diffuse infiltration water.

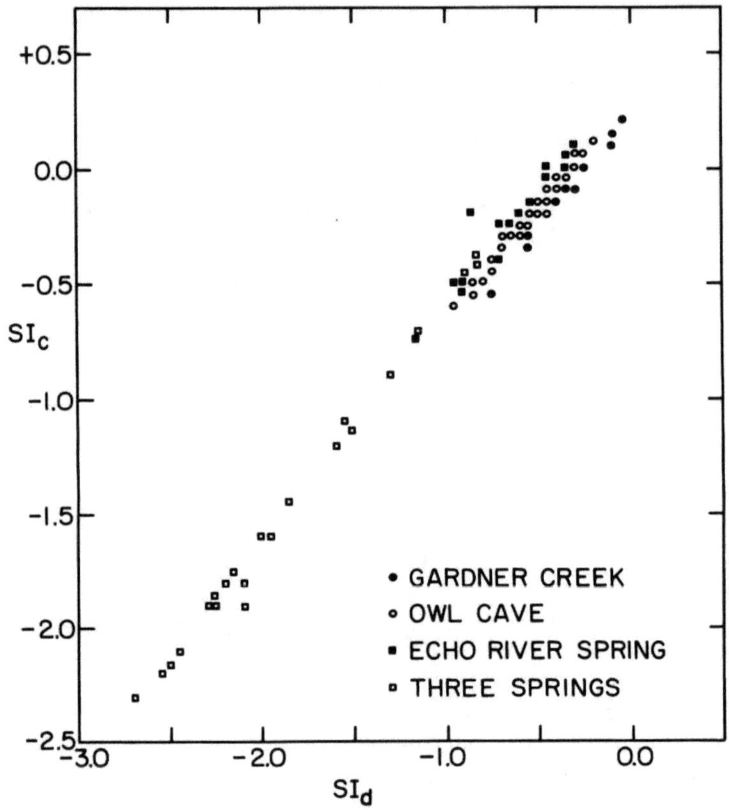

Figure 6.7 Relationship between saturation index for calcite, SI_c, with the calculated saturation index for dolomite, SI_d, for various types of water from the karst aquifer.

The waters are also undersaturated with respect to dolomite and, in general, more highly undersaturated with respect to dolomite than with respect to limestone. Figure 6.7 shows a plot of the SI_c versus SI_d for some examples of all four of the hydrogeologic water types. It can be seen that regardless of the overall hardness, seasonal variations, and other fluctuations, SI_c and SI_d track rather closely. One can derive from simple considerations of the definition of the saturation index that

$$SI_d = \left[\frac{K_c^2}{K_d}\frac{\text{Mg}}{\text{Ca}}\right]^{\frac{1}{2}} SI_c \tag{6.26}$$

This does not take into account the differential kinetics of calcite and dolomite dissolution. Referring to Figure 6.2, the kinetics of dissolution of dolomite are sufficiently sluggish that one would expect SI_d to lag behind SI_c, particularly for the rapidly moving conduit waters. Such is clearly not the case. All data fall neatly into the theoretical prediction. It is believed that most of the magnesium that appears in the analyzed spring-water samples is magnesium from the limestone. Saturation of the spring water with respect to dolomite rock was not found, and indeed the interbeds and pods of dolomite within the Mississippian carbonate section are sufficiently sparse that one would not expect intimate chemical contact between groundwater and the dolomite. Instead the dolomite saturation index is simply a measure of the amount of magnesium in the limestone.

Undersaturation throughout the flow path from any of the localized inputs to the spring outlets is an important point in the interpretation of the evolution of the conduit system. Seepage waters are near saturation; conduit waters are not. The short travel times that allow undersaturated water to emerge from the springs also account for the development of conduits beneath the protective caprock. Models of cave development that require a mixing zone or that require continuous exchange of carbon dioxide with the atmosphere of the vadose zone cannot account for any of the long uniform conduits that developed under many kilometers of caprock. Based on other work, it appears that the critical aperture is on the order of one centimeter. Water moving in fractures and bedding-plane partings with smaller apertures reach near-saturation levels quickly, and solutional enlargement of the fracture or bedding plane can proceed only very slowly. When the aperture reaches the centimeter size, water of the necessary undersaturation can penetrate the entire width of the aquifer, and conduit development takes place rapidly.

Transported Carbonates: Storm Flow and Seasonal Responses

The dissolved carbonate content of the local springs, mostly drainage from the dissected plateau (Fig. 6.8) show a rather large point-to-point variation. The short time scale (high frequency) variation, however, is an oscillation around a mean value that is nearly constant on long time scales. There is little evidence in Figure 6.8 for any sort of annual cycle. In contrast, the hardness curve for the regional springs (Fig. 6.9) exhibits about the same point-to-point variation superimposed on a well-defined seasonal cycle with a pronounced minimum in February through April. The minimum does ndeed coincide with times of high runoff, and the maximum in August through October corresponds to the annual dry period.

It is instructive to examine the calculated carbon dioxide pressures for this same set of samples. For the local springs, which receive much of their flow from the organic-poor soils of the cuesta, the seasonal change in carbon dioxide partial pressure is very small, although there is a substantial point-to-point variability (Fig. 6.10). There is at most a rather vague

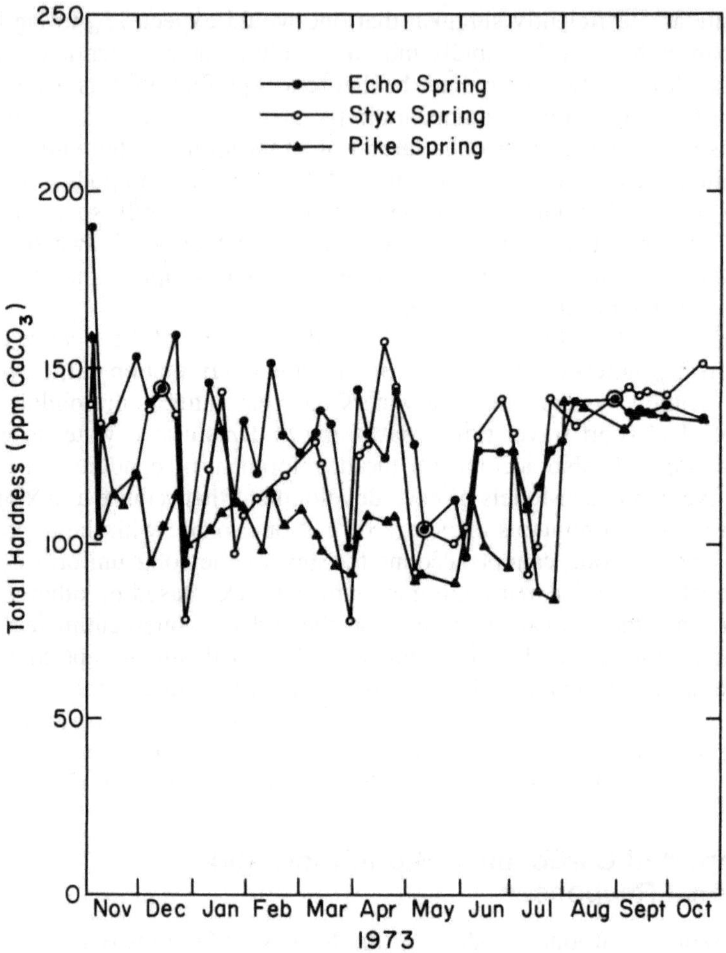

Figure 6.8 Hardness data for three local springs for water year 1972–1973. Each data point represents a single sample and chemical analysis.

seasonal minimum in the period January through March. The annual mean CO_2 pressure is only about one order of magnitude above the atmospheric value. The seasonal cycle is much stronger in the regional springs, which derive substantial parts of their water from the much thicker and more CO_2-productive soils of the Sinkhole Plain (Fig. 6.11). The point-to-point variation is much less, and the average carbon dioxide pressures are higher. There is a pronounced minimum in February through March, which is slightly out of phase with the minimum in the hardness curve. The broad minimum in hardness occurs in March through May and reaches a maximum in August and September. The carbon dioxide curve has a minimum in January through March with a rather broad maximum that spans the summer

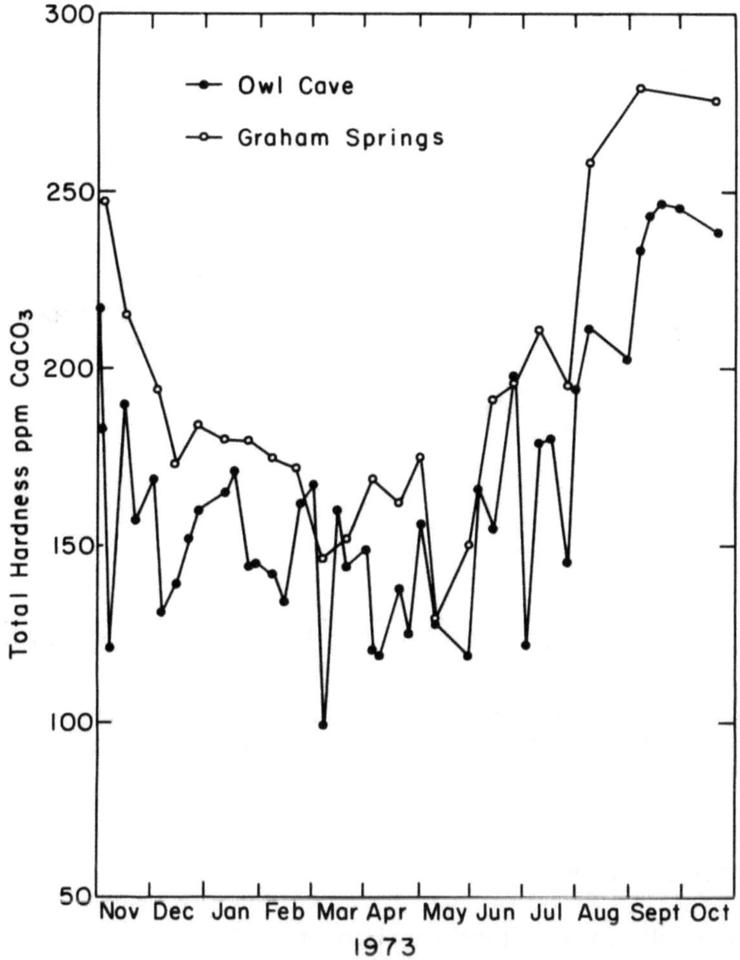

Figure 6.9 Hardness data for two regional springs for water year 1972–1973.

months of July through October. The carbon dioxide curve is controlled mainly by carbon dioxide availability in the recharge area and exactly reflects the growing season. The hardness curve is a superposition of rapid fluctuations due to storm-water flow onto a more slowly varying annual cycle of carbon dioxide production.

The mixing phenomena can be described by the equation

$$C_{\text{outlet}} = \frac{\Sigma C_i Q_i}{\Sigma Q_i} \tag{6.27}$$

Because the differing inputs to the master conduit system have different response times, the chemical response of spring waters to precipitation events becomes very noisy by the time the water reaches the springs.

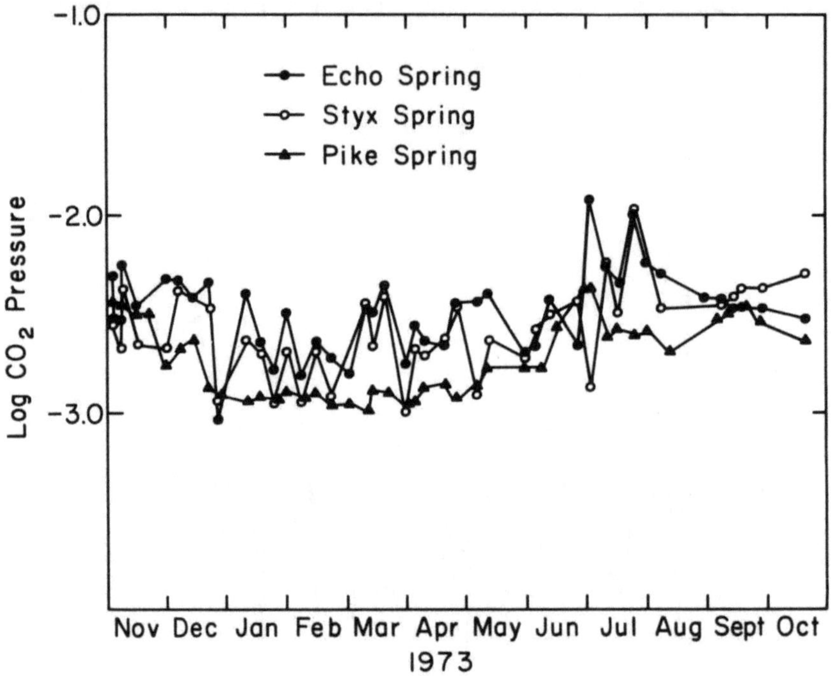

Figure 6.10 Calculated carbon dioxide pressures for three local springs during water year 1972–1973. Each point represents an individual sample and analysis.

Hydrochemical Facies

It was established some time ago (Harmon et al., 1972; Drake and Harmon, 1973) that waters from the various hydrogeologic environments within karst drainage systems carry a characteristic chemical signature. If one looks at the derived parameters of hardness, saturation index, and carbon dioxide partial pressure, these sets of figures are statistically distinct between the various water types. Some notion of what happens within the flow system can be obtained by removing the time variations from the chemical data. The grand averages are plotted in Figures 6.12 and 6.13. All of the data collected during the summer months of June, July, and August and all of the data collected during the winter months of December, January, and February were separately averaged to bring out seasonal effects. Fall and spring samples were omitted from these calculations. Likewise, the data were averaged over all sampling sites in the same hydrogeologic setting. The scatter around these averages is very substantial. The plotted points should

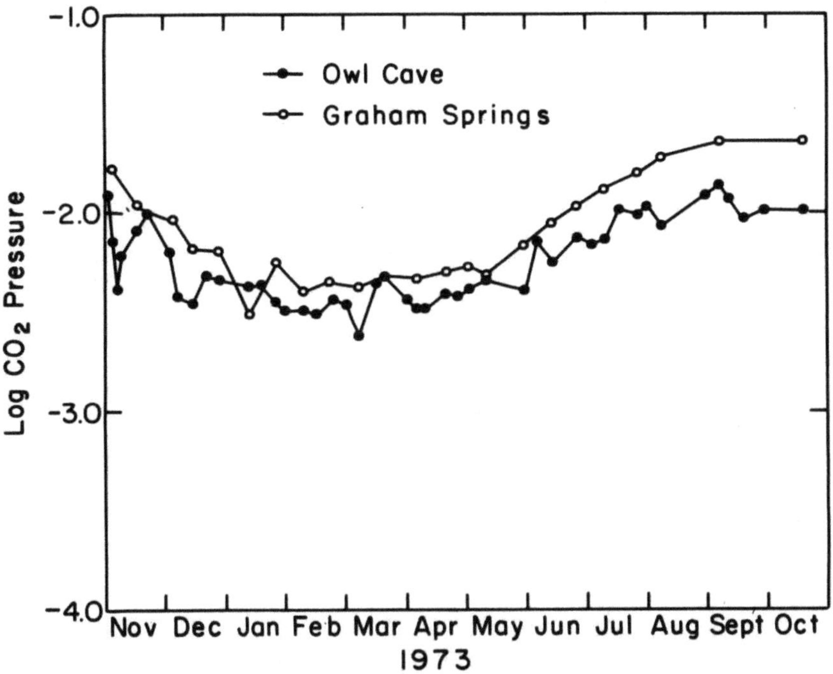

Figure 6.11 Calculated carbon dioxide partial pressures for two regional springs for water year 1972–1973.

be taken only to represent the trends in a large number of individual analyses.

Looking at the uptake of dissolved carbonate, we find that the lowest carbonate concentration is in the Haney Springs. Comparing the Haney Springs with the local springs shows that there is a general increase in dissolved carbonates with little change in CO_2 pressure, although the mean CO_2 pressure is higher in summer than in winter. Between the surface streams and the regional springs, both hardness and P_{CO_2} increase along the flow path. These trends are more or less parallel in winter and summer, but are offset to higher values in the summer. The higher hardness in summer again reflects the high carbon dioxide availability. There is a substantial increase in CO_2 along the flow path reflecting the influence of the diffuse infiltration water from the Sinkhole Plain.

The saturation index likewise increases from Haney Springs to the local springs representing the uptake of dissolved carbonate along the flow path. The saturation index of water from the Sinkhole Plain in winter from the

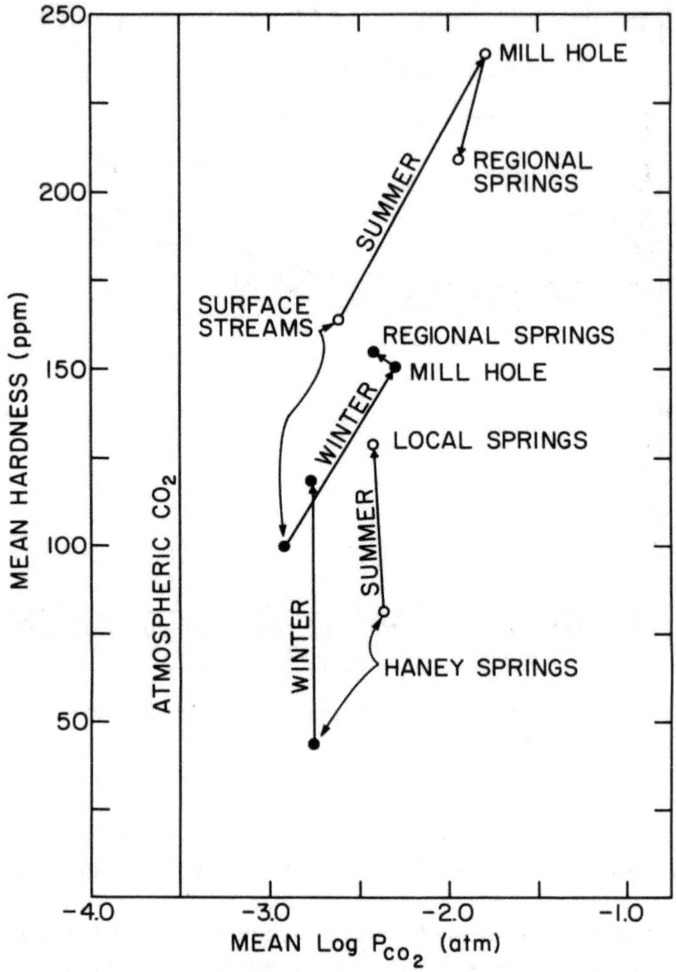

Figure 6.12 Seasonal and hydrogeologic averages of hardness and carbon dioxide partial pressure for various sampling points in the Mammoth Cave area.

surface streams to Mill Hole and the regional springs is of interest. The water remains undersaturated, and there is relatively little difference in the level of saturation between any of the sampling sites, in spite of the increased hardness and increased P_{CO_2}. The result suggests that the waters entering the conduit system are all undersaturated at about the same level, although they could have both a higher carbon dioxide pressure and a higher concentration of dissolved carbonates. The surface streams become

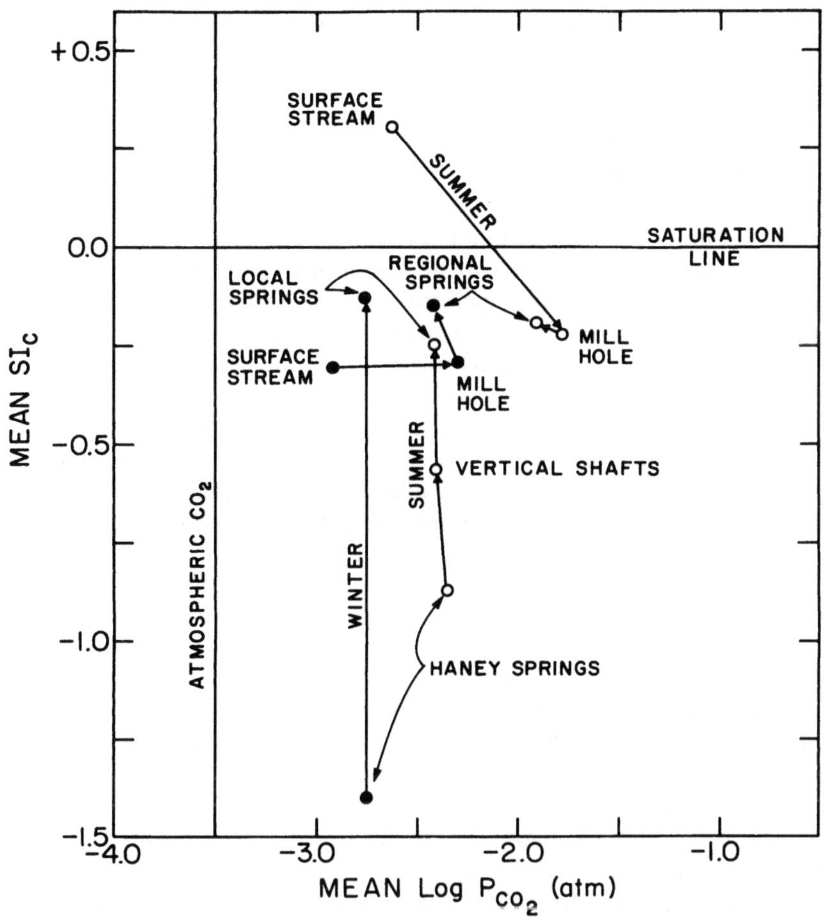

Figure 6.13 Seasonal and hydrogeologic averages of saturation index for calcite and carbon dioxide pressure for various sampling points in the Mammoth Cave area.

supersaturated during the summer months because of the higher temperatures and the relatively lower flows. There is de-gassing of carbon dioxide from the waters seeping into the surface streams from the soils along their banks, leading to supersaturation at the swallets where the surface streams were sampled. Between the swallow holes and the next sampling point at Mill Hole, there is a sufficient replenishment of carbon dioxide in spite of continued carbonate uptake to bring the water that arrives at Mill Hole to the same level of undersaturation as the other points on the plot.

CONCLUSIONS AND FUTURE CHALLENGES

Carbonate Chemistry

From investigations of mostly the input-output chemistry, several hypotheses have been confirmed.

1. Both undersaturated and supersaturated water occur in the vadose zone as Thrailkill (1968) has pointed out. Vertical shaft waters are among the most highly undersaturated to be found anywhere in the system, and seepage waters are the only supersaturated waters found anywhere in the system. These contrasting chemical properties support the concept of treating the infiltration term of classical hydrology as consisting of two terms in karst aquifers: diffuse infiltration and internal runoff.

2. Water discharging from the springs, even springs that drain regional-scale groundwater basins, are undersaturated with respect to calcite and dolomite. Furthermore, the level of undersaturation is roughly in agreement with that required by current laboratory investigations of the rate of dissolution of these minerals as a function of undersaturation.

3. Although dilution and mixing effects dominate the short time scale (days to weeks) fluctuations in spring-water chemistry, these variations are superimposed on an annual cycle controlled by rate of carbon dioxide generation and its seasonal variation.

4. The concept of a "karst water type" or "chemical facies" is confirmed. On the average, there are values of the hardness, saturation index, and carbon dioxide partial pressure that are characteristic of particular hydrogeologic settings.

5. In a very approximate way, the chemical evolution of the conduit waters were traced through the system. Hardness increases along the flow path. In the sequence from the Haney Springs on the ridge top to the vertical shafts, to the regional springs, the carbon dioxide pressure remained constant. In the sequence from the sinking streams at the southern margin of the Sinkhole Plain to the regional springs, the carbon dioxide partial pressure actually increased. Both sets of results imply carbon dioxide uptake along the flow path.

For the future, chemical investigations of the karst system are far from exhausted. We know little or nothing of the detail of the water chemistry within the active drainage system. Aspects that require study include: (1) changes in dissolved carbonates, carbon dioxide, and saturation index along single flow lines; (2) mixing effects as various inputs merge along the flow path; (3) chemical changes over a flood hydrograph for various water types; (4) more information on longer time-scale variations of chemical parameters; and (5) carbon dioxide mass balance between cave waters and cave and soil atmospheres.

Environmental Chemistry

For a relatively rural area, the south-central Kentucky karst has suffered a good deal of abuse. We mention here only a few of the environmental problems.

Sewage contamination of the shallow aquifer beneath the Sinkhole Plain has been a problem for many years. Outhouses, septic tank outfalls, and the direct injection of raw sewage and industrial waste into sinkholes all contribute to massive contamination of the conduit drainage. For this reason, Hidden River Cave, beneath the town of Horse Cave, was abandoned as a show cave and as a public water supply in the early 1940s. In 1961 fecal material could be seen floating on the underground stream. In 1964 the construction of a sewage treatment plant for Horse Cave alleviated some, but not all, of the problems, particularly because the outfall of the plant empties into a sinkhole in the Hidden River drainage (see Chap. 3). At the time of this writing, a new regional sewage system was in the advanced design stages. It should be a substantial improvement; however, in karst terrain, unsewered properties, leaky pipelines, and other malfunctions are likely to maintain some flow of sewage contaminant into the system.

Industrial wastes are a serious pollutant in karst areas. Nickel and chromium contamination in the Hidden River system allowed Quinlan and Rowe (1977) to use the metals as a tracer from Hidden River Cave to the distributary springs on Green River (see Chap. 3). Pollution from industrial waste is a much more serious problem in the city of Bowling Green on the western edge of the karst area. N. C. Crawford's group at the Center for Cave and Karst Studies, Western Kentucky University, has investigated the Lost River system in considerable detail (see Stroud et al., 1986). Hydrocarbons such as gasoline, diesel fuel, chlorinated hydrocarbons, phenolic materials, and other organic compounds float on the karst water table and are transported along the shallow master conduit. When the conduit floods, the organics are floated up with the rising water levels and can enter basements of homes, schools, and businesses. At all times the fumes can migrate upward through solution cavities and dry wells, where they pose both a health and explosion hazard.

Interstate highway 65, which crosses the Sinkhole Plain, is a major north–south transportation route. Accidents involving trucks carrying hazardous materials are surprisingly common, and spillages into sinkholes adjacent to the highway occur very easily (Quinlan, 1986). Because of the high-flow velocities in the conduit system, contamination by highly toxic substances, such as the cyanide spill near Cave City in 1980, requires prompt action by authorities lest the spilled material reach local citizens through their well water.

The chemistry of sewage, agricultural chemicals, industrial wastes, and other toxic substances in the karst groundwater system is an important subject for research. Carbonate waters are an efficient buffer system in most natural environments. Heavy metals should quickly precipitate in karstic conduits, yet Quinlan and Rowe's (1977) results showed that the metals reached the springs. Little is known of the reactions of most organic contaminants in karst groundwater. Most of the research to date has dealt with dispersal pathways and inventories of the contaminants. The next step is to investigate the interactive chemistry of the system.

REFERENCES

Berner, R. A., and J. W. Morse, 1974, Dissolution kinetics of calcium carbonate in sea water, IV, Theory of calcite dissolution, *Am. Jour. Sci.* **274:**108–134.

Buhmann, D., and W. Dreybrodt, 1985*a*, The kinetics of calcite dissolution and precipitation in geologically relevant situations of karst areas, 1: Open system, *Chem. Geology* **48:**189–211.

Buhmann, D., and W. Dreybrodt, 1985*b*, The kinetics of calcite dissolution and precipitation in geologically relevant situations of karst areas, 2: Closed systems, *Chem. Geology* **53:**109–124.

Busenberg, E., and L. N. Plummer, 1982, The kinetics of dissolution of dolomite in CO_2–H_2O systems at 1.5 to 65° C and 0 to 1 atm P_{CO_2}, *Am. Jour. Sci.* **282:**45–78.

Clark, W. C., 1982, *Carbon Dioxide Review: 1982*, New York: Oxford University Press, 469p.

Drake, J. J., and R. S. Harmon, 1973, Hydrochemical environments of carbonate terrains, *Water Resources Research* **9:**949–957.

George, A. I., 1974, Preliminary index of gypsum spelothems in the caves of Kentucky, Indiana, and Tennessee, in *Proceedings of the 4th Conference on Karst Geology and Hydrology,* H. W. Rauch and E. Werner, eds., Morgantown, W. Va.: West Virginia Geological Survey, pp. 169–177.

Harmon, R. S., J. W. Hess, R. W. Jacobson, E. T. Shuster, C. Haygood, and W. B. White, 1972, Chemistry of carbonate denudation in North America, *Cave Research Group Great Britain Trans.* **14**(2):96–103.

Herman, J. S., 1982, The dissolution kinetics of calcite, dolomite, and dolomitic rocks in the CO_2-water system, The Pennsylvania State University, Ph.D. dissertation, 214p.

Herman, J. S., and W. B. White, 1985, Dissolution kinetics of dolomite: effects of lithology and fluid flow velocity, *Geochim. et Cosmochim. Acta* **49:**2017–2026.

Hess, J. W., Jr., 1974, Hydrochemical investigations of the central Kentucky karst aquifer system, The Pennsylvania State University, Ph.D. dissertation, 218p.

Hill, C. A., 1981, Speleogenesis of Carlsbad Caverns and other caves of the Guadalupe Mountains, *Proc. 8th Internat. Congress Speleol. (Bowling Green, Ky.)* **1:**143–144.

Jacobson, R. L., and D. Langmuir, 1970, The chemical history of some spring waters in carbonate rocks, *Ground Water* **8**(3):5–9.

Kern, D. M., 1960, The hydration of carbon dioxide, *Jour. Chem. Ed.* **37:**14–23.

Loewenthal, R. E., and G. v. R. Marais, 1976, *Carbonate Chemistry of Aquatic Systems: Theory and Applications,* Ann Arbor, Mich.: Ann Arbor Science Pub., 433p.

Miotke, F. -D., 1972, Die Messung des CO_2-Gehaltes der Bodenluft mit dem Dräger-Gerät und die beschleunigte Kalklösung durch höhere Flieszgeschwindigkeiten, *Zeitschr. Geomorphologie* **16:**93–102.

Miotke, F. -D., 1974, Carbon dioxide and the soil atmosphere, *Karst u. Höhlenkunde, Abh. Ser. A, No. 9,* pp. 1–49.

Miotke, F. -D., 1975, *Der Karst im zentralen Kentucky bei Mammoth Cave,* Jahrbuch der Geographischen Gesellschaft zu Hannover für 1973, 360p.

Nordstrom, D. K., and J. L. Munoz, 1985, *Geochemical Thermodynamics,* Menlo Park, Calif.: Benjamin/Cummings, 477p.

Nordstrom, D. K., L. N. Plummer, T. M. L. Wigley, T. J. Wolery, J. W. Ball, E. A. Jenne, R. L. Bassett, D. A. Crerar, T. M. Florence, B. Fritz, M. Hoffman, G. R. Holdren, Jr., G. M. Lafon, S. V. Mattigod, R. E. McDuff, F. Morel, M. M. Reddy, G. Sposito, and J. Thrailkill, 1979, A comparison of computerized chemical models for equilibrium calculations in aqueous systems, in *Chemical Modeling in Aqueous Systems,* E. A. Jenne, ed., Washington, D.C.: American Chemical Society Symposium Ser. 93, pp. 857–892.

Plummer, L. N., and E. Busenberg, 1982, The solubilities of calcite, aragonite, and vaterite in CO_2–H_2O solutions between 0 and 90° C, and an evaluation of the aqueous model for the system $CaCO_3$–CO_2–H_2O, *Geochim. et Cosmochim. Acta* **46:**1011–1040.

Plummer, L. N., T. M. L. Wigley, and D. L. Parkhurst, 1978*a*, The kinetics of calcite dissolution in CO_2-water systems at 5 to 60° C and 0.0 to 1.0 atm CO_2, *Am. Jour. Sci.* **278:**179–216.

Plummer, L. N., B. F. Jones, and A. H. Truesdell, 1978*b*, WATEQF—A FORTRAN IV version of WATEQ, a computer program for calculating chemical equilibrium of natural waters, *U.S. Geol. Survey Water Resources Inv. 76–13,* 63p.

Pohl, E. R., and W. B. White, 1965, Sulfate minerals: their origin in the central Kentucky karst, *Am. Mineralogist* **50:**1461–1465.

Quinlan, J. F., 1986, Recommended procedure for evaluating the effects of spills of hazardous materials on ground water quality in karst terranes, in *Proceedings of the Environmental Problems in Karst Terranes and Their Solutions Conference,* Dublin, Ohio: National Water Well Association, pp. 183–196.

Quinlan, J. F., and D. R. Rowe, 1977, Hydrology and water quality in the central Kentucky karst: Phase I, *Kentucky Univ. Water Resources Inst. Research Rept. No. 101,* 93p.

Quinlan, J. F., and D. R. Rowe, 1978, Hydrology and water quality in the central Kentucky karst: Phase II, Part A: Preliminary summary of the hydrogeology of the Mill Hole sub-basin of the Turnhole Spring groundwater basin, *Kentucky Univ. Water Resources Inst. Research Rept. No. 109,* 42p.

Reddy, M. M., L. N. Plummer, and E. Busenberg, 1981, Crystal growth of calcite from calcium bicarbonate solutions at constant P_{CO_2} and 25° C: a test of a calcite dissolution model, *Geochim. et Cosmochim. Acta* **45:**1281–1289.

Stroud, F. B., J. Gilbert, G. W. Powell, N. C. Crawford, M. J. Rigatti, and P. C. Johnston, 1986, U.S. Environmental Protection Agency response to toxic fumes and contaminated ground water in karst topography: Bowling Green, Kentucky, in *Proceedings of the Environmental Problems in Karst Terranes and Their Solutions Conference,* Dublin, Ohio: National Water Well Association, pp. 197–226.

Stumm, W., and J. J. Morgan, 1981, *Aquatic Chemistry,* 2nd edition, New York: John Wiley & Sons, 780p.

Thrailkill, J., 1968, Chemical and hydrologic factors in the excavation of limestone caves, *Geol. Soc. America Bull.* **79:**19–46.

Thrailkill, J., 1972, Carbonate chemistry of aquifer and stream water in Kentucky, *Jour. Hydrology* **16:**93–104.

Thrailkill, J., 1976, Carbonate equilibria in karst waters, in *Karst Hydrology and Water Resources,* V. Yevjevich, ed., Fort Collins, Colo.: Water Resources Pub., pp. 34-1–34-27.

Thrailkill, J., and T. L. Robl, 1981, Carbonate geochemistry of vadose water recharging limestone aquifers, *Jour. Hydrology* **54:**195–208.

Wigley, T. M. L., 1977, WATSPEC: A computer program for determining the equilibrium speciation of aqueous solutions, *British Geomorph. Research Group Tech. Bull. 20,* 48p.

7

CAVE SYSTEMS SOUTH OF THE GREEN RIVER

Thomas A. Brucker

The cave systems in south-central Kentucky present a rather different perspective to the cave explorer than to the hydrogeologist. The enlargement of the conduit system by concentrated groundwater flow took place early in the history of the cave system, often when many of the enlarging conduits were below regional base level and inaccessible to direct observation. Cave explorers came on the scene rather late, after dissection of the surface landscape had broken up many of the conduits so that only fragments remained as caves to be explored. Internally, many other late-stage processes, such as breakdown and passage plugs of flowstone or clastic sediment, truncated the passages; therefore, reconstruction of the original conduit system must be done by systematic mapping and a good deal of conjecture. As a further frustration, there is nothing in the process of conduit development that ensures human-size entrances into the conduits. Inlets might be sinkhole drains that are too small to traverse, and outlets might pass through artesian or alluviated springs. Entrance formation is a stochastic process brought on by random processes of truncation and decay (Curl, 1958); there is no guiding principle that makes entrances in the places an explorer would find most useful.

Exploration, therefore, demands a good deal of persistence. Explorers must first locate entrances, an exercise that often requires excavation of promising sinkholes, stream inlets, and obscure holes on the hillside that are observed to vent cool air in the summertime. A typical entrance gives access only to a local arrangement of passages, and more persistence is required to find connections from these passages into large fragments of the master conduit system.

Cavers are interested in integrating cave systems. Exploration does not follow individual drainage lines; any sort of underground connection through breakdown collapses, up or down vertical shafts, or through the vague shiftings of shaft drains and small distributary and cut-around passages is used to get from one part of the conduit system to another. When the entire complex has been mapped, it constitutes a single cave of some length specified by the survey regardless of how many fragments of different drainage lines it contains.

This chapter offers a description of some of the large cave systems south of the Green River and, in particular, the Mammoth Cave system itself, which, with more than 500 km of surveyed passage contains some of the longest fragments of conduit known anywhere. The presently known caves, which have been physically integrated to form the Mammoth Cave system are shown on Plate 2. Reference should also be made to Figure 2.23, which shows the locations of many of the other caves in the area. The frequency of occurrence of caves with a given length falls off exponentially with increasing length. Table 7.1, which lists the 1987 lengths of the larger caves of the south-central Kentucky karst, shows this idea numerically.

Table 7.1 Major Caves in the Mammoth Cave Area

Cave	*Length (km)*
Mammoth Cave system	523.023
Fisher Ridge Cave system	71.50
Whigpistle Cave	36.21
Hicks Cave (Hidden River system)	31.38
Gradys Cave	19.84
Crumps Spring Cave	17.60
James Cave	16.50
Lee Cave	12.58
Lost River Cave system	10.46
Vinegar Ridge Cave	9.66
Parker Cave	6.92
Coach (Hundred Domes) Cave	5.44
Northtown (Short's Moonshine) Cave	4.90
Smith Valley Cave	4.70
Great Onyx Cave	4.59
Buckner Spring Cave	3.75
Ganter Cave	3.00

Source: All data from Gulden (1988).

HISTORY OF EXPLORATION

Explorers have entered central Kentucky caves for thousands of years. The nineteenth century explorers brought new sophistication in both lighting apparatus and cartographic skills, which enabled both explorers and economic exploiters to venture for kilometers in Mammoth Cave and return. The earliest instrument-grade map of Mammoth Cave was a transit survey by E. F. Lee in 1835 (Quinlan, 1964), but the most complete map until recent time was drafted by Max Kaemper in 1908, with the assistance of guide Ed Bishop. Some 100 km of passage beneath the property of the Mammoth Cave estate were surveyed, yet curiously, passages that extended beyond the property line were omitted. This map was used to open new artificial entrances to Mammoth Cave, and the subsequent "cave wars" between competing commercial enterprises were one result.

The founding of the National Speleological Society (NSS) in 1941 gave focus to those who were interested in the systematic exploration and survey of caves, whether for sport or for scientific research. Survey and exploration techniques were improved, and cavers joined with scientists to interpret a framework of observations and measurements that identified patterns and similarities found not just in adjacent caves, but in those around the world.

In central Kentucky, the early explorers, discoverers, and trail builders included such people as Floyd Collins, George Morrison, Stephen Bishop, Edmund Turner, and the Lee brothers. They gave way to modern cavers whose mark was not just an entrance, trail, or scratched initials in some chamber, but maps that showed with fair precision the features in each cave and their relationship to the land surface above. These maps were so useful to exploration that the connecting of small caves became both a goal and a reality.

The NSS sponsored a week-long underground expedition to systematically map the maze near Floyd's Lost Passage in what was then known as Great Crystal Cave beneath Flint Ridge (Lawrence and Brucker, 1955). Adding to the survey work of Crystal Cave's manager, William T. Austin, Crystal Cave became the region's best documented and fastest growing (in length) cave. In 1954, Austin and Jack Lehrberger explored beyond a pit in tiny Unknown Cave to discover and map more than 25 km of passage beneath Flint Ridge. The connection of Unknown Cave and Crystal Cave the following year led to speculation that the newly named Flint Ridge Cave system might become the world's longest cave.

In 1957 the Cave Research Foundation (CRF) was incorporated with the purpose of sponsoring research in the Flint Ridge caves (Watson, 1981; Brucker, 1955; Smith, 1957). Throughout the 1960s, Salts and Colossal caves were connected to the Flint Ridge Cave system, and it became the world's longest. Passages beneath a karst valley approached Mammoth Cave within 500 m. A high point of CRF efforts was the discovery of the connecting passage between Flint Ridge and Mammoth Cave in 1972 (Brucker and Watson, 1976; Crowther et al., 1984). By this time computers were needed to manage a burgeoning database. In 1981, Hawkins' River in Proctor Cave was linked via the newly discovered Logsdon River in Morrison's Cave to Mammoth Cave (Brucker and Lindsley, 1979; Coons and Engler, 1980). A major connection in 1985 to 85 km of Roppel Cave extended the system to more than 500 km. In the most recent years, exploration was de-emphasized because replacement surveys became necessary to meet improved standards of survey and observation. Extension of the system to 600 km continues as more passage is discovered every year.

A small spring, which fell into a pit near Fisher Ridge, led to the cave known as Crump Spring. Although this cave required much uncomfortable crawling, it was the first significant discovery and mapping effort outside of the national park. After mapping 20 km, interest in the cave waned; however, other groups of cavers discovered significant caves elsewhere in the region.

An entrance on Toohey Ridge led Jim Currens, Jim Borden, and Bill Walter to a dense network of passages far from Mammoth Cave (Borden et al., 1980; Borden, 1982). Roppel Cave had been extended in 1981 to 85 km through discoveries of large conduits and the excavation of a second entrance. The Central Kentucky Karst Coalition (CKKC) was founded in 1974 to document this significant and fast-growing cave system. A connection between Roppel Cave and Mammoth Cave was established through a seasonal sump in Logsdon's River by two teams in wetsuits.

A census of springs, sinking streams, and their relationships to karst springs and cave streams led cavers, under the direction of James F. Quinlan, to two caves. One significant Green River spring near Horse Cave was investigated in 1973, which led to the discovery of Hicks Cave, also known as Hidden River complex. This lengthy base-level conduit was mapped to the limits of human endurance, and it remains one of the few known systems of significant length beneath the Sinkhole Plain. In 1987, a newly completed entrance, drilled into the upstream end of Hicks Cave, provided access to this "rejuvenated" 30-km cave, and the mapping continues. The second discovery, Whigpistle Cave, lies near the edge of the Chester Cuesta (Taylor, 1979). Small passages led to a network of conduits including the largest room in a central Kentucky cave. None of its passages were found to extend beneath the escarpment to the Sinkhole Plain. The discovery of a major river passage in 1985 is yet to be revisited. Its physical difficulty and flood danger make Whigpistle Cave one of the least visited caves in the region, yet it lies astride the Turnhole Spring drainage basin, and has the potential for many more miles of passages.

Fisher Ridge Cave system was discovered by searching a caveless ridge for an entrance (Bean et al., 1981). In a few short years, the surveys of this system have surpassed 70 km. A network of dry, low-gradient trunks and base-level valley drains was found in an area with unusually high passage density. Fisher Ridge itself is in the more highly dissected northeast portion of the Chester Cuesta. One drain passage extends to the east almost to the escarpment, but in a pattern similar to that found in Whigpistle Cave, the passage ended.

New discoveries in the region add an estimated 30 km per year to the surveyed caves in the region. Four organized groups of cavers and several individuals continue to add to our understanding of the patterns and features of central Kentucky caves at a rapid pace. The potential for discovery of more passages might be incalculable.

LOW-GRADIENT TRUNK PASSAGES

Low-gradient trunk passages are found in every part and at every elevation within the limestones south of Green River. The highest trunks are found in the Girkin Limestone near its contact with the Big Clifty Sandstone, only in the proximity of the Green River. The regional dip of the limestone is about 2% toward the northwest. The downdip direction is also toward Green River, where much of the Sinkhole Plain water reaches the surface in karst springs. There is no reason to doubt that the dry trunk passages once served as groundwater conduits just as the feeders of the major springs do today.

Trunk passages seem to have gradients below 1%. This gradient is similar to that measured between the active Turnhole Spring/Mill Hole system. Vertical distribution of known trunks is evidence of the differential between regional dip and trunk passage gradients. Six kilometers from the Green River, trunk passages are found only in the Ste. Genevieve Limestone, some 45 km beneath the sandstone.

As these trunks cut across the various limestone beds, they undergo changes in cross section. Perhaps the only consistent feature found throughout the entire length of a trunk passage is its low gradient. Trunks generally have few tributaries at grade and have somewhat ambiguous orientations. Local dip and strike are often cited as the determinant feature of passage orientation and shape. Palmer (1981) cites the Marion/-Cleaveland Avenue section of Mammoth Cave as an example of strike and dip influences (see Fig. 11.8). The area of the structural trough centered at the Carmichael entrance is crossed by two low-gradient trunks, the lower Cleaveland Avenue and the upper Sandstone Avenue, a component of the Main Cave trunk. These two avenues cross at right angles, suggesting that structural controls alone do not determine passage orientation. Trunk-passage orientation could determine spring locations, yet there are numerous examples of distributary bifurcation, such as Audubon Avenue and Houchins Narrows in Mammoth Cave.

Portions of five known, low-gradient trunks have been sufficiently integrated through exploration to describe their general characteristics: Upper

Salts, Main Cave, Deer Park Avenue, Grand Avenue, and Frost Avenue are the general names proposed for these trunk systems, although each is a composite of several passages. Hidden River Cave, which is beneath Horse Cave, and the Hicks Cave trunk might also belong on this list. There are other low-gradient passage segments within the region; however, only those within the Mammoth Cave system have sufficient "exposure" to be considered in this description, much of which draws on the measurements and observations of Palmer (1981). Although each trunk system is several kilometers in length, none is known to extend from the recharge area on the Sinkhole Plain to a discharge site along the Green River; the known passages cover only one-third of this distance. Detailed measurements of passage gradient have not been made along these trunks; instead, rough approximations can be made by determining the stratigraphic horizon at the known extents of each trunk, and applying what Palmer (1981) has observed about the local dip of individual beds.

THE MAMMOTH CAVE SYSTEM

Spatial relationships in caves are difficult to grasp because one cannot actually see more than a short segment of cave passage at any moment. Maps allow a geometric representation of the overall plan of the cave passages and of their various interconnections. Map making in caves is an arduous and time-consuming undertaking. The CRF/CKKC map of the Mammoth Cave system, shown on Plate 2, is a synopsis of more than 30 years of effort by hundreds of individuals and many tens of thousands of person hours of labor both underground and in data reduction.

Plate 2 shows the layout of the main passages in the Mammoth Cave system in relation to surface topography. Cave descriptions are laced with dozens of place names—every passage and every passage junction gets a name—of which only a few can be shown on a map at the scale of Plate 2. Likewise, it is impossible to give a verbal walking tour of a big cave system without writing a very long and doubtless very tedious treatise. The sections that follow attempt to describe briefly some of the main components of the Mammoth Cave system; they are not detailed passage-by-passage descriptions.

The Upper Salts Avenue Trunk

Upper Salts is the largest passage in the Mammoth Cave system. It is a large canyon, extensively modified by breakdown, and filled with secondary sediments. Dismal Valley, a slump feature caused by the presence of a cross-cutting, stream-carrying passage, exposes 10 m of sediment and 10 m of breakdown in a passage 20 m high. In places, solution features can be seen in the walls within 5 m of the present ceiling. Passageways beneath the floor yield a few known, reliable points to measure total passage height. Total cross-sectional area for Upper Salts might be as much as 400 m^2, with an estimated 25% caused by breakdown enlargement. Numerous meanders and braided loops occur at three levels in the trunk, and at least two occur at grade. The floors of Dismal Valley and the River Map section are at the same approximate elevation. The reason for these meander loops is not known.

From the Salts Cave entrance, this passage follows the dip toward Green River. At its downstream end Upper Salts divides in an apparent distributary network. To the northwest, one section divides several times; however, each branch is truncated by valley wall collapse. Correlations between trunks across valleys can be made by examining relative elevations, stratigraphic horizons, and relative cross sections of passages that oppose each other (Brucker, 1966); however, no information can be gathered about passage character, such as the presence of additional tributaries, across these gaps.

The two levels of Great Onyx Cave might be continuations of the Salts trunk in its distributary network (Palmer, 1981). Edwards Avenue, a 30-m^2 tubular passage, is relatively breakdown free, but it is nearly filled with sediments. This passage joins the slightly lower Cox Avenue, and near the Great Onyx entrance it attains 60 m^2 in cross section. Dyer Avenue in Crystal Cave is another segment of Upper Salts. It is inferred that this passage is a distributary from the northwest end of Salts Cave at the lowest level. Dyer Avenue is aligned perpendicular to the trend of Salts, as it crosses a structural anticline whose center is cut by Green River. This passage is large, approximately 60 m^2, and filled to a great extent by sediments and breakdown. An excavation in a tributary of Dyer Avenue penetrated 20 m of sediment and rock fill without reaching a bedrock floor; therefore, this cross-section estimate might be too low. In 1985, a short piece of Salts trunk was discovered upstream of the Salts entrance sink. Although truncated by a major surface sink, this passage does continue in the trend of Upper Salts, and contains much breakdown in as many as three meanders.

If the assumption is made that Upper Salts has a 1% gradient upstream Salts trunk could correlate to Yahoo Avenue or Downy Avenue in Roppel Cave, 5 km to the southeast. These passages are in the Joppa Member of the Ste. Genevieve Limestone, 40 m lower than the passages at the Salts entrance. Neither avenue is exceptionally large, but the numerous related tributaries suggest that these could be feeders to the Salts trunk. The 5-km gap leaves much room for speculation.

The Main Cave Trunk

The Main Cave trunk includes many of the large passages seen in the developed portions of Mammoth Cave (see Palmer, 1981). Main Cave, and to a greater extent its upstream continuation, Kentucky Avenue, are extensively modified by breakdown and filled with sediments. The sediments were once mined as a source of nitrates for gunpowder near its paleo downstream end at the Historic entrance. Audubon Avenue and Houchins Narrows (and Dixon Cave) are a paleo-distributary network lying quite close to the Green River. Southeast of the Rotunda, Main Cave has a cross section of 250 m^2. Gothic Avenue, which appears as a cross-cutting passage at the ceiling, might actually be a portion of a large meander, from Blue Spring Branch to Blackall Avenue, through Gothic to the Methodist Church. Behind the Methodist Church is a large breakdown-filled room that could be a remnant of a third higher level distributary for Main Cave. Across a short gap at the Violet City entrance chamber, Main Cave continues as Sandstone Avenue. This impressive canyon shows evidence of solution 6 m from the present ceiling, but it is largely filled with sediment, breakdown, and flowstone deposits. A 1-km gap exists between Sandstone Avenue and the beginning of Kentucky Avenue. Kentucky Avenue, at its downstream end, has almost no breakdown, and it exhibits solution features at the ceiling and numerous meanders. Beyond the first kilometer, Kentucky Avenue has 40 m of vertical relief because massive breakdown domes obscure the original level of the passage. At its terminus at the Frozen Niagara entrance, Kentucky Avenue is so modified by breakdown and flowstone mineralization that the historic ceiling elevation is a mystery. There are questionable areas in the Black Chambers, Blue Spring Branch, and in Bismark Dome that suggest significant pieces of Main Cave may yet be discovered.

Deer Park Avenue

Deer Park Avenue is the least exposed low-gradient trunk. More is known about its principal tributary system, Big Avenue/Cleaveland Avenue, a tributary that perhaps had significant recharge in the Chester Cuesta. Deer Park Avenue is 80 m^2 in cross section, and less than 1 km of passage is known. Numerous portions of Deer Park trunk are breakdown-free, however the passage is aligned with a fracture zone (often mapped as a fault) north toward Green River. This fracture zone caused a great deal of truncation by breakdown collapse, creating a jumble of passage fragments. A recent discovery uncovered one passage that might have been a distributary for Deer Park Avenue.

Upstream, Deer Park Avenue extends through a breakdown into one of the largest rooms in the cave, Discovery 70, which is 40 m wide and 20 m high. This room has at its perimeter a network of intersecting passages of a more recent nature. The upstream continuation of this trunk remains undiscovered because the collapse of a karst valley (Deer Park Hollow) terminates Discovery 70. Big Avenue, the major tributary, is aligned along the local dip, which in this area is toward the west. Approximately 1 km from Deer Park Avenue a major portion (the ceiling) of Big Avenue terminates in a sand fill. The remaining portion continues, across a short gap, as Franklin and Cleaveland Avenues. Cleaveland Avenue is a classic tube, with an average volume of 60 m^2. It follows the strike as it passes beneath Mammoth Cave Ridge toward the Snowball Dining Room. At this junction, Marion Avenue joins from the north. Marion Avenue might have drained the early Houchins Valley, the karst valley separating Flint and Mammoth Ridges, rather than the Sinkhole Plain. Continuing as Boone Avenue, the other branch extends from the Snowball Dining Room the entire length of the ridge as Pilgrim, Robertson, and Fox Avenues. These passages have numerous tributaries. Fox Avenue might grade into the floor of Kentucky Avenue at Grand Central Station and near College Heights Avenue. For these reasons, the Cleaveland Avenue tributary might not be representative of a low-gradient trunk.

Grand Avenue

Grand Avenue, beneath Flint Ridge, is perhaps the best exposed of the low-gradient trunks. The upstream component, Grand Avenue in Colossal Cave, also has the greatest volume: 100 m^2. Two primary levels meander from Colossal Dome to Sandstone Tumbledown, the point at which the upper level is terminated by breakdown. Grand Avenue has a canyon shape which, although modified by breakdown, is not modified to the extent of other passages discussed. Sediments in the Pearly Pools Route contain a large pebble fraction, which suggest that high-stream velocities were present in this conduit.

From the Colossal entrance, the downstream terminus of Grand Avenue, a short gap exists to its continuation at Argo Junction. Here, the two components of Grand Avenue continue almost as separate entities, Turner Avenue, the higher, and Mather Avenue, the lower. Turner Avenue is a classic tube throughout its length, averaging 32 m^2 in area (see Fig. 9.12). It extends northward to Brucker Breakdown with little variation in profile or volume. Mather Avenue is the larger, being 60 m^2 and it remains a canyon, oriented along the dip to the northwest.

One-half km beyond Argo Junction, a major distributary, the tubular (6 m^2) Swinnerton Avenue, leaves Mather Avenue at grade, turning westward where it might have met Green River at Floating Mill Hollow. Swinnerton Avenue might have been a piracy or capture by a spring with an hydraulic advantage over springs in the vicinity of Great Onyx Cave or Pike Spring, or it might have been a diversion necessitated by Mather Avenue's collapse somewhere downstream.

Mather Avenue is typically a wide canyon of 60 m^2 cross section, with frequent short meander tubes in the floor or walls. A thin (<1 m) sediment layer is found throughout its length, and this is often covered by breakdown of varying thickness. North Mather Avenue terminates beneath elongated sinkholes, presumably collapsed into Mather, but then it continues as Ralph Stone Hall. At least a portion continues beyond as Huber Trail, a passage that might also intersect a continuation of Turner Avenue; however, the exact relationships remain to be worked out. Turner Avenue intersects Mather Avenue at its midpoint in Albright Junction, a collapse area with a network of small tubes that seem to bypass the base of this breakdown to join North Mather. Turner Avenue terminates in rather abrupt fashion where Brucker Breakdown slices it and fills it with sandstone. Grand Avenue might have drained toward a spring at Three Sister's Hollow or at Great Onyx Cave; no evidence favors either site.

Frost Avenue

Beneath Joppa Ridge lies a network of moderate to large passages that are reached through Proctor Cave. Frost Avenue, which lies within the Joppa Member of thc Stc. Genevieve Limestone near the same elevation as Grand Avenue, has many apparent tributary branches and is often intersected by vertical shafts or modified by canyons. The westernmost 3 km of passage are low gradient. Breakdown terminates Frost Avenue several kilometers from Green River, and no passages have been discovered beyond that might correlate. Likewise, Frost Avenue's upstream branches terminate with no apparent continuation; however, discoveries from passages at lower levels could someday lead up to this level. Longs Cave, a system of very large canyon trunks near the escarpment's edge is a possible relative of Frost Avenue; however, its considerably larger volume (150 m^2) is not accounted for by Frost Avenue.

Other Bits and Pieces

Lee Cave, also in Joppa Ridge, is a low-gradient trunk with a rectangular cross section of 100 m^2 (Freeman et al., 1973). It is partially filled with thick sediments, which, at one location, form unusual tall sediment mounds that resemble the fluvial deposition found inside stream meaders. Lee Cave apparently drained from east to west, roughly along strike, in a manner that seems atypical, when compared to other passages in the area. It is possible that Lee Cave was a subterranean cutoff for Green River. An active river cutoff is known east of Mansfield bend on the Green River.

Longs Cave is enigmatic. Far from the Green River, its enormous sediment-filled trunk(s) does(do) not seem to correlate with other known passages. This 150 m^2 canyon, with two tributaries, has an undulating ceiling. The total cave length is too short to establish a gradient.

Natural Tunnel is much too short to establish a gradient. It is a single, truncated passage segment located above Pike Spring, and although breakdown modified, it appears to have a large cross section (100 m^2). It is mentioned here as a reminder that many low-gradient trunks could be obscured from observation: some have certainly been destroyed by valley solution and collapse; some are filled by sediments, breakdown, or secondary calcite deposits; and some remain undiscovered.

OTHER PASSAGE TYPES AND OTHER CAVES

Shaft Drains

Passages of high gradient, 3% or greater, occur throughout the region. Often these form extensive networks of integrated dendritic passages that are found at all elevations, from the dry upper levels to active, stream-carrying levels within the current Green River flood zone. Many are found to cross above or below other passages without intersection. Frequently, intersections show no evidence of contemporaneous solution, resembling rather the chance intersection of an older passage by a younger one. A great deal of our sense of paleohydrologic history derives from the relationships of these shaft drains to the low-gradient trunks they intersect.

Shaft drains are typically smaller in cross section, with passage height greater than width (classic canyon proportions). Frequently they can be followed upstream to a vertical shaft or shaft complex, often developed beneath the lower contact of the Big Clifty Sandstone at ridge margins. Retreating ridge margins and the collapse of vertical shafts have isolated many shaft drains from their former sources. Often shaft drains can be followed upstream where they intersect other drain passages. Junctions at grade are (were) tributaries, while vertical intersections might be between genetically related passages, or the chance intersection of another shaft drain, with subsequent stream capture.

Downstream, shaft drains will have tributaries at grade, vertical junctions, stream-capture cutoffs, and chance intersections with other passages, or they might suddenly drop into other vertical shafts. Generally, shaft drains increase in cross section downstream and in volume of water flow (for active stream passages) as tributaries coalesce into master drains, or "valley drains." Shorter caves usually contain this type of passage. Crump Spring Cave consists of 18 km of small-shaft drain passages, with only one short segment of "trunk-sized" passage. James Cave, on Bald Knob, is another lengthy system of shafts and drains so intertwined that it resembles a vertical maze. Passage density is typically high in these caves, yet the overall water-carrying capacity is much less than that of a low-gradient trunk.

Valley Drains

Valley drains are the longest continuous passages found in cuesta caves. They have medium gradients (1–4%) and various shapes, from canyon to tube. They can be found perched on less soluble beds of shaly or siliceous limestone, such as the extensive Hawkin's/Logsdon River passage, or grade to base-level trunk streams as the Eaton Valley complex and Bottomless Pit drain to Echo River. Karst valleys typically are floored by numerous shallow sinkholes separated by low divides, perhaps contributing to the number of small-flow tributaries of relatively high frequency found in valley drains as contrasted with the high-flow tributaries found in vertical-shaft drains. Valley drains are intersected by few vertical shafts, and these shafts are smaller than those at ridge margins.

Several significant valley drains are known to span great distances; indeed, most cave connections have been made through valley drain systems. Hawkin's River, in Mammoth Cave, and its upstream continuation, Logsdon River, in Roppel Cave, is a sprawling system of 10 km. From its downstream terminus at a sump 4 km from the Green River, this 15 m^2 passage passes beneath several karst valleys, to an upstream sump less than 2 km from the sinkhole plain. Dye-tracing has linked sinkholes at the escarpment to Logsdon River, but no Sinkhole Plain sinking streams are known to recharge Hawkin's River. Tributary passages, though lengthy, are few. There are numerous examples of capture at the margins of Hawkin's River, especially in Arlie Way in Roppel Cave.

Several other significant valley drains should be noted. Mammoth Cave's Mystic River drains both the north and south flanks of Mammoth Cave Ridge. A recently discovered overflow passage led to a sump 3 m below the elevation of Mystic River. A. T. Leithauser, a diver, has connected Mystic River's downstream sump through submerged distributary passages into Roaring River, another valley drain, and into Echo River, the base-level passage for Mammoth Cave. The P Survey/Elysian Way in Roppel Cave is an example of a recently abandoned valley drain. Although it lies near the top of the flood zone it rarely carries water downstream through its extensive series of rimstone dams and other secondary deposits. This 50-m^2 passage has tributaries that link, or overlap, Hawkin's River and Fisher Ridge Cave, with the Pike Spring drainage.

In this sense "trunk" valley drains establish local base level and influence passage gradients and orientation, causing in some cases updip flow and apparent flow reversals in passages draining to other points. Dye tracing has proven that overflow from one drainage basin to another is common (see Chap. 3). The apparent dynamic nature of valley drains at the basin margins suggests that drainage routes are discrete, but are linked by necessity when their capacity to carry water is exceeded. Although there is no direct evidence that low-gradient trunks had a similar paleohydrologic behavior, the evidence could have been destroyed by the retreating valley walls that isolate the trunks. If it could be established that these types of passages are similar, and not different, the study of extensive valley drains could provide a model for the paleohydrology of the higher and more spectacular passages in the region.

Recognizing and delimiting valley drains is difficult without the ability to trace air-filled passages beyond the limits of human exploration. For this reason identification of the sources for abandoned drains at higher elevations is, at best, conjecture. Several passages appear to have functioned as valley drains, such as Robertson Avenue or Lehrberger Avenue in Mammoth Cave, yet they also have characteristics of low-gradient trunks. More work needs to be done.

It might be assumed that caves found beneath the Sinkhole Plain would most strongly resemble small valley drains because the internal runoff from various sinkholes would drain into the passages at many points. Vertical shafts are small and infrequent in Sinkhole Plain caves such as Parker Cave and Hicks Cave; their large trunk passages with well-defined tributaries more closely resemble the low-gradient trunks found beneath the ridges. Passage trends seem to be closely related upstream to the locations of the terminal swallets of their tributary sinking streams.

Sinkhole Plain caves are more difficult to integrate into lengthy systems. Exploration is often blocked by sediment banks and chert-walled sump pools, or explorers are threatened by the probability of dangerous flash flooding. The caves are strongly influenced by local lithologic variations found within the upper St. Louis, a cherty limestone, which confines, perches, or creates blind sumps in passages within the flood zone. Fewer passage levels are found, in part because the available thickness of limestone above the water table is much less than under the plateau, and in part because the high-gradient passages lead quickly to the flood zone, near base level, within a relatively short horizontal distance.

CONCLUSIONS

The database available from cave exploration and survey has expanded enormously over the past several decades. What 30 years ago, when the systematic surveys began, were a group of large but isolated caves on Mammoth Cave Ridge and Flint Ridge, have now been integrated to a single system that also includes large sections (particularly the Roppel section and the Proctor section), which were unknown at the beginning. Other large caves, such as Fisher Ridge Cave and Whigpistle Cave, have been discovered and surveyed in the Mammoth Cave area. In spite of this impressive progress, it is not enough. The known cave fragments allow only a speculative interpretation of the present and past drainage systems.

REFERENCES

Bean, L., D. Crowl, C. Hopper, K. Ortiz, and P. Quick, 1981, Fisher Ridge Cave system, *Natl. Speleol. Soc. NSS News* **39:**209–214.

Borden, J., 1982, The siege of a super-system, *Caving Internat., No. 14*, pp. 20–31.

Borden, J. D., J. C. Currens, W. G. Walter, and P. W. Crecelius, 1980, Toohey Ridge Cave system, *Natl. Speleol. Soc. NSS News* **38:**79–84.

Brucker, R. W., 1955, Recent explorations in Floyd Collins Crystal Cave, *Natl. Speleol. Soc. Bull.* **17:**42–45.

Brucker, R. W., 1966, Truncated cave passages and terminal breakdown in the central Kentucky karst, *Natl. Speleol. Soc. Bull.* **28:**171–178.

Brucker, R. W., and P. Lindsley, 1979, New Kentucky junction, *Natl. Speleol. Soc. NSS News* **37:**231–236.

Brucker, R. W., and R. A. Watson, 1976, *The Longest Cave*, New York: Knopf, 327p.

Coons, D., and S. Engler, 1980, In Morrison's footsteps, *Natl. Speleol. Soc. NSS News* **38:**127–132.

Crowther, P. P., C. F. Pinnix, R. B. Zopf, T. A. Brucker, P. G. Eller, S. G. Wells, and J. P. Wilcox, 1984, *The Grand Kentucky Junction*, St. Louis, Mo.: Cave Books, 96p.

Curl, R. L., 1958, A statistical theory of cave entrance evolution, *Natl. Speleol. Soc. Bull.* **20:**9–21.

Freeman, J. P., G. L. Smith, T. L. Poulson, P. J. Watson, and W. B. White, 1973, Lee Cave, Mammoth Cave National Park, Kentucky, *Natl. Speleol. Soc. Bull.* **35:**109–126.

Gulden, R. E., 1988, List of the Caves of the U.S.A. over 1 Mile, Odenton, Md.: NSS Committee on Long and Deep Caves, 6p.

Lawrence, J., Jr., and R. W. Brucker, 1955, *The Caves Beyond*, New York: Funk & Wagnalls, 283p. (Reprinted by Zephyrus Press, Teaneck, N.J., 1975.)

Palmer, A. N., 1981, *A Geological Guide to Mammoth Cave National Park*, Teaneck, N.J.: Zephyrus Press, 210p.

Quinlan, J. F., 1964, History and Evolution of the Map of Mammoth Cave, unpublished manuscript.

Smith, P. M., 1957, Discovery in Flint Ridge, 1954–1957, *Natl. Speleol. Soc. Bull.* **19:**1–10.

Taylor, R. L., 1979, Discovery in Whigpistle, *Natl. Speleol. Soc. NSS News* **37:**183–184.

Watson, R. A., 1981, *The Cave Research Foundation: Origins and the First Twelve Years, 1957–1968*, Mammoth Cave, Ky.: Cave Research Foundation, 494p.

8

CAVES AND DRAINAGE NORTH OF THE GREEN RIVER

Angelo I. George

Geomorphology and lithology control the distribution of caves and karst development north of the Green River in Edmonson, Hart, Green, and Taylor counties, Kentucky. Unique to this locality is the presence of pseudokarst and paleokarst. The northern portion of the central Kentucky karst covers approximately 333 km^2 and occupies a much greater area than that defined by White et al. (1970).

Territorially, the northern area is partially bounded by three base-level streams and two major escarpments: Green River on the south, Nolin River to the west, Bacon Creek and the Chester Escarpment on the north, and Little Brush Creek and the Muldraugh Escarpment to the east (Fig. 8.1). The northern frontier is controlled by a major lithologic and surface drainage divide situated between Green River and Bacon Creek. This divide extends along the crest of the Brush Creek Hills from the headwaters of Little Brush Creek on the dip slope of the Muldraugh Cuesta and then westward to Nolin River.

Early work in this part of Kentucky includes discussions on the geomorphic and geographic observations by Burroughs (1923), Sauer (1927), Weller (1927), and Lobeck (1928). Weller (1927) and Lobeck (1928) concentrated on the geomorphology, the cultural geography, and the geologic aspect of speleogenesis. Dicken (1935) differentiated the territorial zonation of the Sinkhole Plain by landform texture. Bretz's (1942) classic monograph on the origin of caves described Bat Cave and Running Branch Cave as

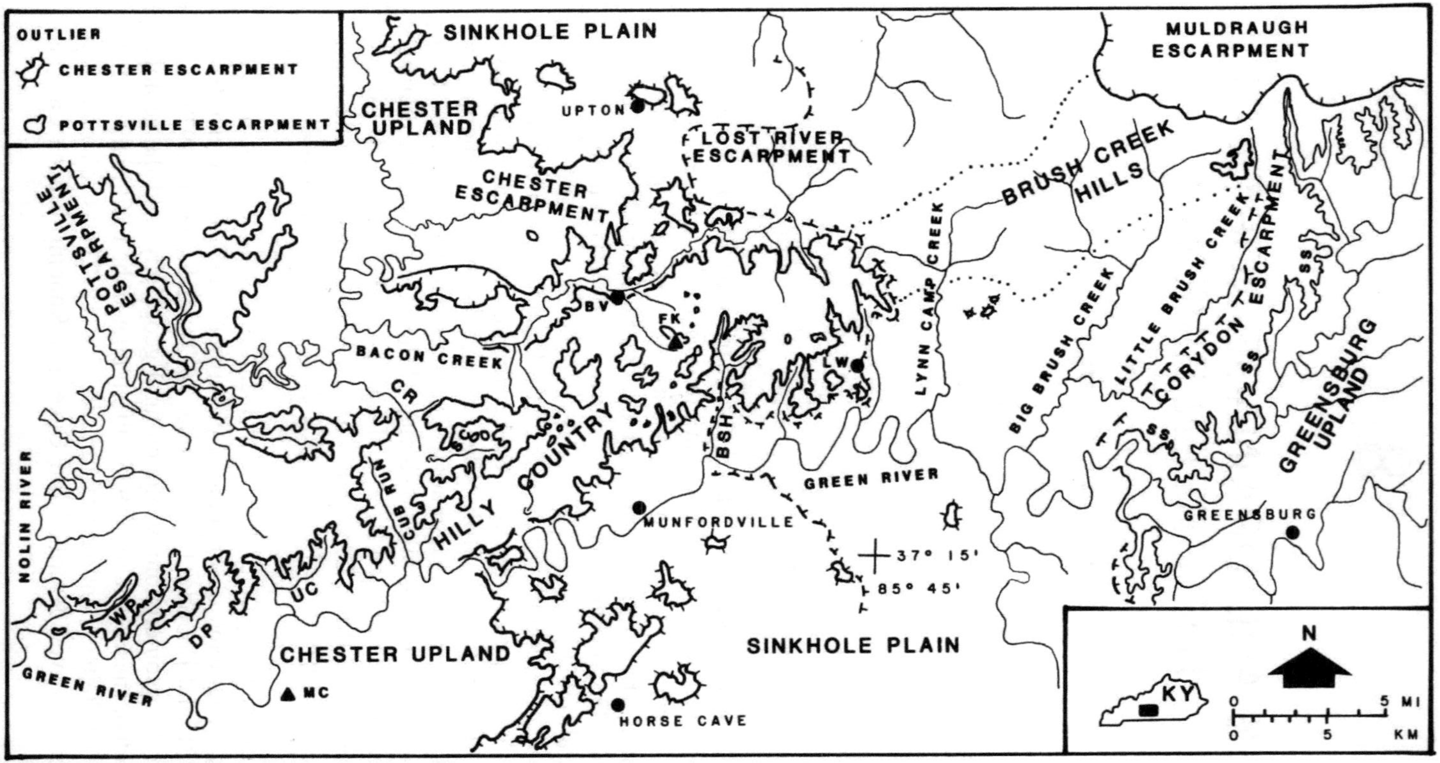

Figure 8.1 Sub-regional geomorphic divisions, north of the Green River in the central Kentucky karst, Mississippian Plateau region, Interior Lowland Province. The following abbreviations are used: BSH = Boiling Springs Hollow; BV = Bonnieville; DP = Dry Prong; SS = Salem–St. Louis contact; FK = Frenchman Knob; LW = Linwood; MC = Mammoth Cave; SC = Sinking Creek; UC = Ugly Creek; and WP = Wet Prong.

examples of cave-passage trends in violation of Gardner's (1935) theory for dip control of cave-passage development. Cushman et al. (1965) examined a number of karst resurgences in the Hilly Country for potential domestic groundwater supplies in Mammoth Cave National Park. Hess et al. (1974) conducted a traverse of Green River to catalog some geochemical parameters of 29 karst resurgences. Quinlan and Ray (1981) subsequently expanded the collection of river-spring and watershed data to encompass the entire central Kentucky karst. George and Schmidt (1977) discussed the origin of caves from the Hilly Country.

GEOMORPHOLOGY

Major geomorphic units consist of the Sinkhole Plain and the Chester Upland (Fig. 8.1). North of the Green River, the Brush Creek Hills, formed from a Pennsylvanian clastic-filled paleo-river valley (Brownsville Channel), bisect the length of the Chester Upland and Sinkhole Plain. Five escarpments are represented: Muldraugh, Corydon, Lost River, Chester, and Pottsville.

Greensburg Upland

Along the eastern limits of the study area is the Greensburg Upland (Fig. 8.1). Landforms contain a prekarst look with coarse dendritic surface drainage, few sinkholes, and a return to surface-stream flow. The locality between Little Brush Creek and the Salem–St. Louis Limestone contact is similar to the New Middleton Upland along the western fringes of the Norman Upland of south-central Indiana. The western terminus of this upland ends with the development of ponors at the base of the Corydon Member (Ash, 1985). Ash (1985) showed that the Indiana counterpart produced a 3-m-high cherty escarpment above the New Middleton Upland.

Sinkhole Plain

The northern complement of the Sinkhole Plain covers about 25% of the study area and is found only adjacent to the Green River and south-draining tributaries. The plain is best revealed as wide meander scars interspersed with numerous clastic-capped outliers, giving a more rugged look than its southern counterpart. This feature is especially apparent from Linwood east to Big Brush Creek. Vertical relief of the Sinkhole Plain averages about 15–18 m except near outliers where the sinkholes tend to be larger in diameter, shallower in depth, and fewer in number.

Outliers are erosional and solutional detached hills from the Hilly Country and Brush Creek Hills. Their height ranges from 15 m to 42.6 m above land surface. The clastic nature of the Brush Creek Hills and their close position to the Green River have crippled the horizontal extent of the Sinkhole Plain. The concept for a fluvial origin of the Sinkhole Plain has been proposed by

Miotke and Palmer (1972), Palmer and Palmer (1975), Quinlan and Ewers (1981), and Palmer (Chap. 11). These concepts are at variance with solution origins proposed by Dicken (1935), and Lehmann (1969), and the stripped structural-surface concept of Quinlan and Pohl (1968).

Chester Upland

Within the central Kentucky karst, the Chester Upland is the type area for large cave development. A complement of this upland is present north of the Green River and extends along two-thirds of the length of the study area (Fig. 8.1). The Chester Upland is a plateaulike landscape divided into two geomorphic units: the Mammoth Cave Plateau (south of the river); and the Hilly Country, locally called The Knobs (north of the river). Despite classic investigations from the Mammoth Cave Plateau, very little investigative attention has been paid to Hilly Country cave and karst development. Only Bretz (1942), George (1975), and George and Schmidt (1977) discussed the cave and karst nature of this rugged upland.

Surface hydrology is the main difference between the north and south side of the river. The Mammoth Cave Plateau has no perennial surface-flowing streams (Davidson and Bishop, 1971). The Hilly Country has perennial free-flowing surface streams in areas of clastic underlayment and intermittent and ephemeral streams along carbonate reaches. During periods of prolonged rainfall, these streams have continuous free flow from headwater divides to their mouths on Green River.

Dry, ancestral stream valleys crossing through the Mammoth Cave Plateau from the Sinkhole Plain, have longer stream lengths, shallower grade, U-shaped cross sections, and much larger catchment areas. Groundwater recharge is from rainfall onto carbonate rocks on the Sinkhole Plain and clastic rocks within the Chester Upland. Efficiency wise, karsting of this landscape has produced a more economical way to transmit water to the Green River through low-gradient cave passages.

Streams in the Hilly Country have steep V-shaped valley cross sections, high-gradient long profiles, and smaller catchment areas. Major aquifer recharge is rainfall runoff from clastic landforms into carbonate valley bottoms. This implies that the water chemistry for the Hilly Country is different from the water chemistry from the south side. Additional work needs to be done on this implication. Most of the Chester Upland as a plateau karst feature is obscured north and east of Munfordville because it is covered with lower and middle Pennsylvanian slump deposits.

Brush Creek Hills

Occupying the northern half of the Hilly Country is a clastic-filled Early Pennsylvanian paleovalley called the Brownsville Channel. Sauer (1927) named this landform the Brush Creek Hills. The southern slope of this resistant feature forms the ragged, low-relief Pottsville Escarpment, where pieces of it are preserved on the Sinkhole Plain. Field identification of the Brush Creek Hills is based on in situ and slumped lower and middle Pennsylvanian conglomerates, sandstones, and clays, producing hilly relief that is interior and peripheral to the Brownsville Channel.

Solution subsidence of the clastic-filled Brownsville Channel (Brush Creek Hills) has truncated the Hilly Country along a line between Munfordville and Bonnieville eastward to and beyond the Muldraugh Escarpment (Figs. 8.1 and 8.11). Immediately west of Munfordville, the Chester Escarpment rises 90 m above the level of the Sinkhole Plain. The elevation of the crest of the Brownsville Channel or Brush Creek Hills is 122 m above the Sinkhole Plain. Mature relief of the Brush Creek Hills produces steep terrain with many low-order, high-gradient ephemeral tributary streams, draining to base level at Green River and Bacon Creek.

Fluvial erosion and interstratal karstification of the Brush Creek Hills have buried portions of the Sinkhole Plain with a thick sequence of slump, alluvium and reworked colluvium, and residuum. This is best seen north of the landform along the southern border of the north-central Kentucky karst. Much of this area east of Upton is a nonkarstlike plain interspersed with islands of mantled carbonate rock with well-developed doline fields. The doline fields are associated with anticlinal domes and anticlinal noses. Nonkarstlike areas are associated with synclinal axes and structural depressions.

INFLUENCE OF LITHOLOGY

Lithologic rock types often control the position of cave passages and texture of karst landforms. Carbonates are more susceptible to solution than clastics; and some carbonates are more soluble than other carbonates. Shale, siltstone, and chert tend to act as aquitards or barriers to the solution of carbonates.

Sinkhole Plain

Development of the present karst plain is influenced by variation in erodibility and solubility of rock types. Elevation of cave entrances, passages, and karstic resurgences throughout the area are controlled by lithology. Lithological differences have perched most subsurface base-level drainage at the Salem Limestone–St. Louis Limestone contact. The size of sinkholes and sites of ponors are often determined by resistant chert intervals.

All large springs and local base-level cave passages are found at or near the Salem Limestone–St. Louis Limestone contact. Examples of caves north of the river and under the Sinkhole Plain are (Fig. 8.2): Buckner Cave Spring, Brush Creek Cave, Aetna (Cushenberry) Cave, and Crump Cave Spring. If the Salem Limestone is covered by more than 7.6 m of St. Louis

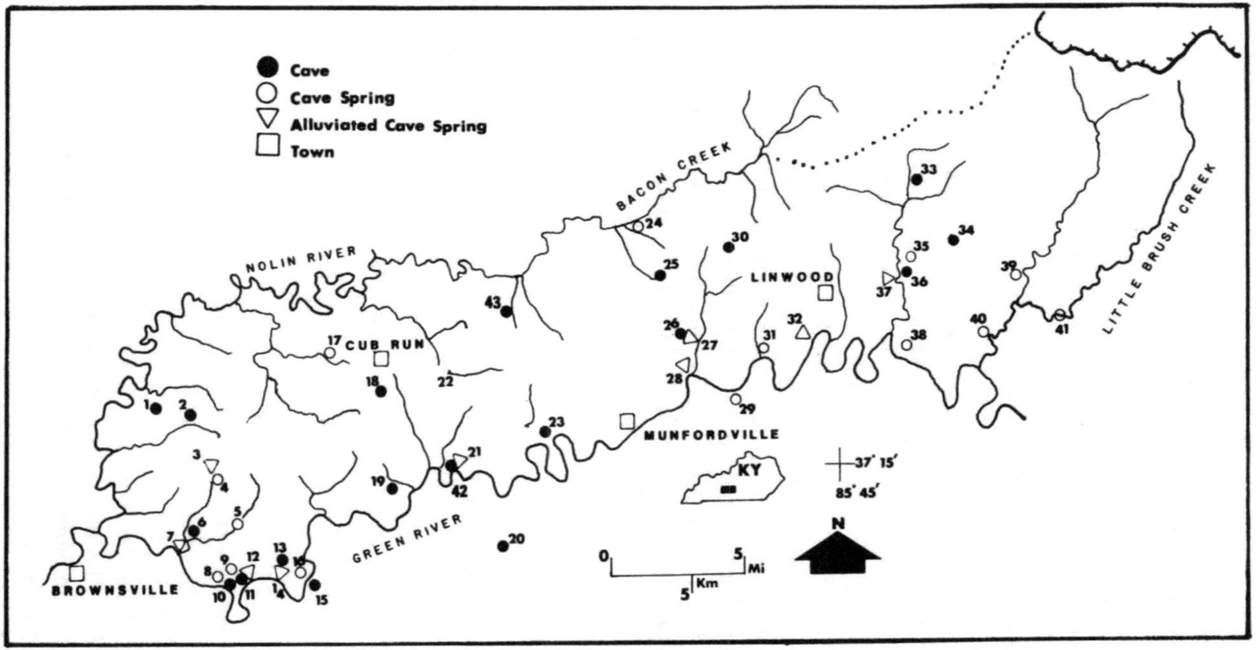

Figure 8.2 Index map of caves and springs referred to in this chapter. The following features are indicated: 1, Holley Cave; 2, Bylew Cave; 3, Boiling Springs; 4, Big Spring Cave; 5, Stephen's Cave; 6, Fort's Funnel Cave; 7, Buffalo Spring; 8, Saunders Cave; 9, Stillhouse Cave; 10, Bat Cave; 11, Ganter Cave; 12, Ganter Blue Hole; 13, Running Branch Cave; 14, Running Branch Blue Hole; 15, Mammoth Cave; 16, Dossey Domes Caverns; 17, Cub Run Cave; 18, Lines Cave; 19, Wilson Cave; 20, Crump's Spring Cave; 21, McCoy Spring; 22, Sinking Creek Ponor; 23, Lawler Pit; 24, Bonnieville Cave; 25, Frenchman Knob Pit; 26, The Big Hole; 27, Boiling Springs; 28, Johnson Spring; 29, Mammoth River (Grady's) Cave; 30, Barnes Smith Maze Cave; 31, Buckner Spring Cave; 32, Rio Spring; 33, Turner Cave; 34, Holley Branch Ponor; 35, Browns Cave; 36, Dry Cave; 37, Powder Mill Spring; 38, Crump Cave Spring; 39, Brush Creek Cave; 40, Cushenberry (Aetna Grove) Cave; 41, Sinks of Little Brush Creek Cave; 42, Forestville Saltpeter Cave; 43, Hester Cave.

Limestone, exploration for groundwater in the Salem will generally produce 1–2 whole orders of magnitude lower water well yields.

One of the physical causes for spring and cave entrance clustering is the presence of a shale facies at the base of the St. Louis Limestone. From east to west (down the depositional slope), the shale thins from 1.5 m (Hudgins quadrangle) to 0.91 m (Canmer quadrangle). With the shale acting as an aquitard, elevation of spring entrances ranges from 186 m to 164 m above sea level. Within the study area few cave passages or springs are found below the Salem–St. Louis contact. Mammoth River (Grady's) Cave, south of the Green River, does breach this shale interval (Joseph Saunders, personal communication, 1976) by vadose canyon development. Where the basal St. Louis Shale is stripped away, caves and springs are developed eastward beneath the Greensburg Upland; and adequate well yields in the Salem Limestone can be developed for domestic supplies. The Salem Limestone tends to contain more calcareous siltstone, shaly siltstone, and much chert mixed with sandstone in the upper part (Pohl, 1970). This composition further helps to restrict cave development.

St. Louis Evaporite

McGrain and Helton (1964) recognized the economic and areal distribution of natural gypsum and anhydrite in the lower one-third interval of the St. Louis Limestone, Mississippian Plateau region. A sebkha environment was needed for the cyclic depositional character of the evaporite sequence during Lower Meramecian time (Pohl, 1970). The basal St. Louis Shale facies is part of the cyclic depositional condition associated with a sebkha environment.

Groundwater solution and removal of sulfates in the subsurface accounts for elevated mineralization in springs and wells, formation of tufa, sheet-like solution cavities, and rock subsidence (George, 1977). The solution of natural gypsum–anhydrite provides for the concordance for some cave passages and spring discharge zones in the lower St. Louis Limestone. This is similar to Gardner's (1935) "static zones" for controlling cave-passage formation within narrow vertical limits of a carbonate aquifer.

Wigley et al. (1973) examined gypsum karst of the Canal Flats in eastern British Columbia, Canada. Formation of caves and geochemical conditions in gypsum karst can be used to show similarities to cave and karst development in Kentucky.

Solution activity will also be enhanced at the lower gypsum contact because the underlying rocks generally will act as a relatively impermeable medium and tend to concentrate flow at this level. If the groundwater has dissolved some limestone (or dolomite) before entering the gypsum bedrock, the addition of gypsum causes a considerable reduction in the water's capabilities to dissolve more limestone. Thus underlying limestones or dolomite beds may function as a completely impermeable barrier to groundwater flow. The consequent flow concentration in the gypsum will be conducive to enhanced solutional activity at the lower gypsum contact (Wigley et al., 1973, p. 127).

Evaporite beds and nodules in the study area are subjected to interstratal karstification, and those rock intervals within the evaporite units become the preferred avenue for groundwater movement, cave-passage development, and position of springs. The thickest and lowest evaporite horizon deposited on a regional scale is positioned just above the top of the Salem Limestone (McGrain and Helton, 1964). The Salem Limestone has scattered gypsum nodules, but does not have continuous beds of gypsum.

Interstratal karsting of the St. Louis evaporites produces: sponge work, breccia zones, and rock collapse or subsidence. Johnson Spring (Fig. 8.2) has excellent examples of "solution looking" voids. Similar voids and in situ gypsum nodules are seen in Squire Boone Cavern, Harrison County, Indiana. Sponge work, gypsum nodules, and sheetlike solution cavities are found together in high-yield water wells at Elizabethtown, Hardin County, Kentucky. Groundwater well yields range from 13 l/sec to 76 l/sec (200–1200 gpm). Interstratal karstification of the St. Louis evaporite has resulted in the regional foundering of overlying rock units. The Brownsville Channel occupies a negative area on the updip edge of the evaporite sequence. The solution edge of the evaporite zone is receded farther westward into the western coal basin than in the central and north-central karst region.

Chert Horizons

Howard (1968), George (1973), Woodson (1981*a*, 1981*b*), Ash (1985), and Hess et al. (1974; Chap. 2), have emphasized the importance of chert horizons in karst as controlling agents in the localization of ponors and sinkholes. Three chert zones are found in the St. Louis Limestone; the Corydon Member contains two zones: (1) St. Louis ball chert (spherical and concentrically banded); and (2) Corydon chert (lenticular, in massive dolomites). This unit is collectively about 9.1–10.7 m thick. About 12 m above this interval is the Lost River chert. A black flint zone is found at the top of the Ste. Genevieve Limestone in the Aux Vases Member. These cherts occur on a regional scale and can be traced through the Mitchell Plain of southern Indiana.

The Lost River chert horizon is found in the upper portion of the Horse Cave Limestone Member, upper St. Louis Limestone Formation (Pohl, 1970). The rock unit is 1.5–3 m thick, fossiliferous, and offers an imposing resistant interval to solution and erosion. It restricts karstic solution at its boundary, perches cave development, and initiates ponor development on the updip side. The Lost River chert has yielded to a partial stripped solution surface in the central Kentucky karst (Quinlan and Pohl, 1968; Quinlan, 1970). A low-relief escarpment can be traced through portions of the Mississippian Plateau region. Usually the escarpment is imperceptible as a topographic rise, yet it can be recognized in the field by observing residual chert occurrence and especially sites of burr piles (Frederick J. Woodson, personal communication, 1981). Burr piles are small stacks of chert that have been collected by farmers.

Clastic Rock Units

Outcrop position of a carbonate-clastic contact intercepts surface runoff at or near this junction. Many vertical shafts develop below a clastic rock cover. Quinlan and Pohl (1968) and Quinlan (1970) have estimated the number of vertical shafts along the Chester Escarpment (to include outliers) at 200–500 shafts per linear mile of caprock. Shafts are responsible for secondary slope retreat of the Chester Escarpment. They are enlarged by groundwater solution and act as a repository for fallen clastic rocks that are mechanically milled and carried off in the bedload of cave streams to base-level resurgences (White et al., 1970). A practical cross section of this operative can be seen along Interstate highway 65 between Munfordville north to Bonnieville. Many Girkin Formation vertical shafts are 23–30 m high and filled with a clastic mixture of Big Clifty Sandstone, Hardinsburg Sandstone, and Kyrock conglomerates. Subsidence dolines often form above these filled shafts. Frenchman Knob Pit is developed at the Girkin Limestone–Big Clifty Sandstone contact and has more than 50 m of relief. Bedload material at the base of the 38.1-m-deep entrance drop are Mississippian–Pennsylvanian clastics and some carbonates.

Sandwiched between clastics, caves in the Haney Limestone and Glen Dean Limestone often encompass the entire vertical carbonate rock mass. Most of the caves are of short lateral extent and display a maze of high, narrow, interconnected canyon passages. Spring caves in these rock units act as recharge zones to lower carbonate rock. Interstratal karstification is operating in this kind of setting (Moore, 1972; Miller, 1969; Quinlan, 1970). The original carbonate rock thickness is probably reduced through the action of groundwater solution just below each clastic caprock. Prototubes often extend laterally away from the cave passage as thin sheetlike solution cavities (bedding-plane partings).

LONGITUDINAL STREAM PROFILES DRAINING KARST TERRAINS

Study of longitudinal stream profiles from karst terrains can yield usable information needed to measure karst modification of surface hydrology to subsurface hydrology (George, 1976, 1982). Average stream profiles in equilibrium will have a concave profile. The profile will be flattest near the mouth and steeper nearest its watershed divide. In carbonate rocks subjected

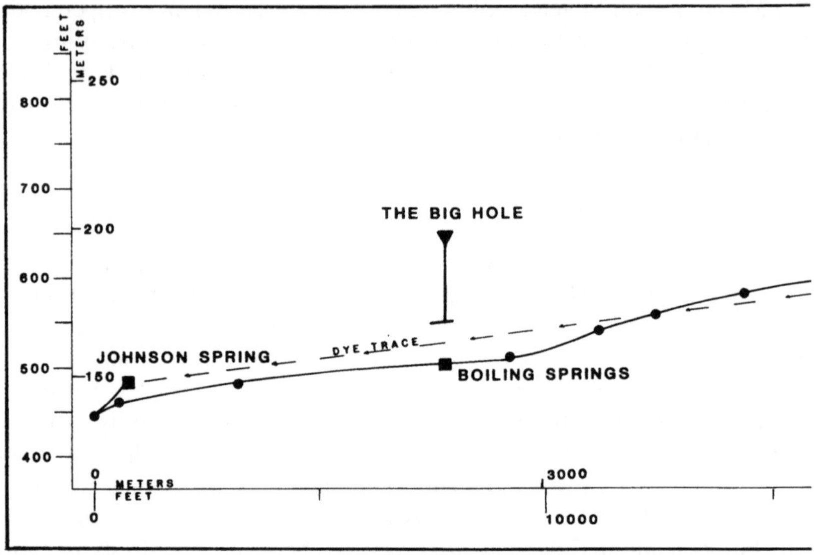

Figure 8.3 Longitudinal stream profile of Boiling Springs Hollow, Hart County, Kentucky. Distances were measured along the channel and include bends. Data extrapolated from the Canmer 7.5-min topographic quadrangle.

to a great amount of karst weathering, a portion of the stream profile could be convex in shape. As stream-bed piracy becomes more complete and better integrated with the subsurface regimen, the convex profile becomes more pronounced as surface stream erosion becomes less effective and shorter lived during rainfall events. Thus, the position of thalweg ponors, hanging valleys, nick points, and lithologic control can be interpreted from longitudinal stream profiles.

All streams in the study area head from clastic catchments associated with the Brush Creek Hills. Table 8.1 summarizes Sinkhole Plain and Hilly Country longitudinal profiles, information that is used in the following discussion.

Boiling Springs Hollow

Boiling Springs Hollow (Fig. 8.3), a third-order stream 9.9 km long, drains from a clastic catchment and generally flows south to Green River. Low Horton numbers in relation to long stream-profile length are typical of streams draining karsted carbonate rocks. West of Boiling Springs Hollow on the Sinkhole Plain are a number of first-order ephemeral sinking streams. These streams once flowed by way of ancestral streams to Boiling Springs Hollow. Their ancestral stream course can be traced by present ephemeral stream trends, saddles, and dendritic nature of contour lines.

Boiling Springs Hollow consists of three hydrologic sections: (1) a lower,

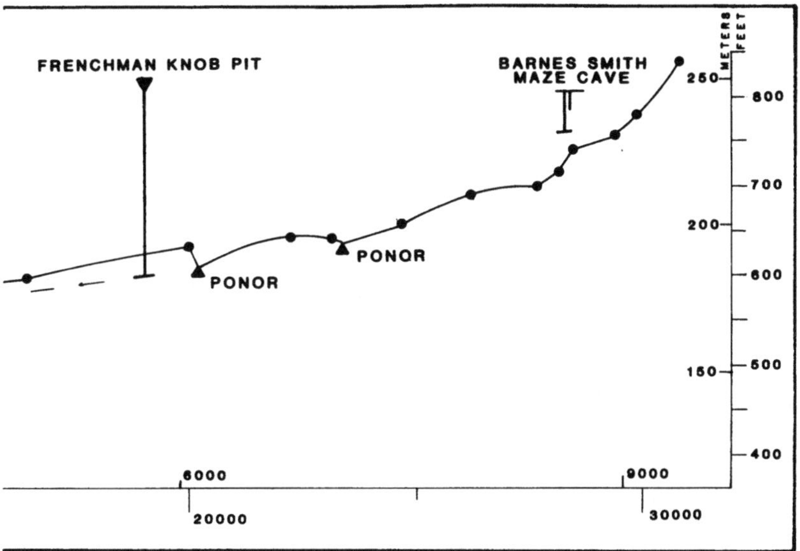

Table 8.1 Basin Parameters for Streams in the Hilly Country

Stream Name	*Stream Length (km)*	*Surface Drainage Area (km^2)*	*Horton Stream Order*	*Head Differential (m)*
Boiling Springs Hollow	10.0	20.9	3	122
Buffalo Creek				
Dry Prong	12.8	12.1	4	134
Wet Prong	8.5	8.1	3	128
Lynn Camp Creek	21.5	60.0	5	172
Stillhouse Hollow	1.8	0.6	2	119
Ugly Creek	10.7	12.1	3	158

spring-fed, 2.4-km-long perennial reach; (2) a middle, 7.6-km-long ephemeral stream channel with convex-upward profile; and (3) an upper ephemeral stream channel with well-developed convex-upward profile in areas of thalweg ponors. These two ponors are found at kilometers 6.5 and 7.5 (miles 4.05 and 4.65), and are silt-filled farm ponds formed by perched water-table conditions. Both of these ponors are situated just above the Lost River chert. Judging from the long stream profile, each ponor probably breaches the chert and connects with Boiling Springs (a blue hole) and Johnson Spring (an occluded tubular bluff spring) resurgences.

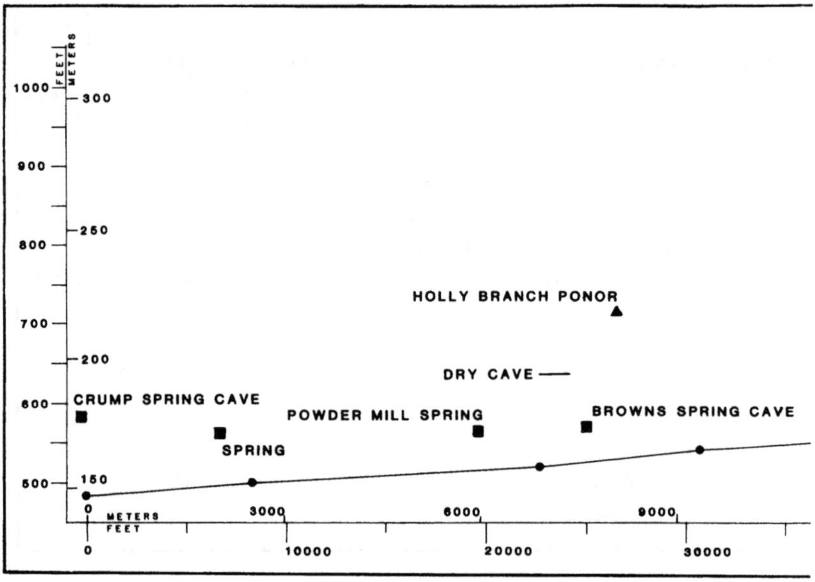

Figure 8.4 Longitudinal stream profile of Lynn Camp Creek, Hart County, Kentucky. Distances were measured along the channel length and include bends. Data were extrapolated from the Hudgins and Magnolia 7.5-min topographic quadrangles.

Vertical profiles of three caves have been projected against the Boiling Springs Hollow longitudinal stream profile (Fig. 8.3): Barnes Smith Maze Cave, Frenchman Knob Pit (Blue Hole of Kentucky), and The Big Hole. Frenchman Knob Pit and The Big Hole owe their origin to two different kinds of karst processes. Both owe their vertical development to vadose flow and base-level lowering. Frenchman Knob Pit is an example of interstratal karstification at the Girkin Limestone–Big Clifty Sandstone contact. The Big Hole is an example of an ancestral stream-bed ponor operating at a higher base-level of Boiling Springs Hollow.

Lynn Camp Creek

Lynn Camp Creek (Fig. 8.4) is a fifth-order stream with 21.5 km of trunk stream-valley drains and approximately 60 km^2. Its Horton number is higher and its long profile is flatter in the lower half. The karst nature of the lower profile is not evident, suggesting better base-level equilibrium and a change in lithology. This is similar to the lower reach in Boiling Springs Hollow. Irregularities observed in the upper third of the Lynn Camp Creek profile correlates to a change in bedrock lithology from carbonate to clastic. Lithology again plays an important role in the localization of springs and

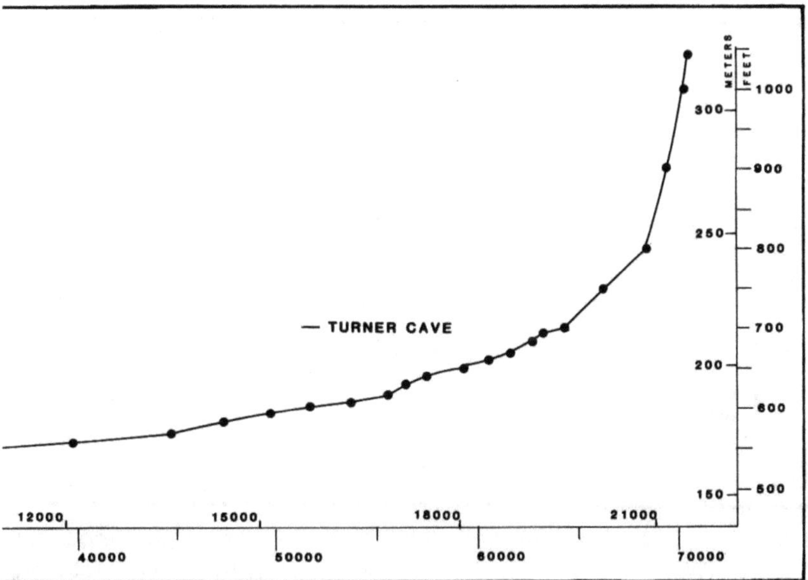

caves at the Salem Limestone–St. Louis Limestone contact above local base level.

Karst topography adjacent to the creek on the Sinkhole Plain is evident. Holly Branch, a 3.4-km-long ephemeral stream east of Lynn Camp Creek sinks into Pennsylvanian slump deposits at the base of Akin Knob. Its groundwater-flow direction is probably west toward Browns Spring. Situated between Lynn Camp and Big Brush Creeks are a suite of karst fens (swamps) and large farm ponds in sinkholes. These features are believed to mark the approximate groundwater divide between the two base-level streams. One of the farm ponds collapsed and its effect was noticed by local residents at Brush Creek Cave on Big Brush Creek, 2.61 km east of the catastrophic collapse (Figs. 8.2 and 8.7). A higher base level is observable between Crump Spring Cave, Dry Cave, and Turner Cave (Fig. 8.4). An intermediate base level is apparent between an unnamed spring, Powder Mill Springs, and Browns Cave Spring.

Hilly Country

Situated within the Mammoth Cave National Park are four long stream profiles from the Hilly Country (Fig. 8.5). Stream-valley slope angle

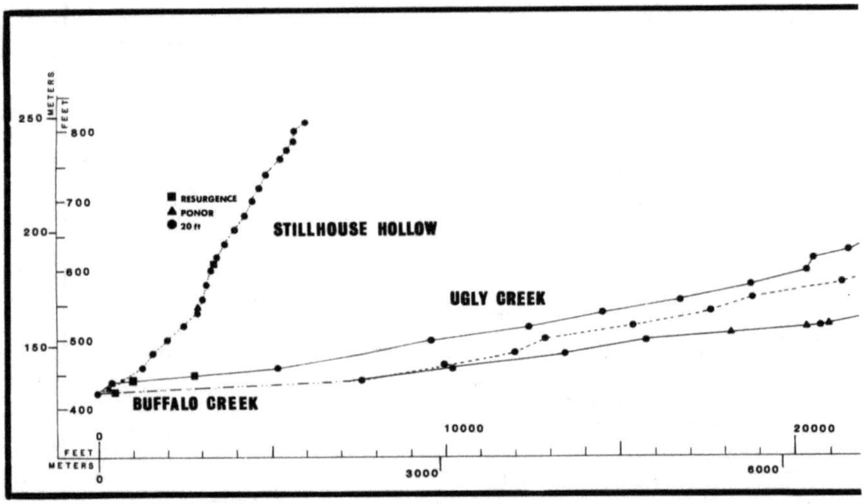

Figure 8.5 Longitudinal stream profile of Stillhouse Hollow, Ugly Creek, and Buffalo Creek (Wet and Dry Prongs), Edmonson County, Kentucky. Distances were measured along the channel length and include bends. Data extrapolated from the Mammoth Cave and Rhoda 7.5-min topographic quadrangles.

for each profile is due to variation in lithology and the karst nature of the valley. Nick points are profile irregularities associated with the Girkin Formation–Big Clifty Sandstone contact. The stream-valley profile then becomes slightly convex along the rest of their reach, a condition indicative of stream-bed piracy. Stillhouse Hollow, Ugly Creek, Dry Prong, and Wet Prong are not in equilibrium with base-level Green River. Sparse cave mapping in Dry Prong seems to point to a more concave subterranean cave stream profile.

Dry Prong is situated in an erosional saddle, flanked by higher-gradient Wet Prong (on the west) and Ugly Creek (on the east). Dry Prong is the master karst valley west of Cub Run Creek. Dry Prong has the lowest gradient and the largest amount of exposed carbonates in its valley bottom. Subsurface stream piracy occurs in all three valleys. Dry Prong and Wet Prong share a portion of their surface flow with ponors and springs situated in each other's catchment area. Headwaters of Dry Prong sink at the ponor in Raymond Hollow, pass under Collie Ridge, and resurge at Big Spring Cave in the headwaters of Wet Prong (Fig. 8.6). An untested cross exchange probably occurs at Fort's Funnel Cave with water mixing from ponors in Wet Prong and Dry Prong.

Valley bottom flooding occurs once subvalley caves become filled to capacity. This effect is sequential as upper valley caves flood first and so on until each ponor is overtaxed and the entire valley has a free-flowing

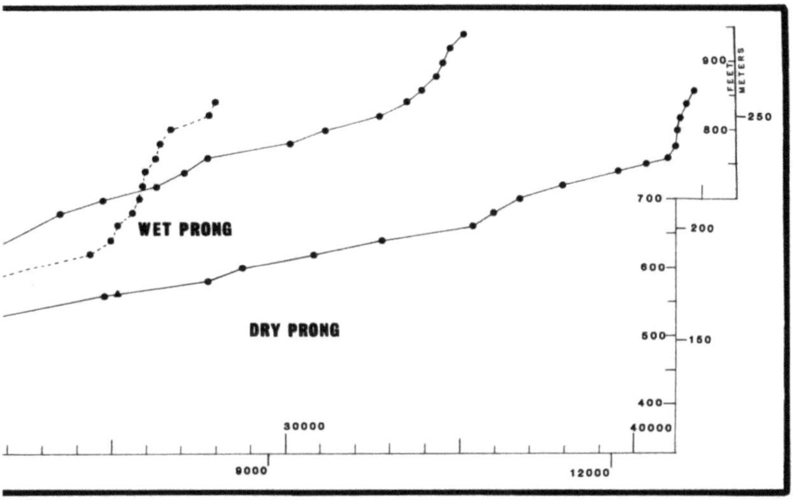

stream. Groundwater-carrying capacity of cave passages is limited and has smaller internal volume and smaller recharge areas.

Smallness of recharge area in the Hilly Country is directly related to the presence of the Kyrock conglomerate-filled Brownsville Channel. Resistance to headward erosion offered by this quartzite conglomerate deposit to stream erosion has resulted in restricted headward valley development, forming hanging valleys above amphitheater-like headwalls and rock shelters. Watersheds are small and provide restricted amounts of water to a developing karst hydrosystem. If the deep Brownsville Channel had not been cut and later filled with conglomeritic clastics, then larger watersheds could have developed with a better integrated cave system and larger-volume passages (George and Schmidt, 1977).

SUBSURFACE TRACER EXPERIMENTS NORTH OF THE GREEN RIVER

Some subsidence tracer experiments have been conducted in the Hilly Country and Sinkhole Plain by James F. Quinlan and his Upland Research Laboratory personnel (Fig. 8.7).

Cane Run Creek is a 15-km-long, third-order tributary to Nolin River. Its upper reach is a half-blind valley developed by Sinking Creek (Figs. 8.7 and 8.8). Sinking Creek sinks into a series of colluvial–alluvial-filled ponors

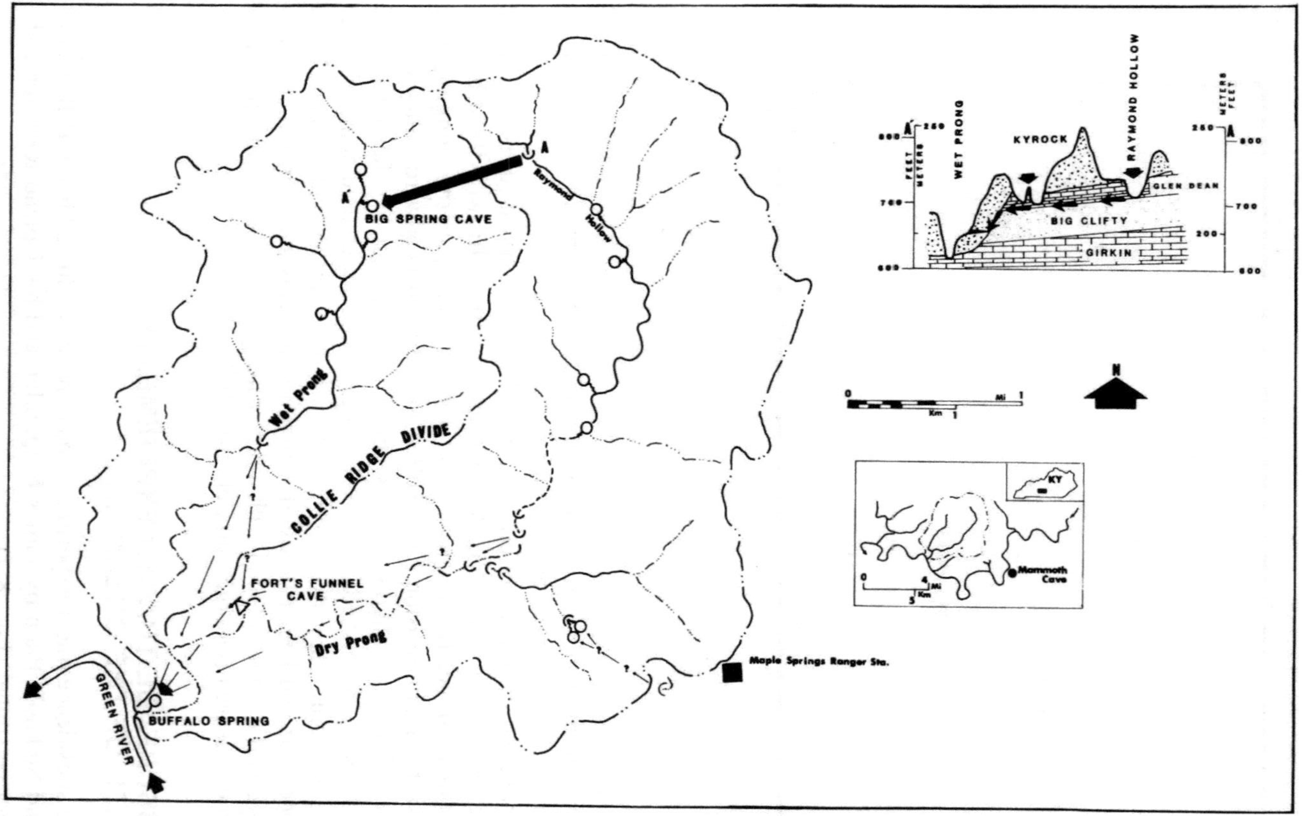

Figure 8.6 Subsurface dye-tracing results from the Hilly Country, Mammoth Cave National Park (based on data provided by James F. Quinlan, personal communication, 1974). Additional subsurface-flow routes are hypothetical.

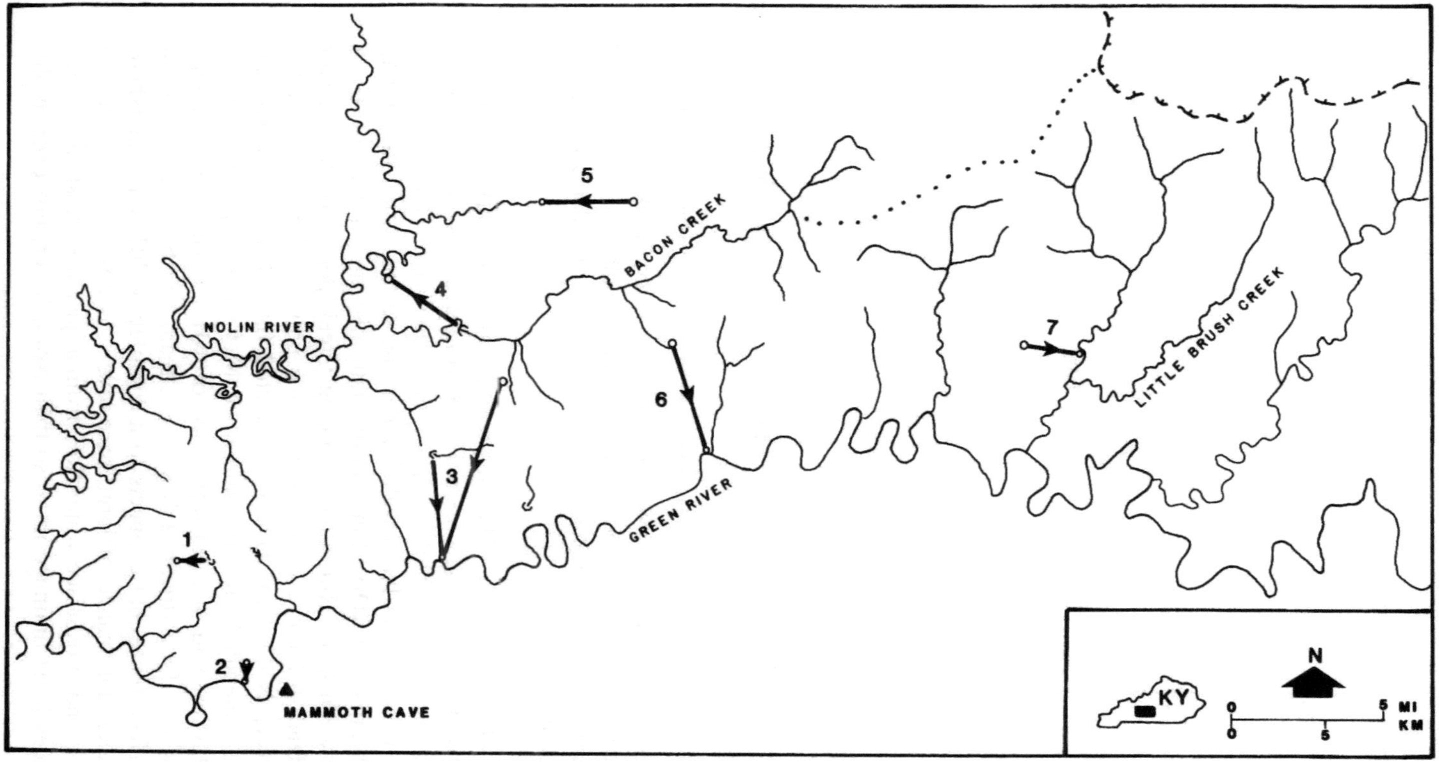

Figure 8.7 Selected subsurface-flow directions, north of the Green River. (Based on data from Quinlan and Ray [1981], James F. Quinlan [personal communication, 1987], and North Shore Task Force [unpublished].) The following selected dye-traces are shown: 1, Raymond Hollow to Big Spring Cave; 2, Running Branch Cave to Running Branch Blue Hole; 3, Sinking Creek and Hester Cave to McCoy Spring; 4, Bacon Creek to Cooch Webb Spring; 5, Vento Sink to Round Stone Spring; 6, Frenchman Knob Pit to Johnson Spring; and 7, farm pond collapse to Brush Creek Cave.

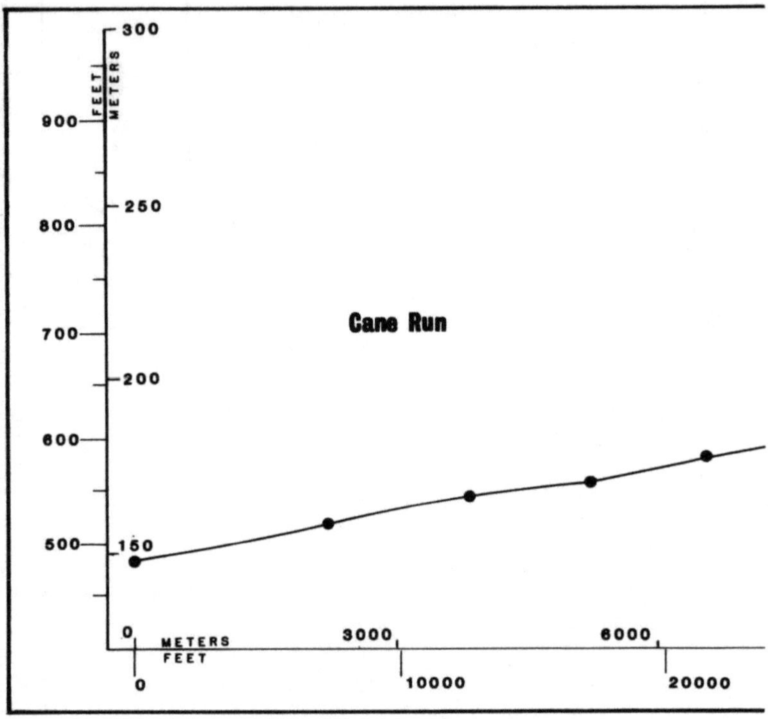

Figure 8.8 Longitudinal stream profile of Cane Run–Sinking Creek, Hart County, Kentucky. Distances were measured along the channel length and include bends. Data extrapolated from the Cub Run, Mammoth Cave and Munfordville 7.5-min topographic quadrangles.

at the bottom of a large uvala. The longitudinal profile is markedly convex downstream from the terminal ponor. Fluorescein dye was injected in the ponor and was recovered at McCoy Spring on Green River. The subsurface hydrology is such that the dye traveled under one surface water divide and one major stream valley (Dry Run), an overland distance of 5.2 km. Head difference between Cane Run Creek–Sinking Creek ponor to Nolin River is 24.4 m. Head difference between the same ponor and McCoy Spring, on Green River, is 42.7 m. Dye was injected into Hester Cave (Bacon Creek surface watershed) and it resurged at McCoy Spring, an overland distance of 9.8 km with a head difference of 101 m. These experiments illustrate how a surface stream and a cave stream can be deranged to an altered subsurface route that connects to a completely different watershed. It also illustrates dominance of large base-level streams in their ability to capture surface and subsurface water flow from adjacent watersheds. Green River influences karst drainage more so than Nolin River; however, Nolin

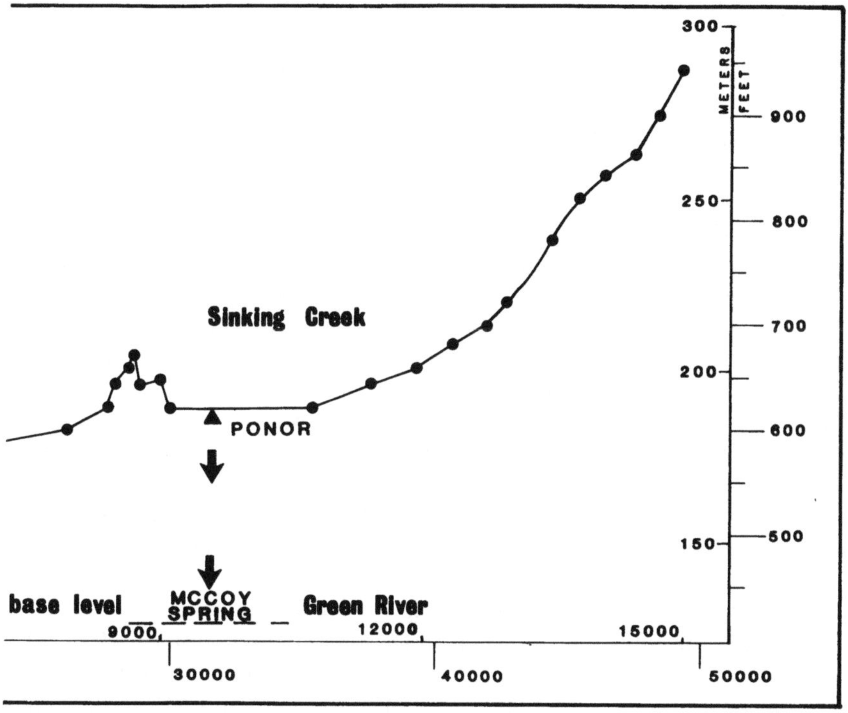

River does impact karst and cave development in the southern portion of the north-central Kentucky karst.

Optical brighteners were used to establish a hydrologic connection between a stream sink ponor in Raymond Hollow and Big Spring Cave (Figs. 8.6 and 8.7). Dye results demonstrate a 1.52-km straight-line connection between a ponor formed in the Glen Dean Limestone, crossing under a major topographic watershed divide (Collie Ridge), penetrating the wall of the Brownsville Channel, and discharging from a cave spring formed high in the Kyrock conglomerate. Presence of springs reveals in part the aquifer potential of the Kyrock conglomerate (Davis et al., 1974; Sedimentation Seminar, 1978).

Groundwater-flow direction from Frenchman Knob Pit has been resolved using water-tracing experiments (Figs. 8.3 and 8.7). The pit is located in the Bacon Creek surface watershed. Yet, its subsurface flow drainage is 5.9 km (3.65 mi) southeast to Johnson Spring on Green River. This gives approximately 102 m (335 ft) of vertical difference between the top of the pit and its discharge point at base level. Johnson Spring is perched above the Salem Limestone–St. Louis Limestone contact.

POST-CHESTERIAN (KASKASKIA) PALEOKARST

At the close of the Mississippian period, the North American cratonic landmass was high enough to support the development of large free-flowing streams. These streams formed a dendritic network of channels draining toward the south and southwest, creating a super delta in the Illinois basin. This was a major erosional event, and streams reached their maximum valley development during mid-Morrowan (mid-Pennsylvanian) time. One of these streams in the study area differentially truncated Chesterian rocks, forming the Brownsville Channel (Bristol and Howard, 1971; Sedimentation Seminar, 1978). This is the farthest southeasterly known channel that drains into the Illinois basin. The valley is uniformly 3–5 km wide and up to 61 m deep. Other deltaic channels occur along the perimeter of the basin. Some are known to be 48 km wide, have 122–152 m of relief, and exhibit complex "scabland-like" topography on their floodplains.

Climatic conditions prevailing during the early Pennsylvanian period are considered to be subtropical to tropical (Sedimentation Seminar, 1978). The locality was subject to torrential and prolonged periods of rainfall onto a lush forest-vegetation community; evapotranspiration and runoff rates must have been very high.

Siever (1951), Shaw and Gildersleeve (1969), and Sedimentation Seminar (1978) showed the existence of two stream terraces formed in the Brownsville Channel. The terraces are preserved on the Haney Limestone and Glen Dean Limestone; they are sharply defined, do not have a weathered zone, and there is not much Chesterian rubble at the terrace contact with the Kyrock conglomerate fill (Sedimentation Seminar, 1978). Sedimentation Seminar (1978) concluded that the river system that eroded the Brownsville Channel was not the same river that deposited two different lithic assemblages. They pointed out a sparsity of fossil terrace outcrops and a way to explain well-developed river terraces on carbonate rocks in a subtropical climate. Terraces should have been developed on resistant clastics rather than more soluble and erodible carbonates. Presence of the terraces did not harmonize well with the presumed warm, humid, subtropical climate (indicated by coal horizons), and the paleolatitude of the study area. Tropical karst areas with exposed carbonates will undergo case hardening on the surface (Sweeting, 1972), and this condition has not been found in the study area.

Perhaps the initial climate during the interim between Mississippian and Pennsylvanian times was temperate in this part of Kentucky. The Brownsville River eroded wide terraces on the Glen Dean Limestone and

Haney Limestone, augmented by a third phase of deep-stage river cutting. Each of these events represents base-level arrestment; therefore, proto cave development in the Glen Dean, Haney, and upper Girkin limestones would have occurred during this time period. Initiation of groundwater flow from the Mammoth Cave area might have occurred and is consistent with the structural geology and shallow phreatic development of that hydrosystem. Absence of case hardening of carbonates on the terraces is explainable for temperate climatic conditions. The climate then changed to subtropical during the deposition of the Kyrock sequence and later Pennsylvanian coal measures.

Quinlan (1972) termed the Mississippian–Pennsylvanian unconformity a fossil karst surface, and called it the post-Kaskaskia paleokarst. Paleokarst is best preserved buried under the thick sequence of Cretaceous sediments in the Jackson Purchase region of Kentucky (Davis et al., 1971; George, 1985*a*). Yet much of the border lands of the Illinois basin exhibit presumed tropical karst landforms on and adjacent to wide alluvial floodplains. Paleogeographic maps (Bristol and Howard, 1971; Davis et al., 1974; Shaw and Gildersleeve, 1969) show numerous steep-sided hills and slump blocks, which appear to be examples of tower karst (peak forest or Fenglin karst) situated on flat-braided floodplains. Upland topography adjacent to this deltaic complex displays low-profile cone karst (peak cluster or Fengcong karst).

Examination of structural contour maps (Gildersleeve, 1965, 1968) drawn at the base of the Kyrock conglomerate and the Hardinsburg Sandstone indicate a number of large, closed depressions. These depressions could represent the presence of sinkholes (Sedimentation Seminar, 1978). These large uvalas or more correctly, glades, are nearly 2 km in length and half that in width; some have 6 m, 30 m, or 43 m of vertical relief. The position of glades to paleotopography suggests that they are half-blind valley features positioned in the middle of former continuous dendritic drainage corridors. They are dendritic in nature and similar to temperate, humid counterparts (i.e., Houchins Valley and Doyle Valley south of the Green River in the Mammoth Cave National Park).

Paleokarst soils formed during the deposition of the Kyrock conglomerate might be present in the study area. McGrain (1960) and McGrain and Hamlin (1962) studied the petrology of three abandoned clay quarries south and southwest of Frenchman Knob. The so-called Haney clay is laminated, white to tan, has nodular gray-blue translucent glassy-appearing clay, iron staining, and might have scattered limonite nodules and stringers. These clays mostly contain halloysite with a lesser amount of gibbsite, kaolin, and silicified fossils. Some of the clays have quartzite pebbles at their base (McGrain, 1960).

Three explanations are offered for the genesis of these clays. Moore (1972) favored interstratal karstification of the Haney Limestone. Whereas Miller (1969) recognized this karstification, he added the interpretation that the Haney Limestone underwent reworking on an erosion surface before deposition of the Caseyville Formation. Understanding the mineralogy of the clays could lead us closer to an answer. Gibbsite is the primary constituent of bauxite, and it weathers from silicates. Occurrence of Gibbsite is rare and virtually unknown as a constituent of sedimentary rocks; it is an important product of certain soils, notably the laterites (Pettijohn, 1957). Laterite is a product of tropical weathering as well as a product of the kaolinites. Kaolinite (an aluminum silicate) is one of the primary cementing clays in the Kyrock conglomerate (Sedimentation Seminar, 1978). Furthermore, limonite nodules and stringers in these clays are probably leached siderite nodules. Large cobbles and boulders of siderite occur in the lower Kyrock conglomerate. The unique occurrence of these Haney clays could have originated on the floor of one of the subtropical karst glades. The clays could be the subtropical, weathered byproduct of the Kyrock conglomerate in sinkholes formed in the Haney Limestone. These aluminum silicates were not derived from the Haney Limestone because the clay mineralogy is not consistent with the lithologic composition of the Haney Limestone.

Much work remains in order to better understand the presence, development, and extent of tropical, fossil karst topography in Kentucky.

PSEUDOKARST

Pseudokarst is the development of karstlike features in noncarbonate rocks. Outcrops of Kyrock conglomerate show good examples of dolines, collapse dolines, caves, vertical shafts, rock shelters, natural bridges, and springs (blue holes and tubular bluff springs). The Kyrock is a quartz pebble conglomerate intermixed with sand. It is cemented together with kaolinite, illite, some chlorite clay, and a minor amount of hematite cement (Sedimentation Seminar, 1978). Hydraulic groundwater removal (clastokarst, suffosion, or piping) of the Kyrock accompanied by mechanical (abrasive) stream bedload action is the key agent needed for the formation of pseudocaves and some rock shelters (Fig. 8.9).

First-order streams head from box-shaped headwalls formed by Pennsylvanian clastics. These streams have steep gradients and drain a recharge territory that has been termed a prekarst area (White et al., 1970). The general surface is prekarst because it lacks large-scale karst features. Close inspection of topographic maps and on-site inspections reveal small-scale karst features formed in the Kyrock conglomerate. A number of suffosion caves are known, but only five have been mapped. These caves are situated along the periphery of the Brownsville Channel (Fig. 8.9), have limited recharge areas, and possess internal karst features normally associated with caves in carbonate rocks. Big Spring Cave (Fig. 8.10) with a proto-tube and a free-flowing stream has formed multilevel erosion wall niches, braided plan, and three entrances, one of which is a pseudokarst window. Holley Cave is, in part, a contact cave developed at the Hardinsburg Sandstone, Glen Dean Limestone, and Kyrock conglomerate interfaces. The main proto-tube and

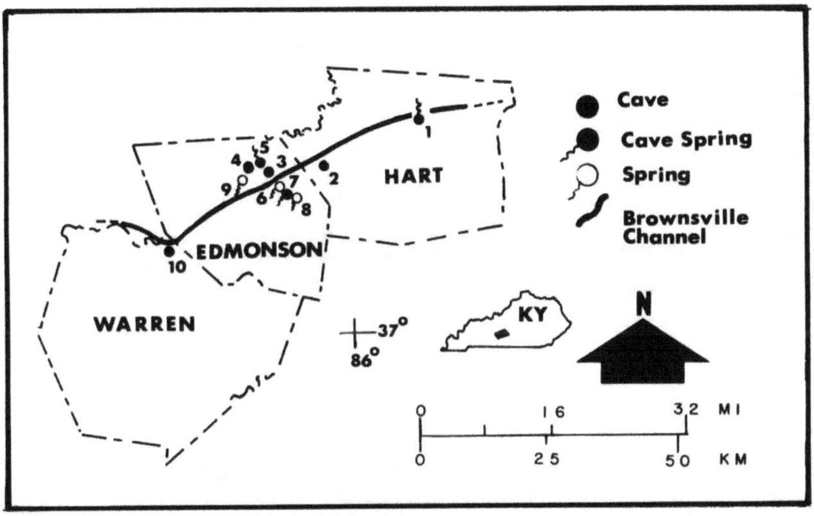

Figure 8.9 Relationship between the Brownsville Channel to suffosional cave and spring occurrences. The following features are known: 1, Frenchman Knob Cave Spring; 2, Lines Cave; 3, Bylew Cave; 4, Holley Cave; 5, Greenbrier Cave; 6, Boiling Spring; 7, Big Spring Cave; 8, unnamed spring; 9, Whistle Mountain Spring; and 10, Deer Hole.

several passages and rooms are formed in the Kyrock conglomerate. The main trunk has anastomosis and scalloped bell holes in the ceiling; stoped, massive breakdown has modified the terminal room in Holley Cave and Lines Cave. Some of this sandstone conglomerate breakdown is believed to have been blasted out and quarried by pre-1815 Holley Cave saltpeter miners (George, 1985*b*). Bylew and Big Spring caves show granular exfoliation activity near each entrance. Solutionlike dolines are found south of Cub Run near Lines Cave. A collapsed doline marks the spectacular entrance to Deer Hole, south of the Green River at Shanty Hollow Lake. Not much work has been carried out on the formation of suffosion caves in consolidated rocks of North America. Most investigations have been restricted to Venezuela (Colveé, 1973; Szczerban and Urbani, 1974).

Pseudokarst and the Destruction of the Brownsville Channel

As the protective Tradewater Formation is stripped away during base-level lowering, the Brownsville Channel and its valley fill are exposed to pseudo and interstratal karst modification. The channel fill is being destroyed internally by suffosion cave development, stoped breakout domes, and vertical shafts formed at the clastic-carbonate interface. The exterior surface of the channel structure is modified by rock shelters, peripheral slump of sandstone

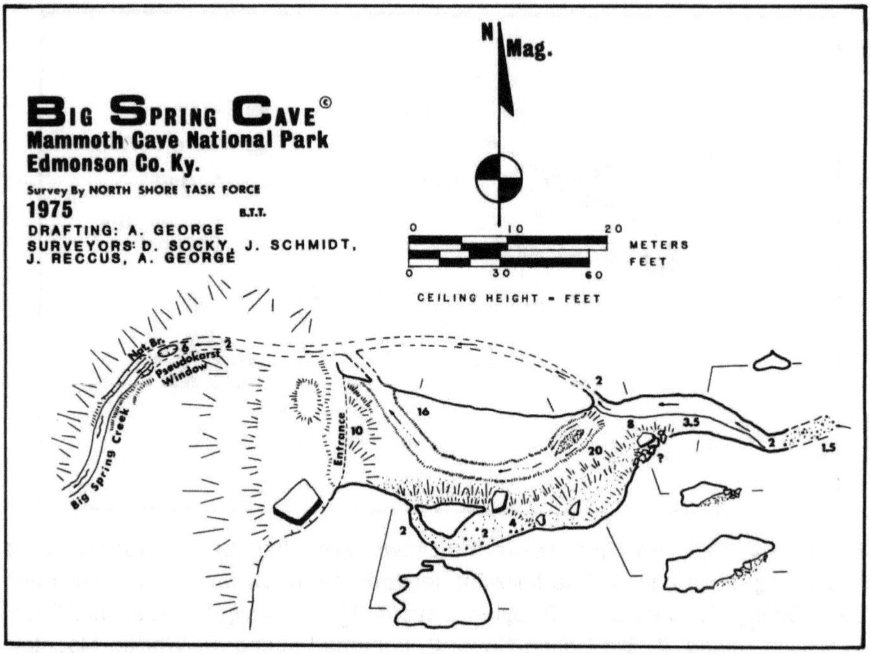

Figure 8.10 Map of Big Spring Cave, Mammoth Cave National Park. Numbers are ceiling heights in feet.

blocks and nonsequential slump into dolines and vertical shafts. East of Munfordville (Fig. 8.11), the Brush Creek Hills represent a major disturbed landform initiated by groundwater weathering and form a discordant feature to the surrounding topography. The resultant terrain displays a chaos of eroded, differentially subsided, collapsed, and faulted blocks of clastics and carbonates. This is similar to the interstratal karst destruction of the Millstone Grit in South Wales, Great Britain (Thomas, 1974). The channel still retains its outward form and stands in a bold inverted relief because the containing mold of the ancestral valley wall has been removed by solution weathering. The contents of the channel fill are scattered along the length and breadth of the Brush Creek Hills and portions of the Sinkhole Plain.

FACTORS THAT INFLUENCE SPELEOGENESIS

The largest caves north of the Green River appear to be subterranean cutoffs associated with a prior and higher base level. These caves are all found near the Green River. Farther back from the river, caves are associated with thalweg ponors and vertical shafts. Interstratal karsting forms some complex, high, narrow canyon caves in the extreme northern part of the study area. A number of phreatic maze caves are found in this part of Kentucky.

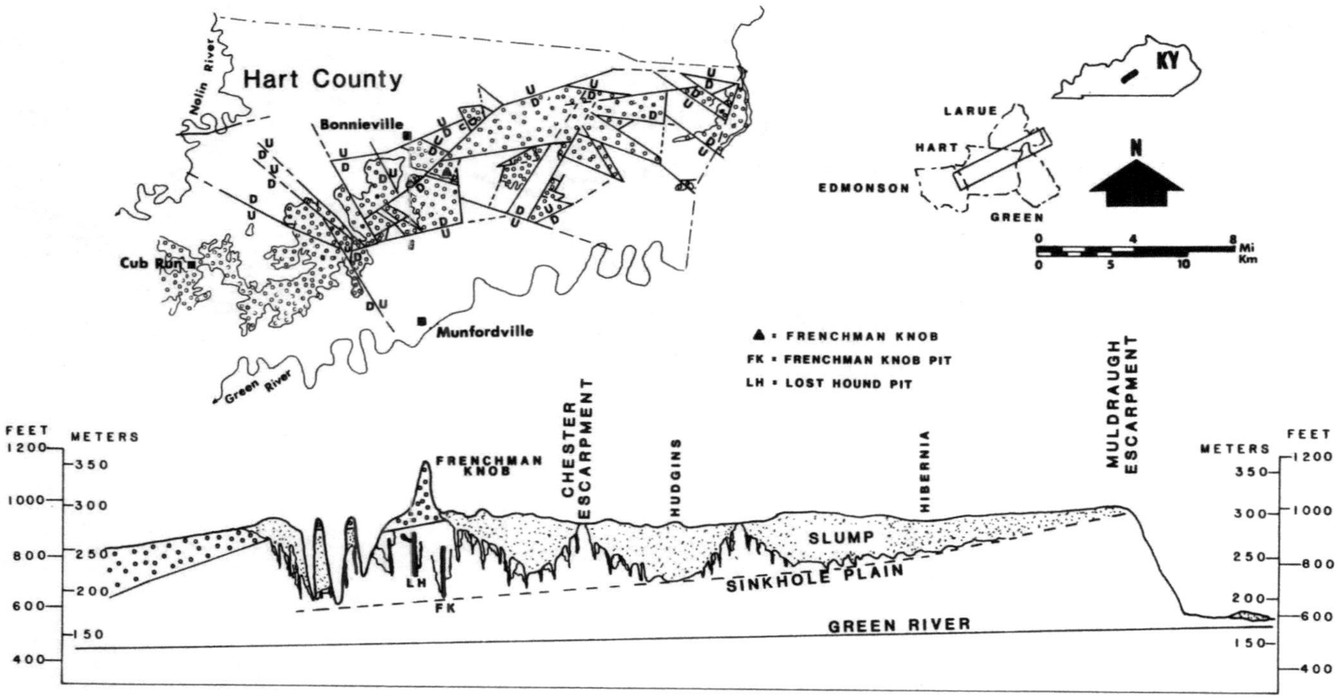

Figure 8.11 Longitudinal sketch map profile and plan view of the pseudokarst weathering of the Brownsville Channel. Once the channel extends east of Munfordville, subsidence on a massive scale ensues as a chaos of fault blocks (based on data from Withers et al., 1931; and Brown and Lambert, 1962). Groundwater solution of carbonates below the Pennsylvanian clastics has caused differential solution subsidence of clastics into vertical shafts and karren.

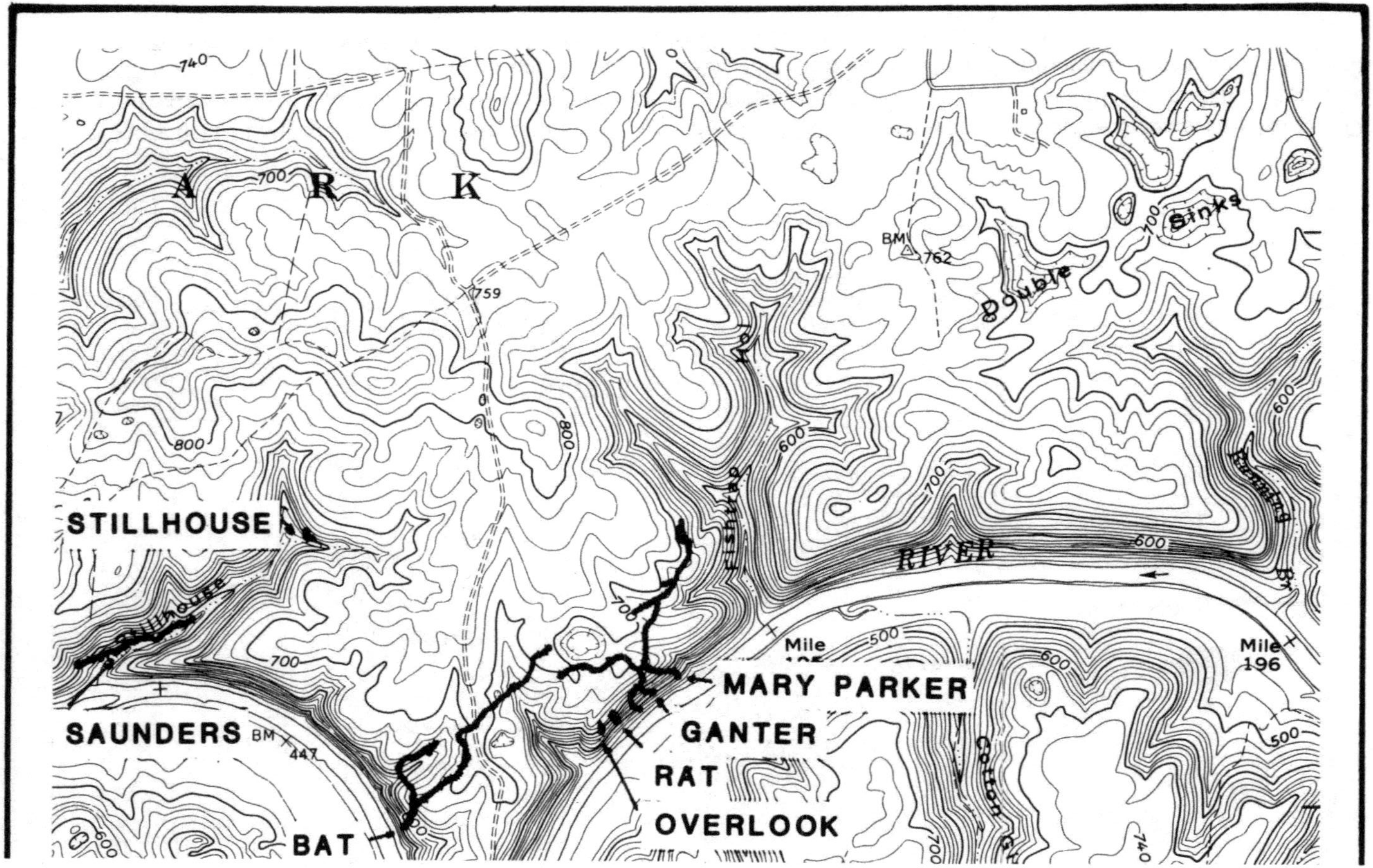

STILLHOUSE
SAUNDERS
BAT
MARY PARKER
GANTER
RAT
OVERLOOK
RIVER
Double Sinks
Fishtrap Hol
Running Br
Cotton
Mile 196
A R K

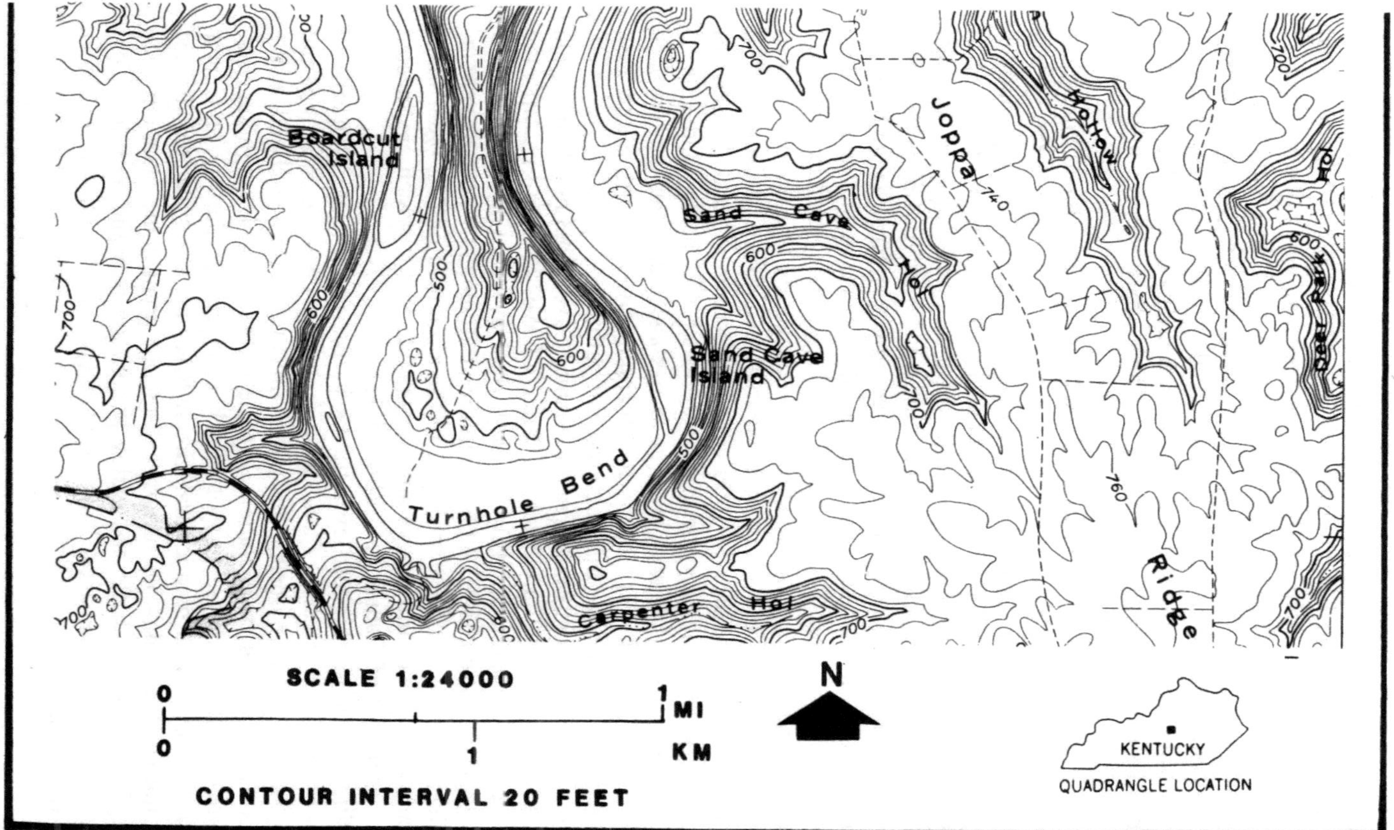

Figure 8.12 Map of the Turnhole Bend Cave system, Rhoda 7.5-min quadrangle, Mammoth Cave National Park. Survey by North Shore Task Force.

No one concise theory can adequately explain cave development on a regional scale. Even within the same drainage basin, caves develop from different hydrogeological conditions. George and Schmidt (1977) discussed the development of some of the smaller caves as being the product of surface-stream piracy (Stephen's Cave, Fort's Funnel Cave, Saunders Cave, and Stillhouse Cave). There is good concordance between master cave passage trends with thalweg and hillside slope positions. Saunders Cave in Stillhouse Hollow is essentially a shaft drain connected to a number of vertical shafts in the cave (Fig. 8.12). These shafts receive intermittent recharge from thalweg ponors.

Dry Prong caves are better integrated and have well-developed trunk-flow systems. Thalweg ponors are the main contributors of water to the system (Fig. 8.6). This is augmented by water input via hillside vertical shafts at clastic-carbonate rock interfaces and connection to stream bed ponor-shaft drains. The regimen is finally combined in a large tubular trunk passage of Buffalo Creek Cave and Fort's Funnel Cave. Probably, this is the same water seen at the Buffalo Spring resurgence on Green River (Fig. 8.6).

Turnhole Bend is historically and geomorphically the most famous feature north of the Green River (Fig. 8.12). It is a classic example of an intrenched meander; its name being derived from a karst vortex (whirlpool) situated upstream and east of the meander bend in Green River (Wilson, 1967). This feature was obliterated in 1906 with the erection of Lock Number 6, just above Brownsville. Apparently, river water was being (and still is) pirated by way of a sub-base level cave passage under the narrow neck of the Turnhole Bend. Exploration and mapping by the North Shore Task Force failed to materialize this subterranean cutoff.

Caves in the Turnhole Bend area are: Ganter Cave, 2993 m (9820 ft); Bat Cave, 1746 m (5727 ft); and Running Branch Cave, 1228 m (4224 ft), the largest mapped caves in the Hilly Country. Bat Cave and Ganter Cave show strong trunk-passage control by being in direct alignment with a lineament. This lineament has surface expression in the alignment of first-order high-gradient tributary stream valleys and the Double Sinks, which are located 1.6 km northeast of Ganter Cave (Fig. 8.12).

Positioning of the highest-level trunks in Bat Cave (Bat Avenue) and Ganter (Moon Dust Passage) Cave suggests that the caves acted as a partial cutoff for Green River during the intrenchment of Turnhole Bend at the 207–210 m (680–690 ft) elevation. Unique to the Moon Dust Passage are the phreatic lift chimneys, two of which connect into the top of Ganter Avenue. Piracy points of ancestral Green River are the truncated dolines called the Double Sinks. The original discharge point is the main entrance to Bat Cave. Vadose shafts and valley ponors are now the only low-flow source of water to these caves. Both caves receive backflooding from higher stages of Green River.

source of water to these caves. Both caves receive backflooding from higher stages of Green River.

Ganter Cave has five well-defined cave levels with numerous vertical shafts and high-gradient drains to the Ganter Avenue trunk passage. The highest shaft in the cave is Dinosaur Dome with more than 49 m of relief. A distributary system of tubes consisting of the Main Entrance, Rat Cave, and Overlook Cave discharged to the valley wall of Green River. By contrast, Bat Cave has at least three levels. The entrance main trunk (Bat Avenue) is believed to be associated with the highest level in Ganter Cave, the Moon Dust Passage. The trunk is augmented with numerous shafts and one long shaft drain called Ellis Avenue.

Developed in the lower St. Louis Limestone, Buckner Spring Cave is another example of an abandoned subterranean cutoff (Fig. 8.13). A steep

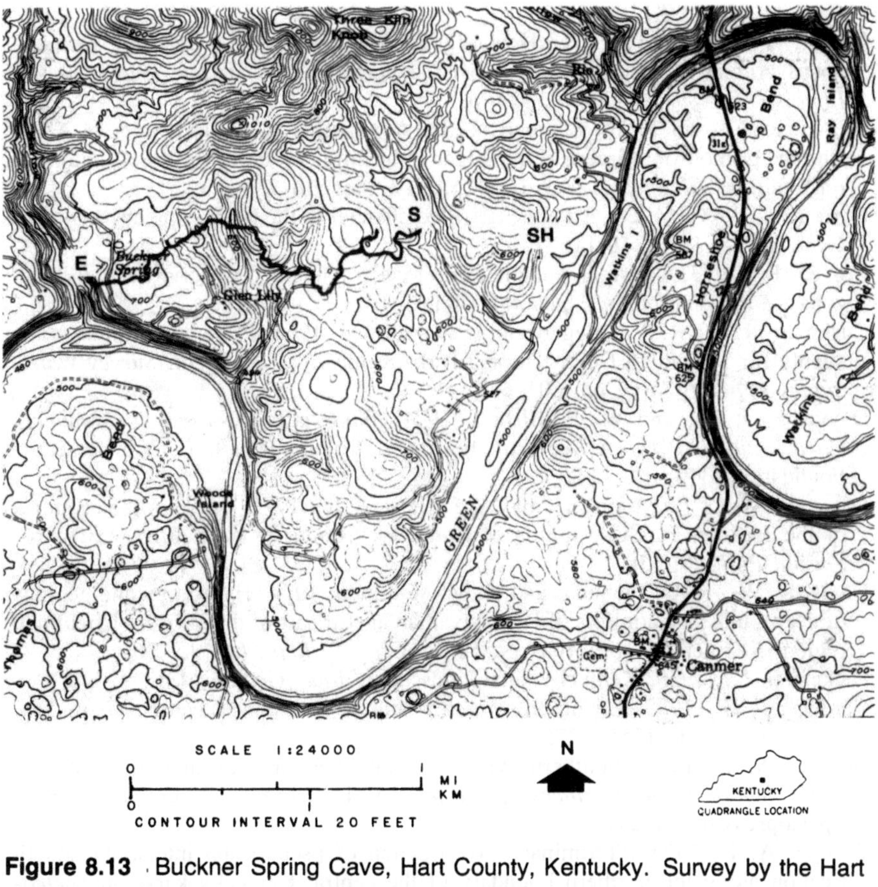

Figure 8.13 Buckner Spring Cave, Hart County, Kentucky. Survey by the Hart Attack Cave Survey. The following abbreviations are used: E, entrance; S, syphon; SH, steep head. *(Adapted from Saunders, 1974)*

head, located at *SH* on Figure 8.13, is believed to be the initial ponor input point. A smaller active cutoff is the Sinks of Little Brush Creek (Fig. 8.2). This is a 70-m-long cutoff through the neck of a meander bend, thereby eliminating 883 m of surface-stream flow. The rest of the cave is a flood route associated with the perennial Little Brush Creek that occupies a lower-level cave passage.

REFERENCES

Ash, D. W., 1985, Geomorphology and karst development in the Mitchell Plain and adjacent Crawford Upland in Harrison and Crawford Counties, Indiana, in *Guidebook to the Kentucky Speleofest, Meade County Fair Grounds, Kentucky*, vol. 14, A. I. George, ed., Louisville, Ky.: Speleopress, pp. 42–59.

Bretz, J H., 1942, Vadose and phreatic features of limestone caverns, *Jour. Geology* **50:**675–811.

Bristol, H. M., and R. H. Howard, 1971, Paleogeologic map of the sub-Pennsylvanian Chesterian (Upper Mississippian) surface in the Illinois Basin, *Illinois Geol. Survey Circ. 458*, 14p.

Brown, R. F., and T. W. Lambert, 1962, Availability of ground water in Allen, Barren, Edmonson, Green, Hart, Logan, Metcalfe, Monroe, Simpson, and Warren Counties, Kentucky, *U.S. Geol. Survey Atlas HA-32*.

Burroughs, W. G., 1923, A Pottsville-filled channel in the Mississippian, *Kentucky Geol. Survey Ser. 6* **10:**115–126.

Colveé, P., 1973, Cueva en cuarcitas en el cerro Autana, Territorio Federal Amazonas, *Soc. Venezolana Espel. Bol.* **4**(1):5–13.

Cushman, R. V., R. A. Krieger, and J. A. McCabe, 1965, Present and future water supply for Mammoth Cave National Park, Kentucky, *U.S. Geol. Survey Water-Supply Paper 1475-Q*, pp. 601–647.

Davidson, J. K., and W. P. Bishop, 1971, *Wilderness resources in Mammoth Cave National Park: A Regional Approach*, Columbus, Ohio: Cave Research Foundation, 34p.

Davis, R. W., T. W. Lambert, and A. J. Hansen, Jr., 1971, Water in the economy of the Jackson Purchase region of Kentucky, *Kentucky Geol. Survey Ser. 10, Special Pub. 20*, 33p.

Davis, R. W., R. O. Plebuch, and H. M. Whitman, 1974, Hydrology and geology of deep sandstone aquifers of Pennsylvanian age in part of the Western Coal Field region, Kentucky, *Kentucky Geol. Survey Ser. 10, Rept. Inv. 15*, 26p.

Dicken, S. N., 1935, Kentucky karst landscapes, *Jour. Geology* **43:**709–728.

Gardner, J. H., 1935, Origin and development of limestone caverns, *Geol. Soc. America Bull.* **46:**1255–1275.

George, A. I., 1973, Southern Warren County karstland excursion, Kentucky, in *Guidebook to the Kentucky Speleo-fest*, A. I. George, ed., Louisville, Ky.: Speleopress, pp. 1–21.

George, A. I., 1975, Preliminary investigation of pseudo and interstratal karstification along the northern boundary of the central Kentucky karst, in *Speleofest Guidebook, Larue County, Kentucky*, A. I. George, ed., Louisville, Ky.: Speleopress, pp. 48–69.

George, A. I., 1976, Karst and cave distribution in north-central Kentucky, *Natl. Speleol. Soc. Bull.* **38**(4):93–98.

George, A. I., 1977, Evaluation of sulfate water quality in the north-central Kentucky karst, in *Hydrologic Problems in Karst Regions,* R. R. Dilamarter and S. C. Csallany, eds., Bowling Green, Ky.: Western Kentucky University, pp. 340–356.

George, A. I., 1982, Peneplains, upland stream terraces, and cave levels in portions of the north-central Kentucky karst, in *Guidebook to the 1982 Kentucky Speleofest, Meade County Fair Grounds, Kentucky*, A. I. George and L. M. McCarty, eds., Louisville, Ky.: Speleopress, pp. 48–68.

George, A. I., 1985*a*, Caves of Kentucky, in *Caves and Karst of Kentucky*, P. H. Dougherty, ed., Kentucky Geol. Survey Special Pub. 12, Ser. IX, pp. 18–27.

George, A. I., 1985*b*, Edmonson County saltpeter sites, *The Karst Window* **21**(5):6–7.

George, A. I., and J. B. Schmidt, 1977, Cave development north of the Green River at Mammoth Cave is strongly influenced by recharge from a filled Pennsylvanian paleo-valley, *Internat. Assoc. of Hydrogeologists Mem. 12*, pp. 415–430.

Gildersleeve, B., 1965, Geology of the Brownsville quadrangle, Kentucky, *U.S. Geol. Survey Geol. Quad. Map GQ-411.*

Gildersleeve, B., 1968, Geologic map of the Bee Spring quadrangle, Edmonson and Grayson Counties, Kentucky, *U.S. Geol. Survey Geol. Quad. Map GQ-757.*

Hess, J. W., S. G. Wells, and T. A. Brucker, 1974, A survey of springs along the Green and Barren Rivers, central Kentucky karst, *Natl. Speleol. Soc. Bull.* **36**(3):1–7.

Howard, A. D., 1968, Stratigraphic and structural controls on landform development in the central Kentucky karst, *Natl. Speleol. Soc. Bull.* **30**:95–114.

Lehmann, H., 1969, On the morphology of the Mitchell Plain and the Pennyroyal Plain of Indiana and Kentucky, *Deutscher Geographentages Bad Godesberg* **36**:359–368. (Trans. by E. Werner, 1975, *Cave Geology* **1**(2):29–39.)

Lobeck, A. K., 1928, The geology and physiography of the Mammoth Cave National Park, *Kentucky Geol. Survey Ser. 6* **31**:227–399.

McGrain, P., 1960, A high-refractory clay in Hart County, Kentucky, *Kentucky Geol. Survey Ser. 10, Info. Circ. 5*, 10p.

McGrain, P., and H. P. Hamlin, 1962, Occurrence of halloysite in Kentucky (abstract), *Geol. Soc. America Symposium Paper 68*, p. 77.

McGrain, P., and W. L. Helton, 1964, Gypsum and anhydrite in the St. Louis Limestone in northwestern Kentucky, *Kentucky Geol. Survey Ser. 10, Info. Circ. 13,* 26p.

Martini, J., 1987, Les Phénomènes Karstiques des Quartzites d'Afrique du Sud., *Karstologie, No. 9,* pp. 45–52.

Miller, R. C., 1969, Geologic map of the Canmer quadrangle, Hart County, Kentucky, *U.S. Geol. Survey Geol. Quad. Map GQ-816.*

Miotke, F.-D., and A. N. Palmer, 1972, *Genetic Relationship Between Caves and Landforms in the Mammoth Cave National Park Area,* Würzburg: Böhler Verlag, 69p.

Moore, F. B., 1972, Geologic map of the Upton quadrangle, central Kentucky, *U.S. Geol. Survey Geol. Quad. Map GQ–1000.*

Palmer, M. V., and A. N. Palmer, 1975, Landform development in the Mitchell

Plain of southern Indiana: Origin of a partially karsted plain, *Zeitschr. Geomorphologie* **19:**1–39.

Pettijohn, F. J., 1957, *Sedimentary Rocks,* New York: Harper & Row, 718p.

Pohl, E. R., 1970, Upper Mississippian deposits of south-central Kentucky, *Kentucky Acad. Sci. Trans.* **31:**1–15.

Quinlan, J. F., 1970, Central Kentucky karst, *Études et Travaux de Méditerranée* **7:**235–253.

Quinlan, J. F., 1972, Karst-related mineral deposits and possible criteria for the recognition of paleokarst: A review of preservable characteristics of Holocene and older karst terranes, *24th Internat. Geol. Congress, Montreal, Proc.* **6:**156–168.

Quinlan, J. F., and R. O. Ewers, 1981, Hydrogeology of the Mammoth Cave region, Kentucky, in *GSA Cincinnati '81 Field Trip Guidebooks*, vol. 3, T. G. Roberts, ed., Falls Church, Va.: American Geological Institute, pp. 457–506.

Quinlan, J. F., and E. R. Pohl, 1968, Effect of vertical shafts on slope retreat and dissection of solution escarpment and Chester Cuesta, central Kentucky karst, *Geol. Soc. America Special Pub. No. 101,* pp. 450–451 (abstract).

Quinlan, J. F., and J. A. Ray, 1981, Groundwater basins in the Mammoth Cave region, Kentucky, *Friends of the Karst, Occasional Pub. No. 1*, Map.

Sauer, C. O., 1927, Geography of the Pennyroyal, *Kentucky Geol. Survey Ser. 6,* **25**, 303p.

Saunders, J., 1974, Exploration of Buckner Cave, *COG Squeaks* **17:**44–47.

Sedimentation Seminar, 1978, Sedimentation of the Kyrock Sandstone (Pennsylvanian) in the Brownsville paleo-valley, Edmonson and Hart Counties, Kentucky, *Kentucky Geol. Survey Ser. 10, Rept. Inv. 21*, 24p.

Shaw, F. R., and B. Gildersleeve, 1969, An anastomosing channel complex at the base of the Pennsylvanian system in western Kentucky, *U.S. Geol. Survey Prof. Paper 650–D*, pp. D206–D209.

Siever, R., 1951, The Mississippian–Pennsylvanian unconformity in southern Illinois, *Am. Assoc. Petroleum Geologists Bull.* **35:**542–581.

Sweeting, M. M., 1972, *Karst Landforms*, London: Macmillan, and New York: Columbia University Press, 362p.

Szczerban, E., and F. Urbani, 1974, Carsos de Venezuela, Part 4, Formas carsicas en areniscas precambricas del Territorio Federal Amazonas y Estado Bolivar, *Soc. Venezolana Espel. Bol.* **5:**27–54.

Thomas, T. M., 1974, The South Wales interstratal karst, *British Cave Research Assoc. Trans.* **1:**131–152.

Weller, J. M., 1927, The geology of Edmonson County, *Kentucky Geol. Survey Ser. 6,* **28,** 246p.

White, W. B., R. A. Watson, E. R. Pohl, and R. Brucker, 1970, The central Kentucky karst, *Geog. Rev.* **60:**88–115.

Wigley, T. M. L., 1973, Chemical evolution of the system calcite–gypsum–water, *Canadian Jour. Earth Sci.* **10:**306–315.

Wigley, T. M. L., J. F. Quinlan, and D. C. Ford, 1973, Geomorphology and geochemistry of a gypsum karst near Canal Flats, British Columbia, *Canadian Jour. Earth Sci.* **10**(2):113–129.

Wilson, G., 1967, *Folkways of the Mammoth Cave Region, No. 2*, Mammoth Cave, Ky.: National Parks Concessions, Inc., 64p.

Withers, F. A., A. H. Sutton, G. R. Wesley, and D. H. Crabb, 1931, Geologic map of Hart County, Kentucky, *Kentucky Geol. Survey Ser. 6*, Map.

Woodson, F. J., 1981*a*, Lithologic and structural controls on karst landforms of the Mitchell Plain, Indiana, and Pennyroyal Plateau, Kentucky, Indiana State University, M.A. thesis, 132p.

Woodson, F. J., 1981*b*, Uppermost St. Louis Limestone (Mississippian): The Horse Cave Member in Indiana, *Indiana Acad. Sci. Proc.* **91**:419–427.

9

HYDRAULIC GEOMETRY OF CAVE PASSAGES

William B. White and George H. Deike, III

There are two competing sets of factors that control the original solutional shapes of the natural drainpipes that make up the conduit system. One set comprises the spatial variations in the rates of solution of the bedrock caused by the distribution and geometry of joints and bedding planes, by the variations in solubility of the limestone, and by the distribution of such lithologic features as shale beds, dolomite beds, sandy layers, and chert nodules. The second set comprises the variations in rates of solution caused by the shifting flow regimes of moving water. If flow velocities are low, the passage tends to be etched into a complex shape controlled by structural and lithologic factors and we speak of an etching geometry or structure-controlled geometry. If the rate of solution varies with flow velocity, the shape of the passages will be modified to accommodate the flow pattern and we speak of a hydraulic geometry. Etching geometries tend to be irregular and angulate; hydraulic geometries tend to be smooth, curvilinear shapes. The forms of solution conduits are further modified by bedrock breakdown, by sediment in-filling, and by the deposition of secondary calcite deposits so that the shapes of cave passages might be quite different from the original solution shapes. Some deduction and interpretation is needed on the part of the karst hydrogeologist.

The aquifer model that has been developed proposes that a system of conduits carries most of the groundwater from the recharge area to springs on Green River. The cave passages are seen as abandoned conduits. The passages can be analyzed for the same sort of properties that are associated with surface stream channels. The comparison of the conduits of limestone aquifers with surface streams must be tempered by the additional complication that many conduits were created by water that filled them completely (pipe flow), whereas others were formed by streams of water with a free-air surface (channel flow). The distinction between pipe flow and open-channel flow is central to the hydraulic geometries of conduit systems.

Some aspects of the hydraulic geometry that might be compared between surface channels and cave conduits include: (1) channel width/channel depth characteristics; (2) sinuosity; (3) braiding; (4) ordered branching ratios; and (5) distinct catchment area/discharge relationships. These comparisons are developed in some detail in the sections that follow.

Channels are usually classified as rigid or erodable. Rigid channels have fixed walls that are not affected by the fluid flow. Concrete or metal flumes and spillways are examples. Erodable channels have walls of unconsolidated materials that can be moved and redeposited by the moving fluid. Most natural channels (creeks and rivers) that typically have beds cut in their own alluvium are of this type. Cave passages are natural examples of rigid channels, at least on the time scale of individual flow events.

Complete cave systems, the complete set of abandoned fragments of conduit beneath residual uplands, can be compared with surface valleys. The nature of surface downcutting is such that most of the earlier record of the channel and its associated valley is destroyed as the valley deepens. In general, all that remains to interpret are a few terraces. Because the surface channel is erodable, it constantly changes its form with time and with major flood events. All that one sees is the current form of the channel plus perhaps a few abandoned segments to indicate where it has been in the recent past. In contrast, the cave system preserves an incredible wealth of detail. The channel is rigid so all details of its evolution are preserved. There is no equivalent to slope retreat and bank caving in the cave conduits. Superimposed conduits and downcutting canyons provide a record of flow regimes far back into the Pleistocene or even to late Tertiary time. Indeed, there is so much recorded detail that there is difficulty in interpretation. The hydraulic geometry of the big cave systems of central Kentucky provide a comprehensive picture of the evolutionary history of the conduit aquifer system.

SOLUTIONAL SCULPTURING

The uneven dissolution of cave passage walls creates a number of small solutional features. Many of these were originally described by Bretz (1942) and used as evidence for the deep phreatic theory of cave origin. Others have been added to the list and although there is no comprehensive catalog of such features, Renault (1958) and Sweeting (1972) briefly described many of them. Anastomosis channels, scallops, and vertical flutes are the most important forms in the central Kentucky caves, although examples can be found of nearly all solutional features.

Anastomosis Channels

Bedding-plane anastomoses are braided, freely interconnected networks of solution tubes found on distinct bedding planes at many elevations within the central Kentucky caves. They appear to be truncated by the main cave passages and therefore usually appear as rows of small openings in the passage walls (Fig. 9.1). Sometimes they appear as half-tubes crossing a cave ceiling. The walls of the anastomosis tubes are generally smooth, rounded, and show little evidence of any additional solutional sculpturing. Where seen in passage ceilings or where beds have fallen, the tubes appear as a braided, highly interconnected network revealing on a small scale the kind of pattern described by Palmer (1975) as an "anastomotic maze." The

Figure 9.1 Anastomosis channels on side wall of Mammoth Cave near Wooden Bowl Room. Note orientation along bedding planes.

cross-sectional diameter ranges from a few centimeters to a maximum of 1 m. Very few of the anastomosis tubes are large enough to permit human exploration and as a result little is known about the details of their pattern. In particular, it is not known how far the anastomosis tubes extend back away from the main cave passage.

There appears to be an evolutionary scheme for the tubes. In early stages diameters are smaller and the tubes are well separated. As they grow, they maintain the same circular cross section, expand out, and sometimes weaken the bedding plane to the point where they become a causative factor in cave breakdown. Although uncommon, anastomotic mazes in such places as Rickwood Caverns in central Alabama have expanded to a point where entire cave rooms appear to be essentially an enlarged and collapsed anastomotic maze.

Figure 9.2 Anastomosis channels with grooving along Fossil Avenue in the New Discovery section of Mammoth Cave.

The anastomosis channels form an additional volume for bank storage during the time in which the cave passages are in the floodwater zone. Groove patterns are sometimes found beneath the openings to the anastomosis channels (Fig. 9.2). What apparently happens is that during floods, water is driven back into the anastomosis tubes and stored there. When the waters recede, the passages drain and water running out of the anastomosis tubes dissolves these small grooves.

Bretz interpreted the anastomosis channels as being primeval pathways that had formed in the limestone before the initiation of cave development by random circulation of groundwater at considerable depths. He envisioned the cave as having formed along one particular pathway through the anastomotic maze. Ewers (1966), in a series of laboratory experiments, showed that the anastomosis tubes are contemporaneous with the main-passage development and essentially represent subsidiary flow lines. They appear in passages whose primary structural control is a bedding-plane parting. One central pathway along the bedding becomes the optimum hydraulic path along which the development of the main cave passages takes place. Other pathways divert from the main one, move off through the bedding plane for short distances and then return to it. These are indeed braided channels in the same sense that braided channels occur in surface streams.

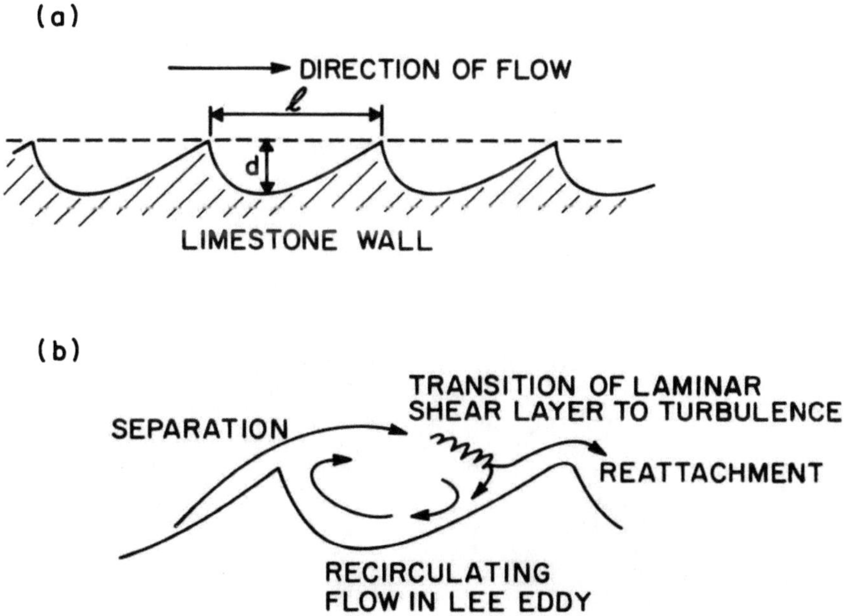

Figure 9.3 *A*, Cross-section sketch of a scalloped wall showing characteristic dimensions and shape in relation to direction of flow. *B*, Mechanism of scallop formation according to Curl (1974).

Scallops

Scallops are shallow cuspate pockets dissolved into cave walls. They are asymmetric with a steep wall and a shallow wall (Fig. 9.3). The axis of the scallop is generally oriented parallel to the axis of the passage. Scallops form in a range of sizes from a few centimeters to more than 1 m. Indeed, the upper limits to scallop size are difficult to define because the largest scallops are comparable to the size of the passage itself and it becomes difficult to distinguish between a scallop and a bend in the passage.

It has long been known that scallops are indicators of flow direction (Coleman, 1949). The steep cusp of the scallop is on the upstream side and the smooth slope is on the downstream side. Scallop markings are also indicators of flow velocity and thus form the basis for calculations of paleo-flow and paleo-discharge. High-velocity flows produce small scallops and low-velocity flows produce large scallops.

Scallops appear in the south-central Kentucky cave systems in three main situations. Some of the large conduits are uniformly scalloped on walls and ceilings (the floor is usually obscured by fill; Fig. 9.4). The large tubular

Figure 9.4 Uniformly scalloped walls and ceiling of a large-tube passage: Grand Avenue in the Flint Ridge Cave section of the cave system.

Figure 9.5 Scalloped walls of a large canyon passage: Upper Salts Avenue in the Great Salts Cave section of the cave system.

conduits generally contain large scallops, from 20 cm to more than 1 m. Secondly, canyon passages are also frequently scalloped uniformly from floor to ceiling (Fig. 9.5). The scallops that occur in canyon passages are often smaller, generally on the order of a few centimeters to 10 cm. The third situation (Fig. 9.6) is the large conduit passage with large scallops or no scallops in the higher parts of the passage but with a zone near the floor with small scallops. This is interpreted to mean that there were two stages in the hydrologic regime of the passage: one in which the conduit was pipe full, followed by a later stage in which the passage contained a higher velocity free-surface stream. Scalloped walls tend to be scalloped uniformly with scallops of approximately the same size (Fig. 9.7). The distribution in scallop sizes is illustrated in Figure 9.8 for three types of cave passage.

It has been demonstrated that scallops are a strictly hydraulic phenomenon (Curl, 1966). Their characteristic size depends on fluid flow and not, to a

Figure 9.6 Passage wall with large (low-flow velocity) scallops on the upper portion and small (high-flow velocity) scallops near the floor. Note hard hat for scale.

first approximation, on the characteristics of the bedrock. Scallops formed on gypsum blocks in the laboratory (Goodchild and Ford, 1971) and scallops formed on ice and snow by moving wind (Curl, 1966) have the same hydraulic properties. Scallop length and fluid flow parameters are related by a dimensionless Reynolds number

$$N_R = \frac{\overline{v}\overline{L}\rho}{\eta} \tag{9.1}$$

where $\overline{v}$ is the velocity of fluid moving past the scallop, ρ is the density of the fluid, and η is the fluid viscosity. The Reynolds number has a numerical

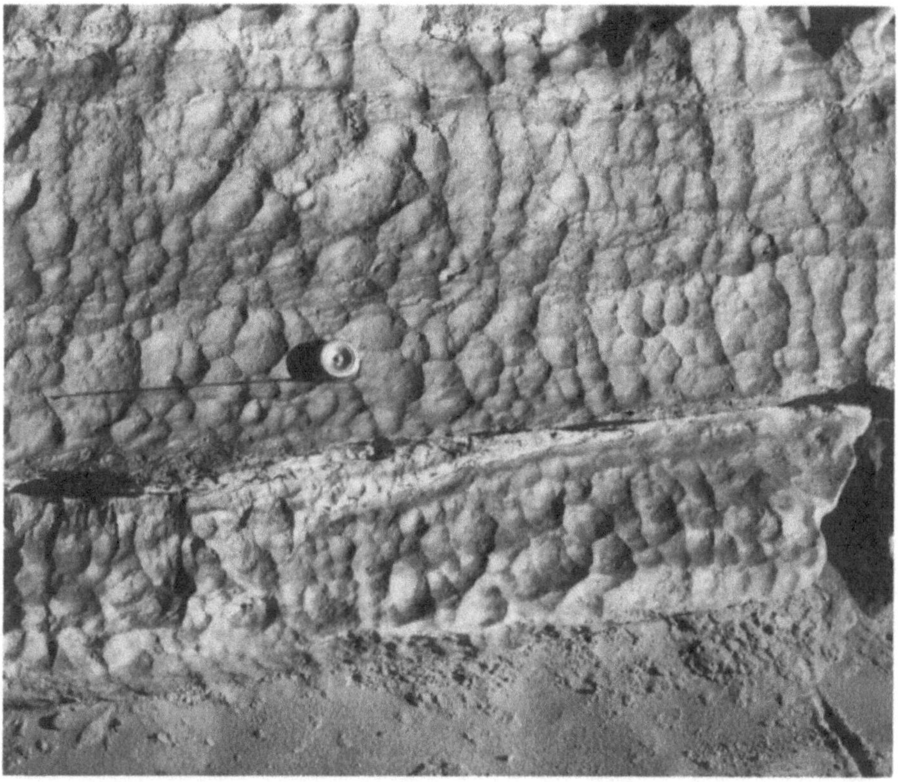

Figure 9.7 Small scallops, some asymmetric, and some with a range of scallop sizes on the wall of a small canyon: Bicycle Avenue in the Flint Ridge section of the cave system. Scale is 1 ft long.

value of 22,500 for the very regular forms known as flutes. $\overline{L}$ is the mean scallop length that Curl (1974) argued should be the Sauter mean

$$\overline{L}_{32} = \frac{l_i^3}{l_i^2} \tag{9.2}$$

rather than the usual arithmetic mean.

By dimensional analysis it was shown that

$$N_R^* = \frac{\overline{L}_{32} v^* \rho}{\eta} \tag{9.3}$$

where v^*, a friction velocity, equals $\sqrt{\tau\rho}$, where τ is the average shear stress at the wall and ρ is the fluid density. The scallop Reynolds number, N_R^*, based on friction velocity, is a universal constant for scallop formation

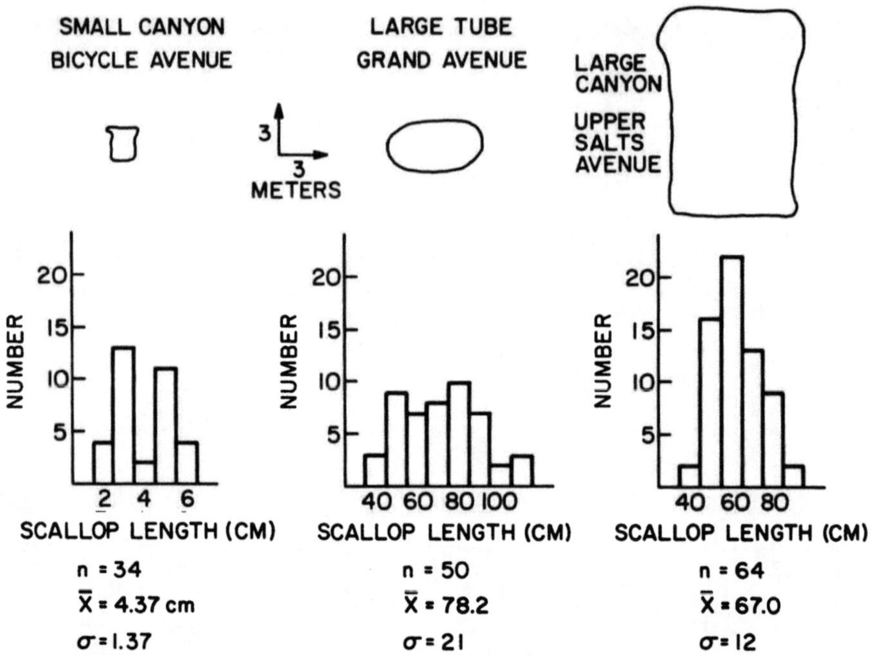

Figure 9.8 Distribution of scallop sizes for three types of passage shown in Figures 9.7 (small canyon), 9.4 (large tube), and 9.5 (large canyon). *(After White and Deike, 1976)*

and was found from model experiments to have a numerical value of 2200 (Blumberg and Curl, 1974). Curl (1974) assumed two limiting geometries to calculate an actual channel velocity from the friction velocity Reynolds number. For a circular conduit

$$N_R = N_R^*\left[2.5\left(\ln\frac{D}{2L_{32}} - \frac{3}{2}\right) + B_L\right] \tag{9.4}$$

For a rectangular conduit (canyon cross section)

$$N_R = N_R^*\left[2.5\left(\ln\frac{D}{2L_{32}} - 1\right) + B_L\right] \tag{9.5}$$

where D is the diameter of the circular conduit or the distance between the walls of the rectangular conduit. The quantity B_L is determined only by the wall roughness and was found from model studies to be equal to 9.4. At the time this research was done, Curl reduced the solution of the above equations to a calculator program; the graphical relationships shown in Figure 9.9 were obtained with this program.

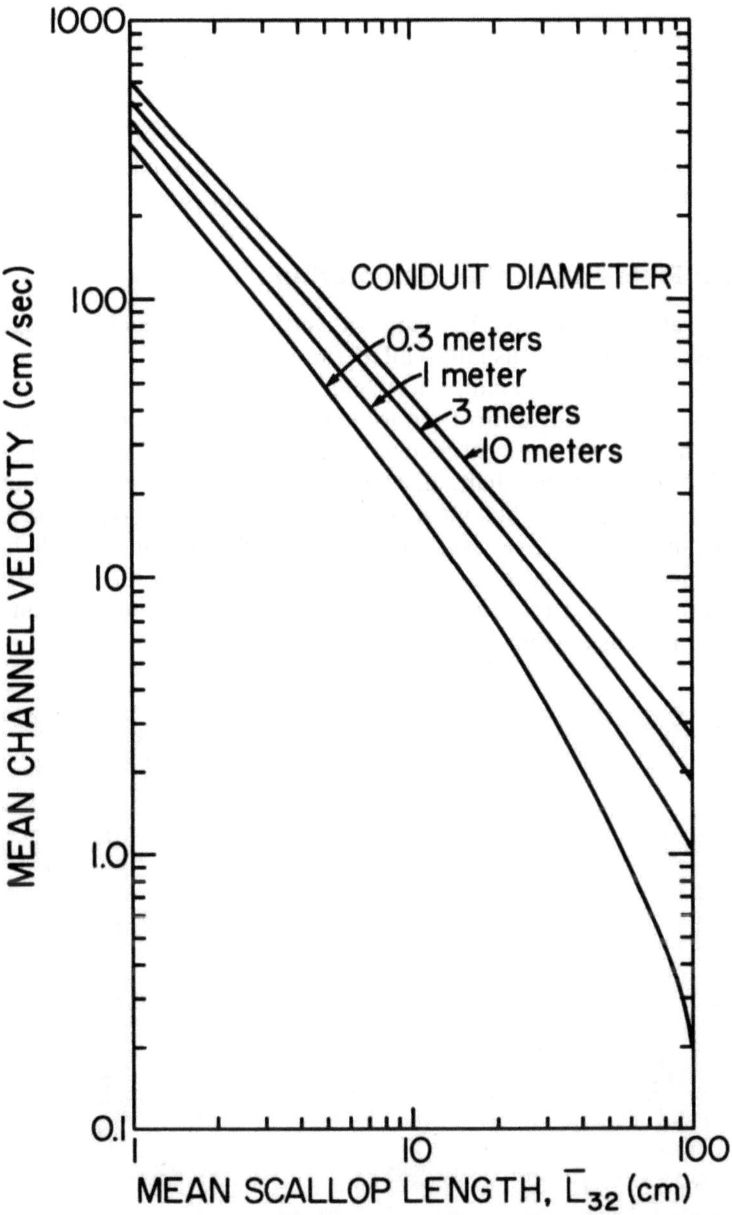

Figure 9.9 Relation of the scallop length (Sauter mean) to mean channel flow velocity for conduits of several diameters. Calculator program devised by R. L. Curl (personal communication, 1976).

The discharge and thus the channel velocity through cave conduits varies with the rate of recharge at the surface. It is not known with certainty which velocity is calculated from scallop measurements. It has been assumed, based on the observation that the rate of dissolution of limestone is not strongly velocity dependent, that the scallop pattern should record the most probable flow, which is then taken to be the mean flow.

Vertical Flutes and Other Sculpturing

Vadose water descending the walls of vertical shafts incises long parallel grooves, which are usually called flutes (not to be confused with the scallop-related fluting found in horizontal passages with very uniform flow velocities). Flutes vary from 1 cm to many centimeters in depth and from a few centimeters to many centimeters wide. Vertical flutes frequently extend the entire height of the shaft, distances as much as 50 m. The flutes are rounded at the bottom but meet at a sharp crest thus giving the shaft a crenulate pattern in cross section. When shafts are cut through beds of greatly varying resistance to solution the resistant beds form ledges thrust outward in relief. In this situation the flutes are discontinuously cut across the outthrust ledges (Fig. 9.10).

Figure 9.10 Vertical fluting near the base of a vertical shaft: Colossal Dome in the Colossal Cave section of the cave system. Iron pipe on the floor is approximately 1 m in length.

Figure 9.11 Rough sculptured walls and ledges derived from high-velocity free-surface flow in a high-gradient passage: El Ghor in Mammoth Cave.

The other environment in which rough and jagged sculpturing of the bedrock occurs is in high-gradient passages. Most passages have smoothly sculptured walls except where the smooth surfaces are broken by breakdown. In high-gradient passages, high stream velocities reach a totally rough-flow regime, and the walls and floor of the passages are carved into jagged pinnacles and deep-grooved potholes (Fig. 9.11).

CHANNEL GEOMETRY

Tubes and Canyons

The original solutionally carved cross-sectional shape of cave passages is not always evident because a clastic fill covers the floor of many passages to depths of from a few decimeters (tubes in Long, Mammoth, and Flint Ridge caves) to perhaps 15 m (Long Cave), and breakdown often alters the

Figure 9.12 Cave passage with ideal tubular geometry: Turner Avenue in the Flint Ridge Cave section of the cave system.

walls and ceiling. The observed simple and complex solution conduits are composed of two end-member shapes: tubes and canyons. Many of the tubular passages in the Mammoth Cave area are remarkably uniform elliptical pipes with the long axis of the ellipse parallel to bedding (Fig. 9.12).

The cross sections in Figure 9.13 illustrate a continuity of shapes from various sizes of tubes to tubes with canyons to canyons with only faint traces of a tubular component at the top. Cross sections 1–5 (Fig. 9.13) illustrate some of the sizes and shapes of tubular passages. Low, wide ellipses are most common; size ranges from 1 m in smallest dimension to 15 m in largest dimension. In cross section 2 the low extension to the left is a stream undercut.

Cross sections 6 and 7 (Fig. 9.13) show tubes with small canyons. Cross sections 8, 9, and 10 illustrate proportionately larger or deeper canyons below the tubes. Rose's Pass and Boone Avenue are also good examples. In cross sections 11–16 the canyon is wide compared to the evident tubular

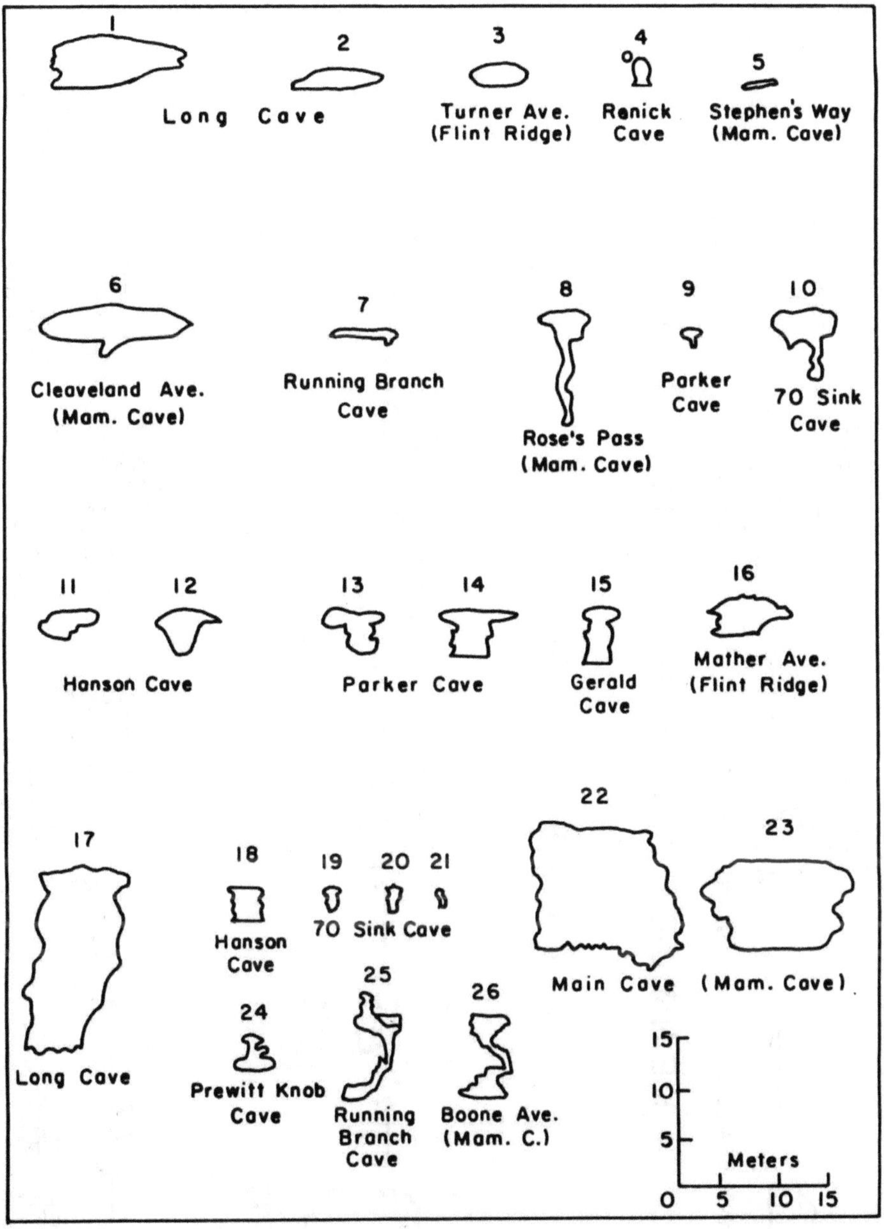

Figure 9.13 Cross-sectional shapes of cave passages for various caves in the Mammoth Cave area.

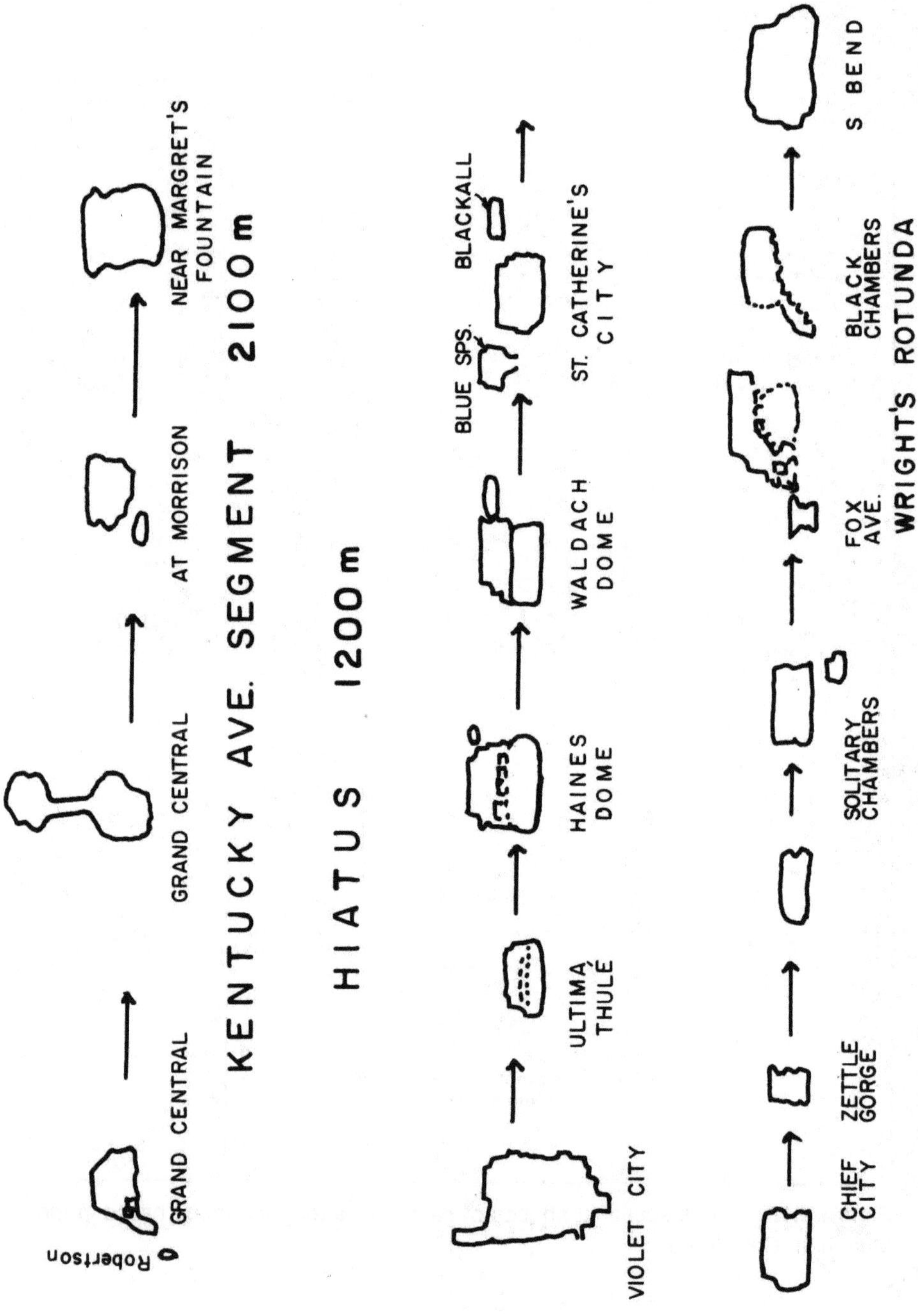
Robertson
GRAND CENTRAL
GRAND CENTRAL
GRAND CENTRAL
AT MORRISON
NEAR MARGRET'S FOUNTAIN
KENTUCKY AVE. SEGMENT 2100 m
HIATUS 1200 m
VIOLET CITY
ULTIMA THULE
HAINES DOME
WALDACH DOME
BLUE SPS.
ST. CATHERINE'S CITY
BLACKALL
CHIEF CITY
ZETTLE GORGE
SOLITARY CHAMBERS
FOX AVE.
BLACK CHAMBERS
S BEND
WRIGHT'S ROTUNDA

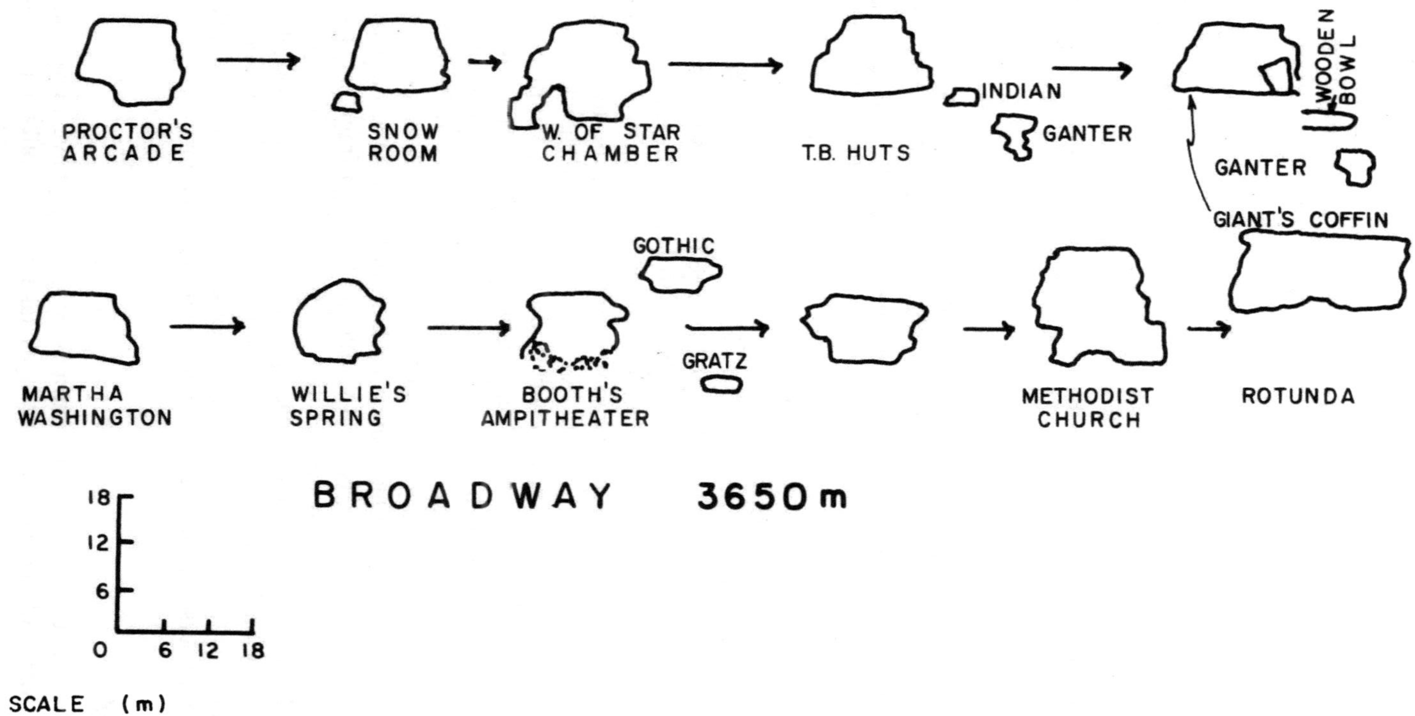

Figure 9.14 Sketch showing passage cross sections along the upper-level master trunk of Mammoth Cave. Place names are mostly on the Mammoth Cave tourist trails.

component above. Cross sections 17–21 illustrate passages that are almost all canyons. Small passages of this type are the common form seen in shaft drains. In some parts of the length of such passages a widening at the top suggests a tubular component. The canyon in the ceiling of Bransford–Nicholson Avenue has a very small trace of a tube at its top. The size of canyons or canyon components of passages ranges from a few centimeters to 15–20 m. The walls of the canyons are approximately parallel through the height in most cases. Flow markings are common on the walls. Cross sections 22–26 are complex. In cross section 22 the trapezoidal passage is altered by breakdown, probably from a very large tube or tubes. Cross section 23 could be a large tube with a broad canyon, but might be two superimposed tubes. Cross sections 24–26 each show two or three tubular components, connected vertically by a canyon.

The widths of canyons are approximately constant along their lengths. In three-dimensional view, canyon plans change continuously through their height, and shape (except for width and height) changes continuously along the length. Tubular passages have relatively invariant sizes and shapes. The height varies no more than 15% from average along lengths up to 400 m, and width varies only 20% from average in lengths up to 1500 m. These numbers suggest the reason that passage components, defined by shape, can be followed for long distances.

One of the longest continuous fossil conduits is the upper-level main trunk passage of Mammoth Cave (Fig. 9.14). The passage is first seen near Grand Central Station near the New Entrance to Mammoth Cave. East of this point the trunk is filled with breakdown from the collapse of an upper-level segment of passage. From Grand Central Station, the trunk can be followed downstream for 2.1 km as Kentucky Avenue. Kentucky Avenue terminates against a hillside breakdown, and there is a 1.2-km hiatus. The trunk is seen again as a short segment known as Croghan Hall, which is blocked by breakdown at both ends. A short distance west of the breakdown the trunk can be entered again at the Violet City Entrance from which it can be followed continuously for 3.65 km to the Rotunda where there is a downstream bifurcation. One branch leads to the collapse at the Historic Entrance to Mammoth Cave, beyond which another short fragment, known as Dixon Cave, is close to the location of the discharge point onto the ancestral Green River. Details of passage geometry vary because of later modification by breakdown and because of cut-arounds and other braiding of the conduit. There are only a few tributaries and some residual drainways, and the main course of the master conduit is never in doubt.

Sinuosity

Many of the conduits in the south-central Kentucky karst have a pronounced sinuosity. Many of the sinuous passages take on a regular meandering pattern not unlike the meandering patterns seen in surface rivers where the bend spacing of the meanders scales with the channel width according to a power law (Leopold and Wolman, 1960). Figure 9.15 shows the relationship between meander bend spacing and passage width for some cave passages, mostly in Mammoth Cave. Comparison of a large set of data from other caves in the United States and elsewhere in the world (Deike and White, 1969) shows that many cave passages follow similar relationships and that the coefficients in the power function are very similar to those describing surface rivers.

Three types of sinuous passages are found. First, small canyons originating in residual streams cutting down through the sediment-covered floors of large tubular passages often exhibit incised meanders. The residual stream established a meandering pattern on the clastic material flooring the large conduit, and this pattern was maintained when the passage cut down into the underlying bedrock. Second, many of the larger canyons, not obviously related to large tubes, also meander. Third, many of the large elliptical conduits, such as Cleaveland Avenue in Mammoth Cave, have a pronounced sinuous pattern.

Meandering patterns in surface rivers arise because these streams have erodable beds and the mobile alluvial sediments can be rearranged to form the channel pattern demanded by energy balance and flow hydraulics. Such explanations are difficult to apply to meandering bedrock tubes guided mainly by bedding planes. A detailed statistical analysis of Serpentine Cave in New South Wales (Ongley, 1968) and observations of sinuous passages in some New York caves (Baker, 1973) gave generally the same results. However, Smart and Brown (1981) plotted bend length/passage width relationships for a series of stream passages in Ireland and New Zealand and arrived at a regression line that is almost at right angles to all previous results.

Channel Form and Discharge

In general, wide canyons have larger scallops and slower flow velocities than small canyons. Velocities computed from the scallop measurements show a weak inverse trend with channel width. An earlier plot showing velocity varying with the -0.7 power of the passage width (White and White, 1970) had less statistical significance when more data points were added.

Tubular passages are taken to indicate flow under pipe-full conditions, whereas canyons are taken to represent open-channel flow. Robertson Avenue in Mammoth Cave (Fig. 9.16) is a remarkable example of a 1000-m segment of conduit that alternates between canyon and tube geometries. The profile (Fig. 9.17) shows that the original tube was undulating in the vertical plane. The high parts of the undulation were downcut to produce canyon cross sections, while the low parts of the undulations enlarged as

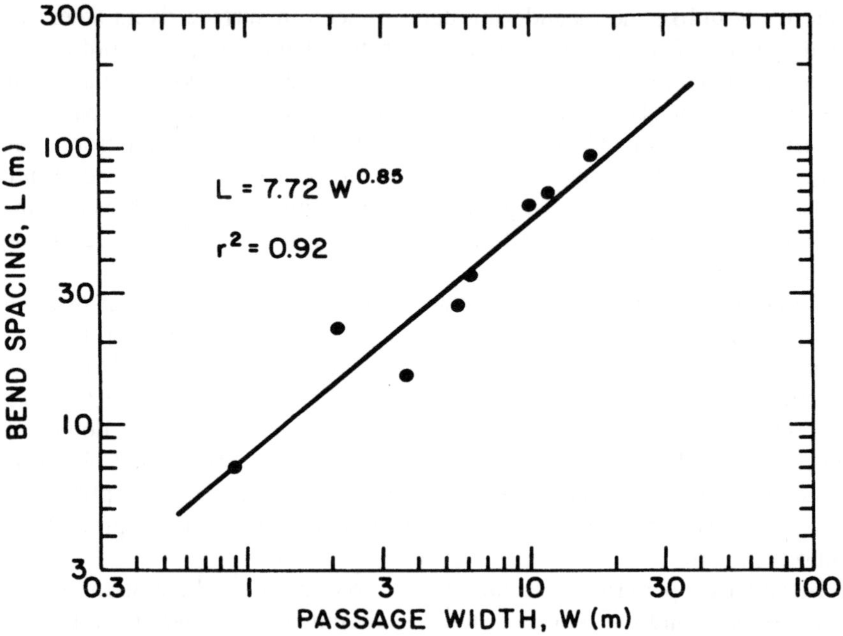

Figure 9.15 Relationship between meander bend spacing and passage width for sinuous passages in the Mammoth Cave area. Solid line is the least-squares regression fit to the data.

tubes. This is the pattern predicted by Ford and Ewers (1978) for a shallow phreatic cave leveled along an old water-table surface. The walls of Robertson Avenue are scalloped with large (L_{32} = 48 cm) scallops in the tubular part and smaller (L_{32} = 16 cm) scallops in the canyon part. Assuming that the sediment veneer in the tubular passage is thin, one can substitute the passage dimensions and scallop sizes into equations (9.1–9.5) and calculate a mean flow velocity of 4.8 cm/sec and a discharge of 0.42 m^3/sec in the tube. Corresponding calculations for the canyon segment yielded a velocity of 19 cm/sec. The flow volume must be the same in both canyon and tube. A flow depth of 0.7 m was calculated in the 3-m wide by 2.5-m high canyon passage. When Robertson Avenue was an active conduit, a visitor would have seen a knee-deep underground stream, flowing with noticeable current, that welled up from a flooded conduit at the upstream end and sumped into a flooded conduit at the downstream end. Given the size of the conduit and the magnitude of the discharge, it is likely that the hypothetical visitor would have been standing knee-deep in the water table. The paleo-water table is recorded by the elevation of the transition from tubular to canyon passage morphology.

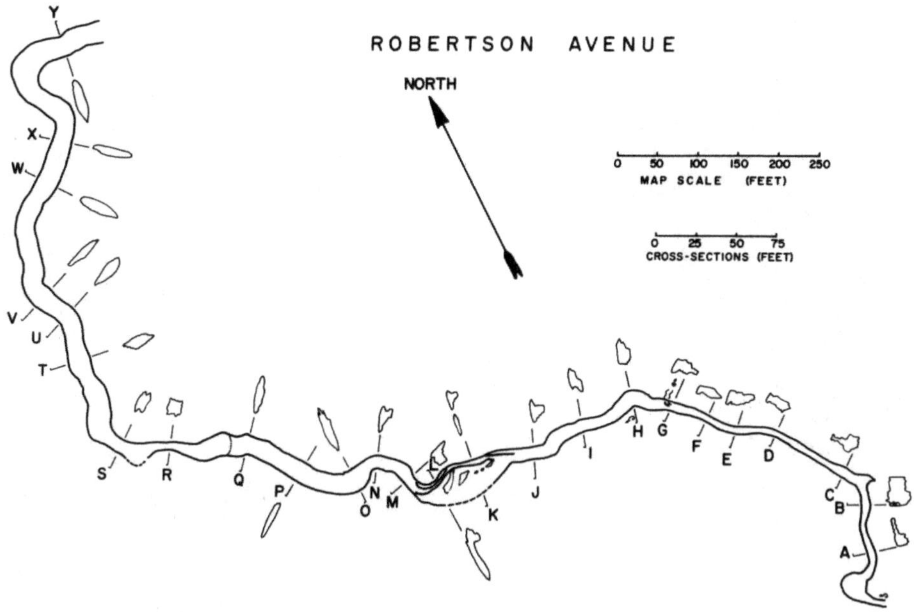

Figure 9.16 Map of Robertson Avenue, a one-time tourist route in the eastern portion of Mammoth Cave.

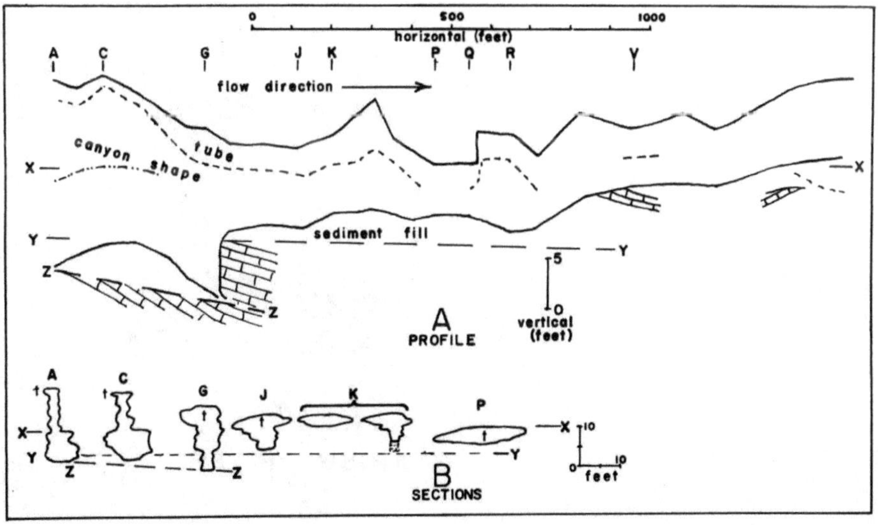

Figure 9.17 Profile and cross sections of Robertson Avenue showing transitions between canyon and tube morphology.

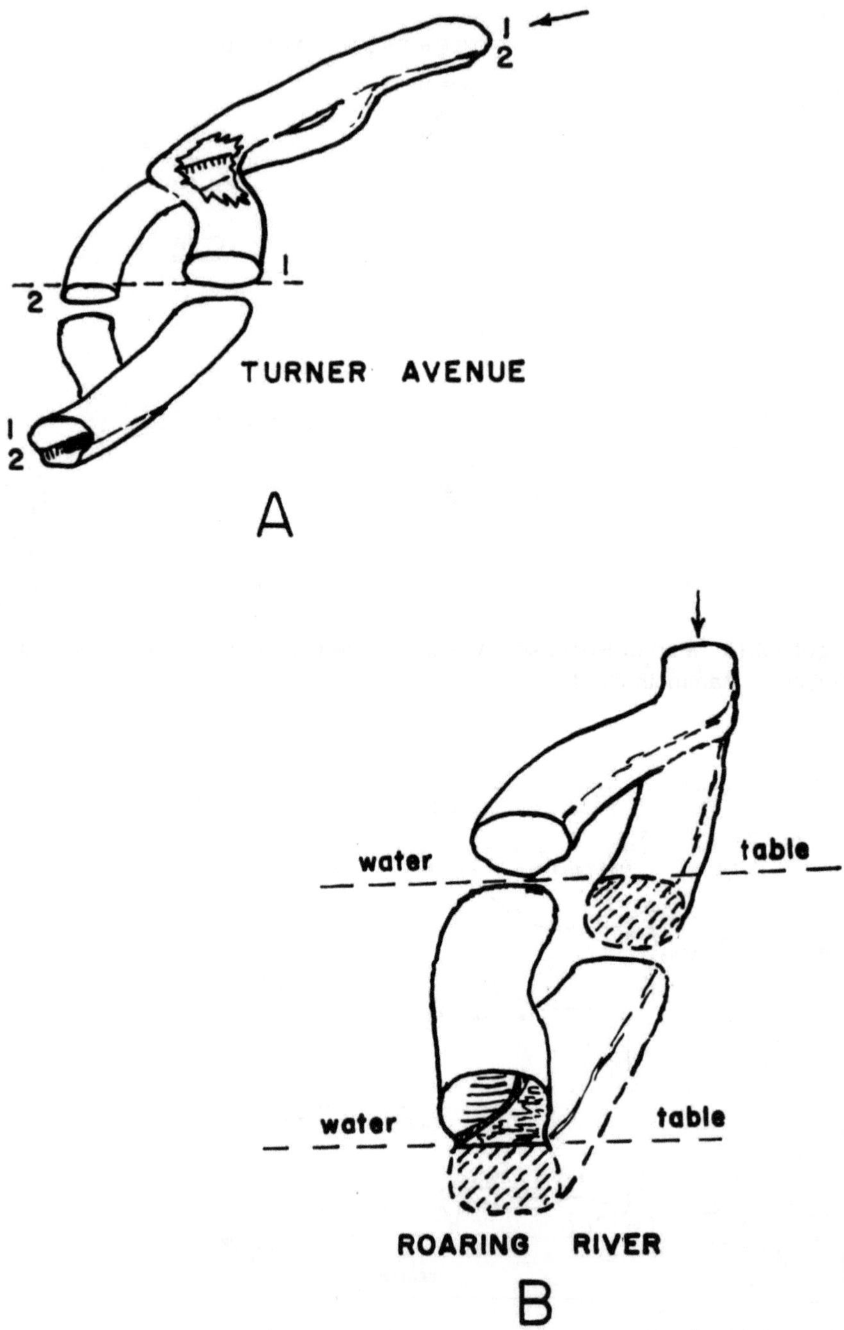

Figure 9.18 Sketches of two examples of braided tubes.

A braided surface stream is one with an interweaving network of channels with intermediate islands. Braided streams are usually found in low-gradient rivers with large sediment loads. Passage forms equivalent to braiding occur in conduit systems. Most nearly analogous to surface stream braiding is the network maze or anastomotic maze pattern (Palmer, 1975) in which water moves through all passages in the maze simultaneously and the intervening pillars become the "islands." Such maze patterns form in hydrogeologic settings where the infiltration of water into the carbonate rock is restricted either by percolation through overlying noncarbonate rocks or by constrictions in outflow or both. In this situation, flow velocities are controlled by geologic barriers, and the early aggregation of the flow path into a single conduit does not occur. Some complex maze caves occur high up near the margins of the Chester Cuesta of which James Cave is the best example. Network mazes of the sort described by Palmer (1975) are not common in the south-central Kentucky karst.

A second, and very common, form of braiding occurs in the evolution of the tubular conduits. Conduits tend to form along bedding planes with a sinuous ground plan. Either contemporaneously or as base levels are lowered, a second conduit forms a few meters below the first, often with a sinuosity that is out of phase with the conduit above. The result, as seen in ground plan, is a braided pair of tubes that loop in and out of contact (Fig. 9.18). Where the tubes are coincident there is a single high passage. In other reaches are offset, separate passages known to cavers as "cut-arounds."

Conduit Continuity and Conduit Junctions

Conduits join and merge in several different ways. One can identify the following:

1. Tributary branches (passages forking, at grade, in the upstream direction).
2. Distributary branches (passages forking, at grade, in the downstream direction).
3. Independent passages that crossed and later enlarged to a size where they intersected.
4. Merging passages resulting from the downward evolution of the conduit system.
5. Diversionary passages formed by a conduit being diverted into an alternate, usually lower, flow path.

The examples that follow are drawn in large part from the tourist passages of Mammoth Cave where they can be readily inspected by visitors. The place names referred to can be found on various maps elsewhere in this volume.

An example of an upstream-graded tributary can be seen at the Snowball Dining Room in Mammoth Cave where Marion Avenue and an intermediate-level passage that represents the downstream part of Boone Avenue join at grade to form Cleaveland Avenue, one of the lowest gradient tubular passages found in the cave system. See Figure 11.8 for a map of this junction, where both ceiling and floor join smoothly. This intersection appears to be a main groundwater junction of two tributaries that were active contemporaneously at shallow depth below what was the base level at the time. There are many examples of minor tributaries entering the main trunks at many locations in the cave system where later-developed valley drains and shaft drains have flowed into the main trunks.

Downstream distributary systems are also common in the cave system. One such distributary occurs at the Rotunda near the Historic Entrance to Mammoth Cave (see Fig. 11.6). The main downstream drainage line splits into two parts, one following Houchins Narrows past the present location of the Historic Entrance through Dixon Cave and out to a paleospring location on Green River, and the second following Audubon Avenue past another bifurcation and a terminal breakdown to another paleospring location on Green River. Something similar must also occur at present-day base level. The flow through the base-level passages of Mammoth Cave bifurcates in the vicinity of Echo River. Part of the water flows through 1000 m of submerged conduit to discharge at Echo River Spring. The other, slightly higher route follows a series of passages to emerge at Styx River Spring, nearly 1 km upriver from the Echo River Spring.

Many converging passages, however, are offset vertically and, to an amazing extent, maintain their hydrologic continuity. Figure 9.19*A* shows the convergence of three passages in the historic section of Mammoth Cave. Pensico Avenue, Buchanan's Way, and Bunyan's Way merge at three slightly different elevations, labeled 1, 2, and 3 on the diagram, to form a single tube. The passage width and shape are maintained in the downstream direction. Figure 9.19*B* is a junction in the Main Cave of Mammoth Cave where a major conduit, now mostly obscured by breakdown merges with the Main Cave itself, forms two superimposed tubes with a hydraulic form that can be traced a substantial distance downstream from the junction point.

In contrast, some tubes intersect but apparently did not originally represent a passage junction (Fig. 9.19*C*). Figure 9.19*C* shows Gothic Avenue where it crosses the Main Cave along the tourist trail. Gothic Avenue is a higher level passage than the Main Cave, but it has incised itself downward and accidentally cut through the ceiling of the passage beneath (or the Main Cave enlarged upward to make the junction). Gothic Avenue apparently converges with the Main Cave some distance downstream, but the present intersection appears to be a fortuitous one.

The common characteristic of all of these passage intersections is the maintaining of the hydraulic geometry as the conduits enlarge, downcut,

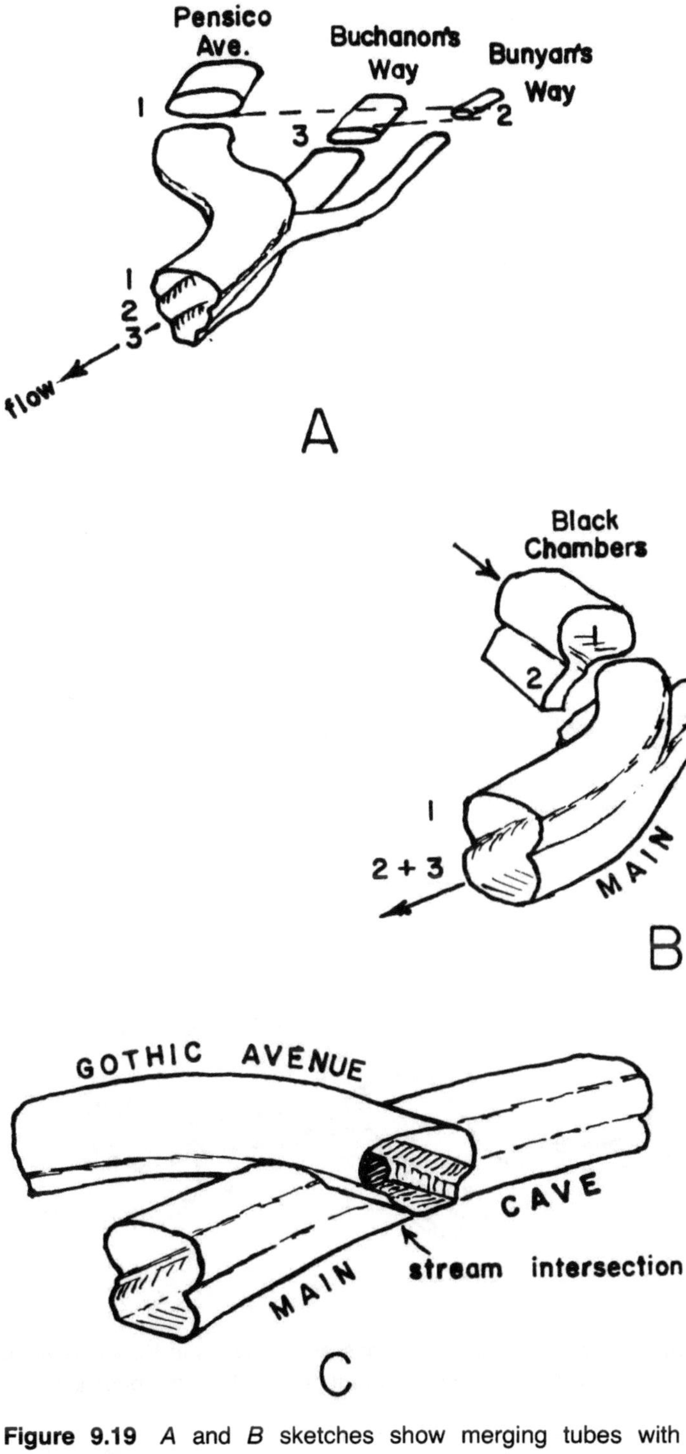

Figure 9.19 *A* and *B* sketches show merging tubes with maintenance of hydraulic form and *C* sketch is an example of the fortuitous intersection of two crossing tubes.

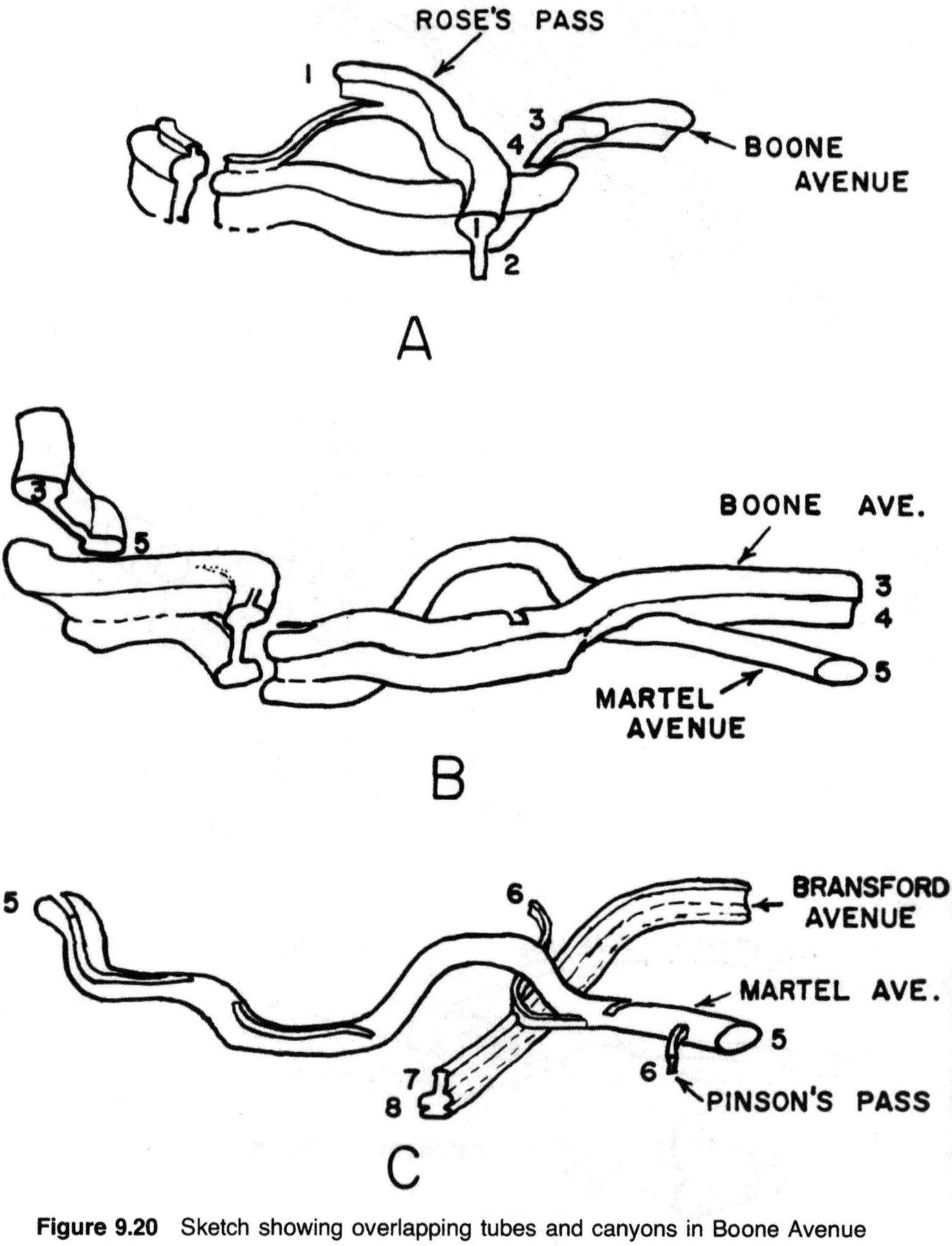

Figure 9.20 Sketch showing overlapping tubes and canyons in Boone Avenue area of Mammoth Cave. *A, B,* and *C* are different segments of the passage.

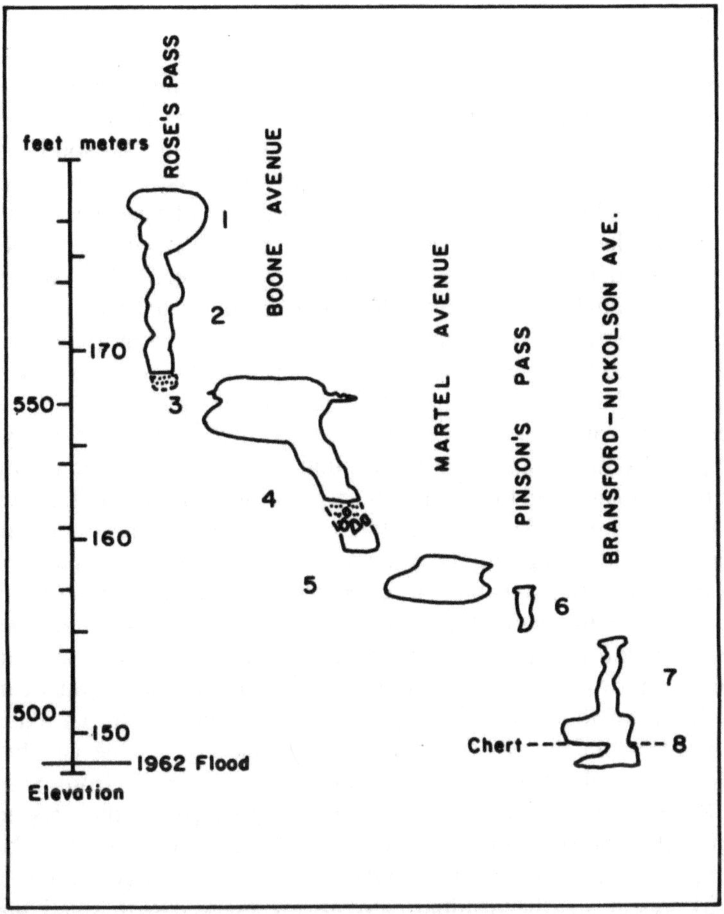

Figure 9.21 Elevational relationships for the various tubes and canyons in Boone Avenue area of Mammoth Cave.

diverge, bifurcate, and merge in a constantly changing pattern. This geometry is particularly well illustrated in the middle part of the Mammoth Cave along what is known as Boone Avenue on the present-day tourist route where a series of superimposed intermediate-sized passages of both tubular and canyon geometry criss-cross over each other (see Fig. 10.14). Figure 9.20 illustrates the evolving sequence of conduits. These passages alternate between tubes and canyons and appear to be sequential in time (Fig. 9.21). Palmer (Chap. 11) has interpreted this sequence as a result of vadose downcutting of a tributary to Cleaveland Avenue. The uppermost component, Rose's Pass, is a distributary passage from Kentucky Avenue, the high-level major trunk (see Fig. 9.14). Kentucky Avenue is blocked by breakdown at the Forks of the Cave and this damming effect induced the diversion passages. The intermediate and lower components might represent diversions that took place farther and farther upstream. The lowest component, Bransford Avenue, extends for a long distance to the southeast and might indicate that the diversions migrate upstream as base level recedes. What is most striking about this complex sequence of passages over more than 40 m of elevation is the degree to which the individual components maintain their identity. The passages merge and diverge, but can still be traced as individual drainage lines for long distances until they are either plugged with clastic fill or lost in breakdown. These are not graded passage junctions, but are the superimposition of independent flow lines that record a long period of downcutting in south-central Kentucky.

Diversionary routes are extremely common throughout the cave system. What appears to have happened in the trunk passages is that the main high-volume flows abandoned the trunks for lower routes somewhere far upstream as base levels lowered. This process left smaller flow volumes, derived from local sources, still occupying the trunk as small, free-surface streams. These streams become perched as base level is lowered because the bed of clastic material insulated the bedrock floor of the trunk from solution, and the flow volumes were too low to rearrange the clastic material. As local gradients increased with lower base levels, however, the streams cut residual canyons into the floors of the tubular passages. Residual canyons are an extremely common feature in the floors of many trunk passages. One can be seen crossing beneath the floor of Cleaveland Avenue not far from Snowball Dining Room, which both illustrates the morphology of the canyon and also shows that the sedimentary cover over the bedrock floor of the tube is only a thin veneer. Whereas the tubular conduits have large scallops or no flow markings at all, the residual canyons are marked with small scallops indicating much higher flow velocities but their narrow widths indicate low-flow volumes.

Two other diversionary routes are illustrated in Figure 9.22. Pohl Avenue in the Flint Ridge portion of the system (Fig. 9.22*A*) is a lower level trunk located at the top of the floodwater zone. Near the downstream end of the presently accessible part of the passage a canyon is developed in the floor of the tube that gradually deepens in the downstream direction. The canyon veers off to the right to become an independent passage loop, which then crosses the tubular passage again. The canyon continues to deepen as it loops to the left and then to the right to cross the tube a second time. On the second crossing, the canyon is deep enough to pass beneath the tube with no breakdown, and the passages do not intersect. At about this depth, the canyon transforms into a tube, known as Columbian Avenue, which becomes a completely independent conduit draining to the present Pike Spring. Pohl Avenue apparently drained to a paleospring loca-

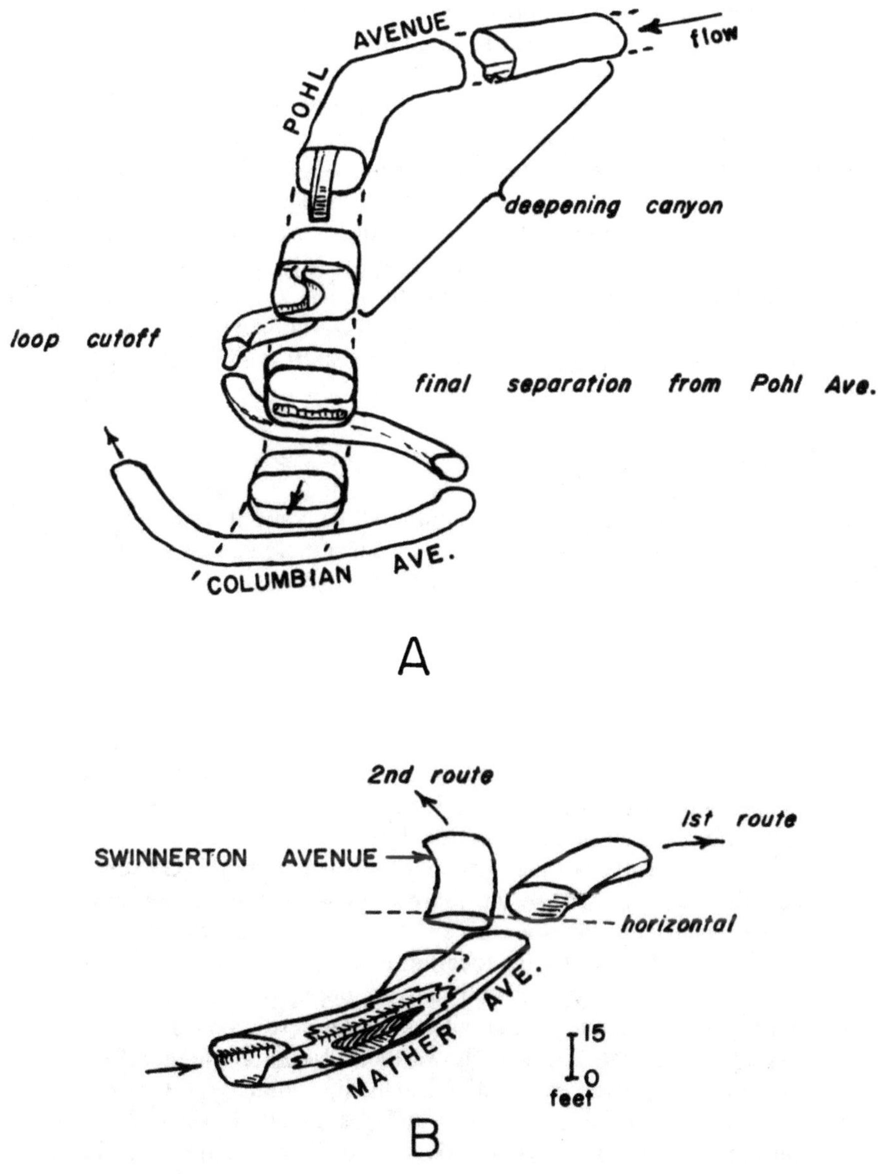

Figure 9.22 Sketches showing two examples of graded diversionary passages.

tion several kilometers downstream from Pike Spring. The second example (Fig. 9.22*B*) shows a major diversionary route in which a trunk carrying water to the north, Mather Avenue in the Flint Ridge Cave system, was diverted to the west by the development of Swinnerton Avenue. The original discharge was to a paleospring near the present location of Pike Spring; the diversionary route must have reached the river several kilometers downstream from Pike Spring.

The degree of hydraulic continuity in the various passage junctions, of which there are hundreds in the cave system, is both remarkable and cautionary. The passage junctions are remarkable in the amount of detail that has been preserved compared with the evidence of previous channel behavior in surface valleys. Passage junctions are cautionary in that the superimposed conduits are often separated by only a few meters, whereas the annual fluctuation in water level might be 10–20 m. Deduction of exact flow routing in long-abandoned passages is hazardous when it is known that detailed flow paths through the active system shift considerably depending on river stage and volume of recharge.

Conduit Gradients

Because the recharge area is in the south and east and the springs on Green River are to the northwest of the Chester Cuesta there cannot be much doubt that the overall regional gradient is to the northwest, the same direction as the regional dip. Questions can be raised, however, about the local gradients of individual segments of conduit and about the actual magnitude of the regional gradient as recorded by the cave passages. Canyon passages have floors that slope continuously in the downstream direction but tubular passages are observed to be undulatory. Indeed there is no reason why tubular conduits could not extend considerable distances below regional base level and the local slope of a passage floor or ceiling might have little to do with the regional or even the local hydraulic gradient.

In 1936 H. D. Walker of the U.S. Geological Survey (USGS) ran a third-order leveling survey through the main tourist routes of Mammoth Cave. The Walker data identifies the survey stations by detailed description and gives elevations to 0.001 ft. By locating these stations within the cave and eliminating those that were established on breakdown piles or otherwise were not in the solutional conduit, it was possible to construct a profile of Mammoth Cave (Fig. 9.23) that shows the regional gradients. The evolutional sequence of passages is taken up in this volume by Palmer (Chap. 12) with a discussion of cave levels and their relationship to Pleistocene downcutting. For present purposes, five "levels" can be identified in Mammoth Cave: (1) a fragment of high-level conduit much modified by breakdown near the Frozen Niagara Entrance at 670 ft (205 m) elevation; (2) the "main cave" including Kentucky Avenue, Croghan Hall, and Sandstone Avenue and the Main Cave to the Historic Entrance at approximately 600 ft (183 m); (3) Cleaveland Avenue and its continuation in the New Discovery part of Mammoth Cave at 550 ft (168 m); (4) River Hall, Roaring River, and other large trunks at the top of the presently active flood zone at 450

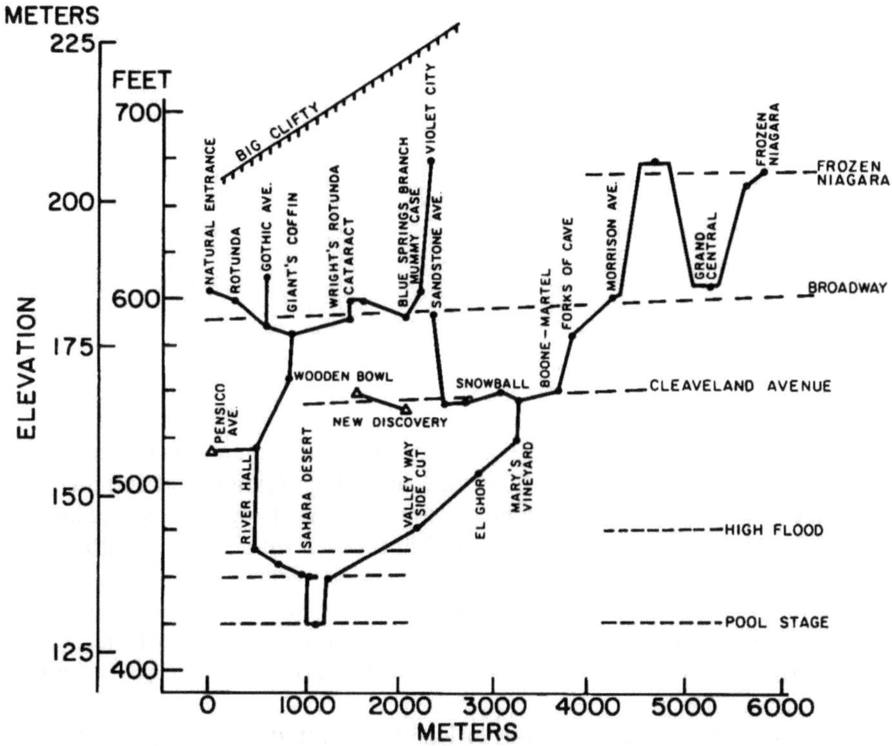

Figure 9.23 Profile of Mammoth Cave. Points are elevations determined by Walker's (1936) leveling survey of the cave.

ft (137 m); and (5) a flooded series of active trunk conduits approximately 7–10 m below pool stage of Green River. These "levels" are roughly parallel and have regional slopes as indicated by the dashed lines on Figure 9.23 of 0.0009 (4.7 ft/mi), roughly one-tenth of the regional dip. Silliman Avenue is identified as a steep-gradient passage, approximately parallel to the regional dip. In the upper reaches of Silliman Avenue the solutional sculpturing (Fig. 9.11) suggests velocities high enough to place the conduit in a regime of totally rough flow.

If canyon passages indeed are evidence for open-channel flow, the relationship between slope (S), hydraulic radius (R), and mean velocity ($\overline{v}$) are related by Manning's equation

$$\overline{v} = \frac{1}{n} R^{2/3} S^{1/2} \tag{9.6}$$

The roughness factor, Manning's n, for limestone-walled canyon passages is taken as $n = 0.02$, the handbook value (Chow, 1959) for a rectangu-

Table 9.1 Hydraulic Characteristics of Some Canyon Passages

Canyon	*Width (m)*	*Scallop Length (cm)*	*Flow Velocity (cm/sec)*	*Hydraulic Radius (m)*	*Slope (m/m)*	*Slope (ft/mi)*
Dismal Valley	10.4	51	6.1	2.13	5.4×10^{-7}	0.0029
Upper Salts Avenue	7.0	56	3.8	1.84	2.6×10^{-7}	0.0014
Robertson Avenue	3.0	16	19	0.49	3.8×10^{-5}	0.20
Bicycle Avenue	3.2	10.8	30	0.44	1.1×10^{-4}	0.57
Indian Avenue	0.9	15	15	0.18	8.9×10^{-5}	0.47
Becky's Alley	0.3	5	44	0.06	3.3×10^{-3}	17

lar concrete channel with a gravel bottom and for a gunnite-coated channel. These materials are most like limestone in surface texture and roughness. Velocities can be calculated from scallop patterns using Curl's equations (9.1–9.5). Hydraulic radius is more difficult to determine because although one can measure the channel width, there is no record of the flow depth except in a few cases such as Robertson Avenue discussed above. Based on such evidence as we have, flow depth was taken to be one-third the width in absence of any other depth indicators. The results are shown in Table 9.1.

The calculated slopes span four orders of magnitude. The large canyons, represented by Upper Salts Avenue and Dismal Valley, have extremely low gradients. The intermediate-sized canyons, Robertson Avenue and Bicycle Avenue, have intermediate gradients. The small canyon, Becky's Alley, has a moderate gradient.

Becky's Alley is taken as representative of many small canyons found throughout the cave system. Many of these are downcutting shaft drains and others appear to be pathways for steep-gradient vadose waters. The gradient for Becky's Alley is about what would be expected for a free-surface stream that is a direct gravity drain. The calculated flow velocity is also typical of what is observed in small, free-surface streams.

Although the intermediate and large canyons contained moderate flow velocities according to the scallop calculations, the low gradients suggest that these conduits are ponded. The ponding could result from the undulating profile of the conduit as was observed in Robertson Avenue and which is predicted by the Ford–Ewers model for the development of shallow phreatic cave passages. The low gradient in the large canyons could also result from base-level control. Large, open conduits such as Upper Salts Avenue have a negligible hydraulic resistance; if the canyon downcuts to maintain grade with lowering base level, it would contain ponded water very nearly at base level. Water levels in the conduits would rise and fall with river stage.

CHANNEL HYDROLOGY

Hydrologic Function of the Conduit System

Casual examination of regional-scale cave maps reveals a bewildering tangle of passages with only one distinguishing characteristic: a general northwest-trend direction. Detailed examination of the passage morphologies and interrelationships suggests at least the following four hydrologic categories:

1. Master trunk drains that carried water beneath the Chester Cuesta from catchment areas on the Sinkhole Plain.
2. A complex of smaller passages that acted as evolutionary, diversionary, and distributary routes related to the master trunks.
3. Valley drains that carry the water from the ephemeral streams in the karst valleys to the master trunks.
4. Shaft drains and other small passages related to the movement of vadose water draining from the ridge tops.

Upper-level passages of the first two categories could have formed before the breaching of the caprock on the ridge tops. However, shafts and valley drains formed as part of the evolving karst valley system would intersect and modify the pre-existing passages. Vertical shafts are frequently observed to intersect, partially intersect, or nearly miss pre-existing horizontal passages. In the lower levels of the cave, master trunks and the other categories of passage have formed or are forming together.

There are no passage characteristics that act as specific hydrologic markers for passage function. There is only passage shape, size, and, in some cases, an indication of past flow velocities as calculated from the scallops. In order to deduce the hydrologic function of a group of passages, it is necessary to map in greater detail than is customary in cave mapping and to pay strict attention to passage intersections and changing cross-sectional size and shape.

Passage Catchment Areas

Elliptical tube conduits with thin sediment deposits have a well-defined cross-sectional area. If the tubes are scalloped, the flow velocity calculated from the scallops and the cross-sectional area permit a calculation of the flow volume. The mean annual runoff from the karst region under present climatic conditions is known (see Chap. 4). If this runoff figure is normalized to catchment area, the resulting runoff intensity factor can be used with the calculated flow volume to estimate the catchment area. This estimate gives

Table 9.2 Catchment Areas for Some Conduits

Conduit	*Cross-sectional Area (m^2)*	*Scallop Length[a] (cm)*	*Flow Velocity[b] (cm/sec)*	*Flow Volume (m^3/sec)*	*Catchment[c] Area (km^2)*
Austin Avenue (C)	2.8	77	1.67	0.047	2.4
Lehrberger Avenue (T)	12.7	76	2.79	0.35	18
Great Onyx Avenue (T)	17.7	63	3.90	0.69	35
Grand Avenue (T)	20.8	89	2.55	0.53	27
Upper Salts Avenue (C)	28.8	71	3.85	1.11	57
Robertson Avenue (T)	8.8	48	4.82	0.42	22

Notes: (C) = canyon morphology; (T) = tubular morphology.
[a]Sauter mean.
[b]Calculated from equations (9.1–9.4).
[c]Based on a mean runoff intensity of 19.5 l sec^{-1} km^{-2} (see Table 4.5).

the paleo-catchment required to produce the amount of water estimated to have moved through the conduit at the time that it was an active drain.

The results of such calculations for several intermediate sized and large conduits are given in Table 9.2. It should be clearly understood that in spite of their apparent precision, the estimated catchment areas are the end result of a long series of assumptions, each of which is subject to considerable uncertainty. Two assumptions are of special importance. Use of present-day flow intensity to calculate catchment area from flow volume assumes that runoff in the past was similar to that at present. Yet we know that southern Kentucky probably underwent substantial climatic excursions during the Pleistocene. Calculation of mean velocity from scallops assumes that the calculated velocities are close to the calculated means. Gale (1984) and Lauritzen et al. (1985) both argued that scallops form in exceptionally high-velocity regimes, the peak 5% of the flow according to Lauritzen et al. (1985). Scalloping is a dissolutional rather than an abrasional process. The rate of development of scallops depends on time of contact between the bedrock and water moving at a certain velocity. Therefore, scallops should record the most probable flow. The most probable flow was assumed to be the mean flow for the figures given in Table 9.2, but it might very well be the low flow.

The values for the catchment areas are interesting. They are of the right order of magnitude to represent the areas of groundwater sub-basins, but they are a little low. Present-day catchments draining to the large regional springs (see Plate 1) have areas of several hundred square kilometers. Upper Salts Avenue, which has the physical appearance of a regional master drain, has a calculated catchment of only 57 km^2. However, the present-day catchment area for Echo River Spring is 33 km^2, and other of the intermediate-sized springs have comparable areas (see Table 4.8). If the scallop markings recorded only extreme flows, the calculated catchment areas would be much too large, but, if anything, the calculated areas are a little small. We conclude that paleohydrologic calculations based on scallop markings give reasonable results although improved accuracy would be desired.

Certainly, these results support the initial hypothesis stated in this chapter. The cave system provides access to fragments of the abandoned conduit system. The conduits have hydraulic geometries that are comparable, but not identical, to those of surface streams. They do act as master drain systems and only a few of them need be active at any one time to provide the necessary flow path from large catchment areas to the springs.

REFERENCES

Baker, V. R., 1973, Geomorphology and hydrology of karst drainage basins and cave channel networks in east central New York, *Water Resources Research* **9**:695–706.

Blumberg, P. N., and R. L. Curl, 1974, Experimental and theoretical studies of dissolution roughness, *Jour. Fluid Mechanics* **65**:735–751.

Bretz, J H., 1942, Vadose and phreatic features of limestone caverns, *Jour. Geology* **50**:675–811.

Chow, V. T., 1959, *Open-Channel Hydraulics*, New York: McGraw-Hill, 680p.

Coleman, J. C., 1949, An indicator of water-flow in caves, *Bristol Univ. Speleol. Soc. Proc.* **6**:57–67.

Curl, R. L., 1966, Scallops and flutes, *Cave Research Group Great Britain Trans.* **7**:121–160.

Curl, R. L., 1974, Deducing flow velocity in cave conduits from scallops, *Natl. Speleol. Soc. Bull.* **36**:1–5.

Deike, G. H., III, and W. B. White, 1969, Sinuosity in limestone solution conduits, *Am. Jour. Sci.* **267**:230–241.

Ewers, R. O., 1966, Bedding-plane anastomoses and their relation to cavern passages, *Natl. Speleol. Soc. Bull.* **28**:133–140.

Ford, D. C., and R. O. Ewers, 1978, The development of limestone cave systems in the dimensions of length and depth, *Canadian Jour. Earth Sci.* **15**:1783–1798.

Gale, S. J., 1984, The hydraulics of conduit flow in carbonate aquifers, *Jour. Hydrology* **70**:309–327.

Goodchild, M. F., and D. C. Ford, 1971, Analysis of scallop patterns by simulation under controlled conditions, *Jour. Geology* **79**:52–62.

Lauritzen, S. -E., J. Abbot, R. Arnesen, G. Crossley, D. Grepperud, A. Ive, and S. Johnson, 1985, Morphology and hydraulics of an active phreatic conduit, *Cave Science* **12**:139–146.

Leopold, L. B., and M. G. Wolman, 1960, River meanders, *Geol. Soc. America Bull.* **71**:769–794.

Ongley, E. D., 1968, An analysis of the meandering tendency of Serpentine Cave, N.S.W., *Jour. Hydrology* **6**:15–32.

Palmer, A. N., 1975, The origin of maze caves, *Natl. Speleol. Soc. Bull.* **37**:57–78.

Renault, P., 1958, Eléments de spéléomorphologie karstique, *Annales Spéléol. 3rd ser.* **13**:23–48.
Smart, C. C., and M. C. Brown, 1981, Some results and limitations in the application of hydraulic geometry to vadose stream passages, *Proc. 8th Internat. Congress Speleol. (Bowling Green, Ky.)* **2**:724–727.
Sweeting, M. M., 1972, *Karst Landforms,* London: Macmillan, and New York: Columbia University Press, 362p.
Walker, H. D., 1936, *Mammoth Cave, Kentucky, Third-Order Leveling,* Mammoth Cave, Ky.: U.S. Geological Survey, Survey Book B6708, 8p.
White, W. B., and E. L. White, 1970, Channel hydraulics of free-surface streams in caves, *Caves and Karst* **12**:41–48.
White, W. B., and G. H. Deike, III, 1976, Hydraulic geometry of solution conduits, in *Proceedings of the 1976 National Speleological Society Annual Convention,* E. Werner, ed., Huntsville, Ala.: National Speleological Society, pp. 57–60.

10

FRACTURE CONTROLS ON CONDUIT DEVELOPMENT

George H. Deike, III

ROLE OF STRUCTURE

In combination with the stratigraphy, the geologic structure has profound influence on the hydrology of karst regions. In the Mammoth Cave region the major aquifer is 90 m of Girkin, Ste. Genevieve, and St. Louis limestones with varied lithology, underlain by shaly beds in the lower St. Louis and upper Salem formations, and capped by the Big Clifty Sandstone, which has basal shale locally. The cap supports perched aquifers, and protects the underlying limestone from vertical infiltration of surface water. Because the soluble limestone is protected from erosion by the cap, an extensive series of solution conduits that date far back in the history of groundwater movement in the aquifer has been preserved. The shales underneath limit downward movement of water in the aquifer.

In many karst areas the joints, with their regional pattern, are the most important avenue of groundwater movement. In combination with bedding partings the joints permit continuous flow paths with rectilinear patterns to develop.

No detailed studies of the influence of structural factors on hydrology in the Mammoth Cave region were made before the 1960s. In October of 1850 Benjamin Silliman visited the cave and reported, "Here we have the dry beds of subterranean rivers—water, following the almost invisible lines of structure in the rocks, began to hollow out these winding paths—"

(Silliman, 1851). An extensive descriptive literature contained no maps good enough to give geologists any idea of the actual nature of the cave. Bretz (1942) expressed the belief that bedding-plane anastomoses were the primitive openings from which Mammoth Cave took its start. Ewers (1966) pointed out that anastomoses form in strata where bedding planes provide the most important avenues for groundwater percolation, in areas of poorly jointed limestone. He observed that anastomoses were directed in part by short joints.

In contrast, Blanc (1958) surmised, on the basis of the poor maps available, that "the complex network of Mammoth Cave in the United States can only be explained by the network of brachyclases," these being defined as joints that cut a single bed, as opposed to true diaclases, which cut a series of beds.

The role of anastomoses in the Mammoth Cave region is actually minor. They are rare in outcrop and in the caves, are associated with larger conduits, and are not found in isolation. Bedding planes are the most important planes of permeability in the aquifer, but only locally did solution produce anastomoses. Usually the anastomoses appear to have developed outward from existing conduits.

The jointing in the Mammoth Cave area does consist largely of what Blanc (1958) would call brachyclases, but the cave "network" is in fact a complex historical sequence of branching tributaries and main conduits whose pattern is not rigorously joint controlled.

This chapter is based on studies conducted from 1961 to 1964 by Deike (1967). The work was supported by the Cave Research Foundation, which sponsored exploration and mapping that made the first detailed studies possible. Subsequent observations on the influence of fractures on the hydrology of the region appear in Palmer (1977).

REGIONAL STRUCTURAL PATTERNS

The regional structure was mapped and published as part of the U.S. Geological Survey's (USGS) geologic quadrangle map series in 1964–1966.

General Structure

The regional dip is northwest toward the Illinois basin at about 1:150. In the Sinkhole Plain the dip direction varies from north to west. Several areas of steeper dip that trend east–west might reflect structures in the basement. Under the Chester Upland there are gentle folds with axes trending northwest that are superimposed on the regional dip. While overall pattern of groundwater flow, and thus cave development, is related to the general structure, Palmer (1977) has shown that finer details of structure, not visible on regional maps, have the greatest influence on the flow paths taken by the groundwater.

Joint Pattern

The orientation, extent, and characteristics of joints were measured and observed in outcrops scattered throughout the area (Deike, 1967). Scarcity of outcrops, particularly in the Sinkhole Plain, prohibited a rigorous grid-sampling plan. At each outcrop, fractures were sampled at one-half outcrop height at four equally spaced sites along the length of the outcrop. Every fracture was measured within one-twentieth of the outcrop length in both directions from the sampling sites. Outcrops of the Mississippian limestone aquifer account for 82 of the 157 outcrops studied and 1232 of the 2307 joints.

The orientation of the measured joints is shown in Figures 10.1 and 10.2. The most numerous joints are in a very prominent set at 10–50° azimuth, while a less prominent set is centered around 110–120°. Minor sets can be observed at about 150–160° and 70–80° on some histograms.

The orientation data were subjected to a Chi-square test against a rectangular distribution, with the results shown in Table 10.1. The hypothesis that joints are randomly distributed with respect to direction can be rejected. The same test was run on the joints in each quarter of the USGS 7.5-min quadrangles, and results were significant at the 0.05 level in every area where more than 75 joints were measured, and at the 0.001 level in 11 of 14 such areas, including all quarters of the Mammoth Cave and Horse Cave quadrangles, which cover the Chester Upland in the cave area.

At 21 outcrops of the Mississippian Limestone, where 20 or more joints were measured, data were also collected on extent and character of joints (Fig. 10.3, on page 267). Binomial tests were run on the data for 10° sectors, and on larger categories based on joint sets defined by orientation data (Table 10.2). The expected values were based on the actual probabilities observed for all directions combined. Significance level is shown by asterisk in the relevant direction categories. There are significant differences in roughness, spacing, and extent of joints between the major joint sets as defined in Figure 10.2.

The data for these two peak directions of joints were subjected to Chi-square two-sample tests (Table 10.3, on page 266) with the following results:

1. Joint roughness is not independent of joint set.
2. Joint spacing is not independent of joint set.
3. The horizontal extent of joints is not independent of joint set.
4. The vertical extent of joints has less than a 10% chance of being independent of joint set.

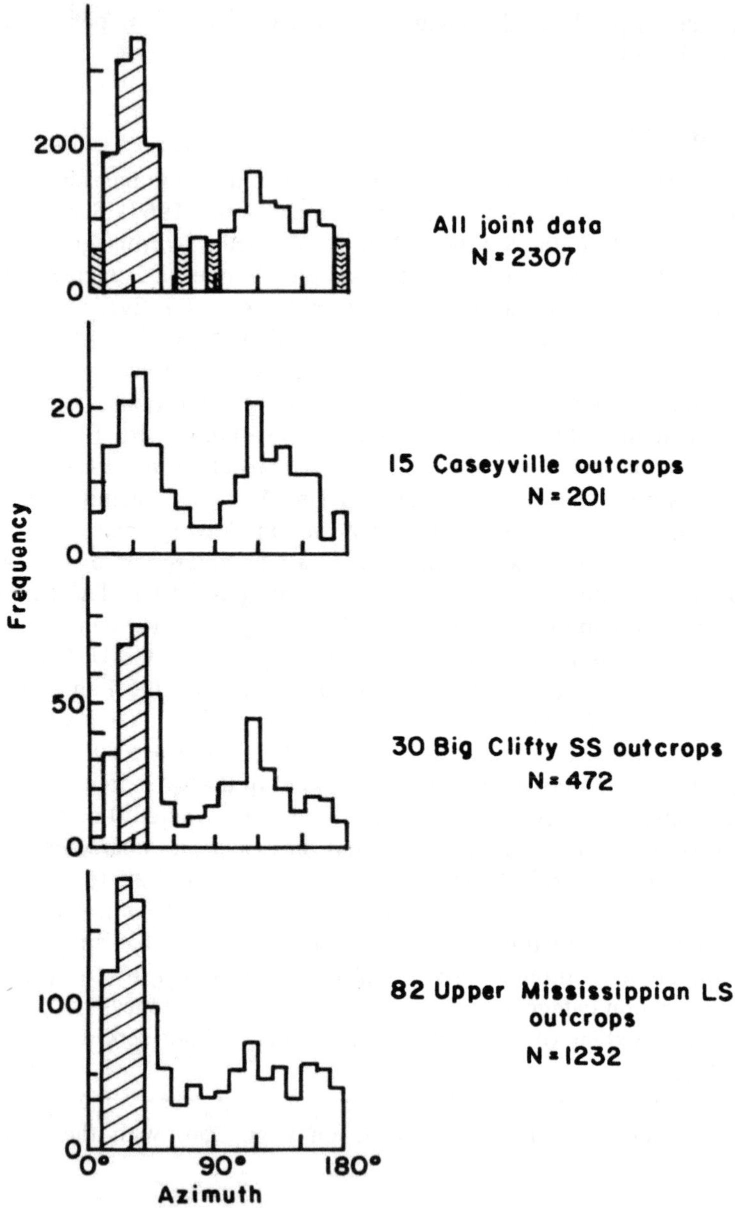

Figure 10.1 Joint direction histograms for most-sampled formations and for all joint data. Prominent sets are at 10–50°; minor sets are at 70–80° and 150–160°.

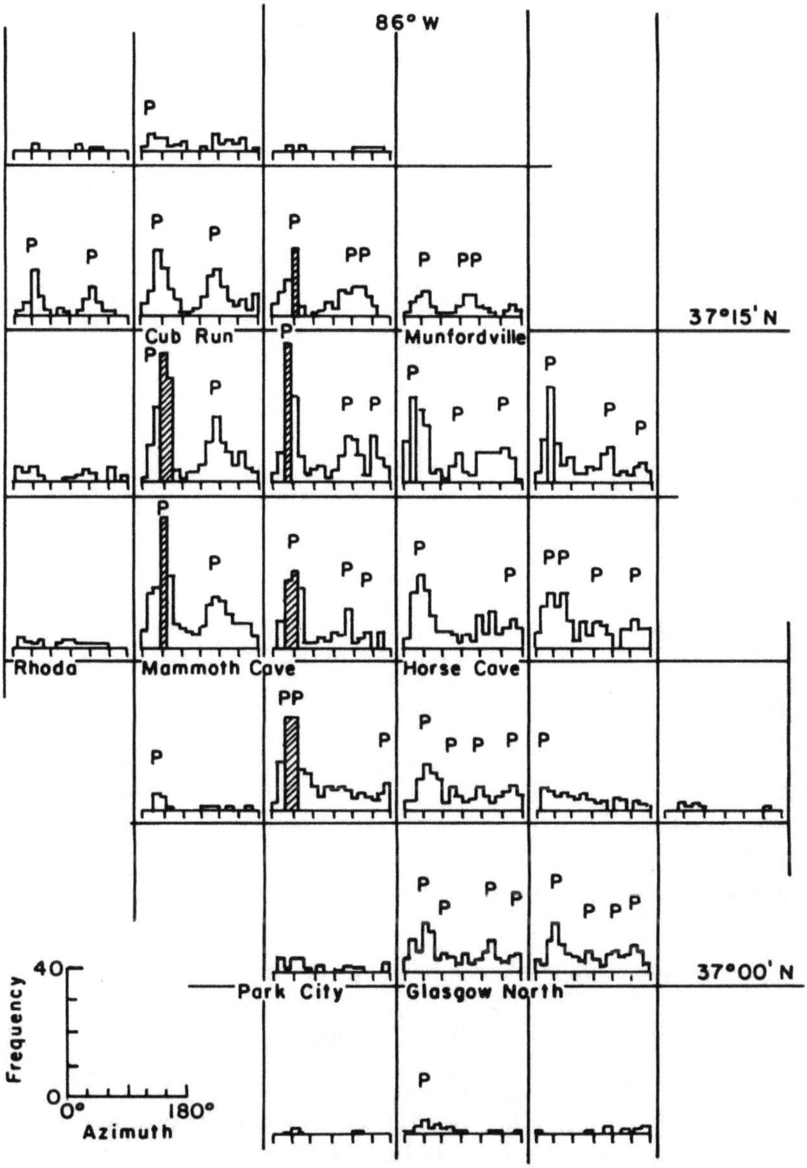

Figure 10.2 Joint orientation histograms. Prominent sets are at 10–50°; minor sets are at 70–80° and 150–160°.

Table 10.1 Chi-square Tests Against Rectangular Distribution and Joint Orientation, by Areas

One-quarter Quadrangle	*Number of Observations*	*Degrees of Freedom*	x^2	*P*
Glasgow South (NW)	21	2	8.0*	
Glasgow North (NE)	55	8	9.63	0.25
Glasgow North (NW)	100	17	35.7**	
Glasgow South (SE)	110	17	31.35*	
Glasgow North (SW)	114	17	36.65**	
Park City (NE)	162	17	110.44***	
Park City (NW)	15	2	7.6*	
Park City (SE)	24	2	6.75*	
Horse Cave (NE)	127	17	99.3***	
Horse Cave (NW)	158	17	94.0***	
Horse Cave (SE)	137	17	48.3***	
Horse Cave (SW)	141	17	71.7***	
Mammoth Cave (NE)	175	17	177.2***	
Mammoth Cave (NW)	193	17	180.4***	
Mammoth Cave (SE)	114	17	136.2***	
Mammoth Cave (SW)	204	17	115.2***	
Rhoda (NE)	33	5	7.18	0.25
Rhoda (SE)	18	2	2.33	0.4
Munfordville (SW)	57	8	22.12**	
Cub Run (NW)	38	5	4.32	0.6
Cub Run (SE)	98	17	83.2***	
Cub Run (SW)	124	17	67.8***	
Dickeys Mills (SE)	53	8	52.1***	
All joint data	2307	17	964.***	
Mammoth Cave Limestone	1232	17	535.***	
Big Clifty Sandstone	472	17	298.5***	
Caseyville Sandstone	201	17	64.6***	

Note: Level of significance: * = 0.05; ** = 0.01; *** = 0.001.

The two dominant joint sets can be characterized as follows:

Northeast set: centered at 10–50° azimuth

- Planar—smooth to the eye, few minor steps
- Close-spaced—spacing less than 15 cm common, in swarms
- Short—seldom more than 2.4 m long as observed
- Often confined to one bed

Southeast (northwest) set: centered around 100–120° azimuth

- Rough—irregular surfaces with relief of 2.5 cm or more
- Wide-spaced—often up to 0.61 m apart

Table 10.2 Chi-square or Binomial Tests on Joint Roughness and Extent, by Direction

	Roughness, Binomial Test		*Spacing, Chi-square Test*		*Horizontal Extent, Binomial Test*		*Vertical Extent, Binomial Test*	
	Planar	*Rough*	*Close*	*Wide*	*< 2.4 m*	*> 2.4 m*	*1 Bed*	*Several Beds*
Class Interval								
0–10	*				0.1			
10–20	*				0.1		0.06	
20–30	**		0.1		0.2		***	
30–40	***		*		**		**	
40–50	0.05		**				*	
50–60	0.4							
60–70	0.4				0.1			
70–80	0.4							
80–90		0.3						
90–100		0.1						
100–110		**						
110–120		**						
120–130		**						
130–140		**						
140–150		0.4						
150–160		0.05			*		0.1	
160–170		0.3						
170–180		0.6						
Combined Classes								
0–80					0.07			
40–80	0.12							
80–100		***						
80–150						**		
100–120				0.1				
100–130						***		*
140–180		***						

Note: Level of significance: * = 0.05; ** = 0.01; *** = 0.001. For each set of data, in each direction where sufficient data exists, the probability or significance level is shown under the predominant or significant effect by data in that direction.

- Long—often more than 2.4 m long, with one example exposed for 13.7 m
- More often cuts more than one bed; observed vertical extent up to 6.1 m

The more extensive character of the rough northwest joint set raises the possibility that the joint component of aquifer permeability is anisotropic, permitting easier water movement parallel to this joint set. Palmer (1981)

Table 10.3 Chi-square Two Sample Tests on Joint Roughness and Extent Data for Major Joint Sets

A–Roughness

	10–50°	100–120°
Planar	115	11
Rough	29	28

$N = 183$; $\chi^2 = 37.8$***

B–Spacing

	10–50°	100–120°
< 0.3 m	80	5
> 0.3 m	58	15

$N = 158$; $\chi^2 = 6.37$*

C–Horizontal extent

	10–50°	100–120°
< 2.5 m	56	9
> 2.5 m	12	14

$N = 91$; $\chi^2 = 6.03$*

D–Vertical extent

	10–50°	100–120°
1 Bed	62	6
Several beds	16	5

$N = 89$; $\chi^2 = 3.31$; $P = 0.1–0.05$

observed that the thickest beds have the most extensive and conspicuous joints.

Fracture Trace Pattern

U.S. Department of Agriculture (USDA) stereoscopic aerial photography of the area, which was taken in 1958 and 1960, was used to map photogeologic fracture traces (Lattman, 1958). Fracture traces are defined as less than 1.6 km long. Longer aligned features, or linears, were not observed, except perhaps for the valley of Cub Run north of Green River, which follows a fault for several miles and is straight on a large scale. No fracture traces were detected along Cub Run. Fracture traces are considered to be manifestations of zones of joints, which might influence groundwater movement more effectively than individual joints.

North of the Chester Escarpment, vegetative, soil tonal, and stream alignments were the features most commonly mapped. The patchwork of small fields and woodlots makes tracing a continuous feature difficult, especially in Mammoth Cave National Park where abandoned farmland is returning to forest. On the Sinkhole Plain tonal lines and aligned dolines were the features mapped in most cases. In all, 4025 fracture traces were mapped, an average of about 6/km^2 (15/mi^2).

The pattern of orientation of the fracture traces is not as clear as that of the joints (Figs. 10.4 and 10.5). Concentrations at 30–40°, 70–80°, and particularly at 90–120° are superimposed on a broad maximum. Some histograms prepared for one-quarter quadrangle areas show a strong concentration at

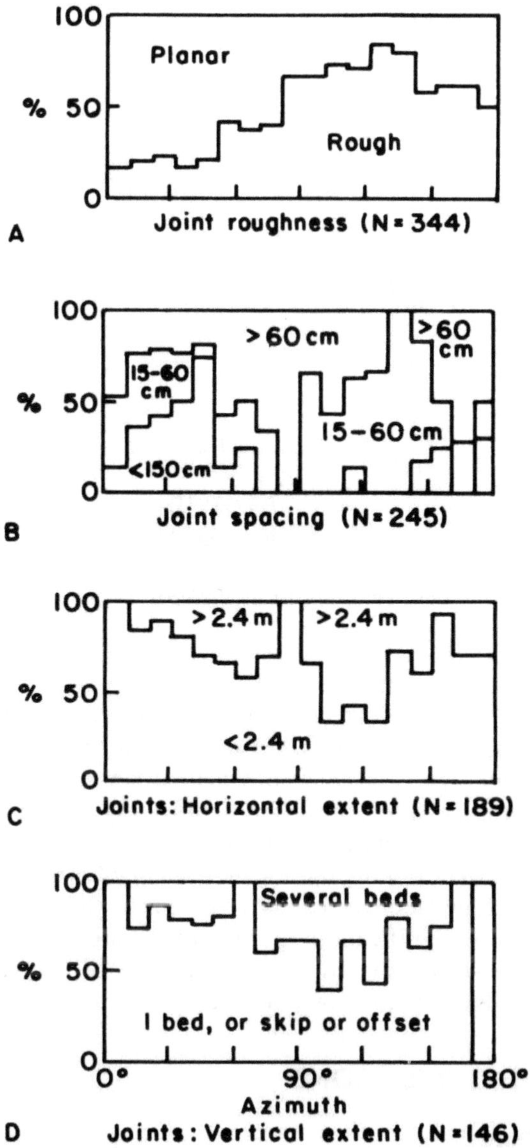

Figure 10.3 Joint surface characteristics and extent (Upper Mississippian limestone).

30–50°, parallel to the most common joints, but the common direction is parallel to the more extensive, rougher joints in the northwest–southeast set.

A Chi-square test against a rectangular distribution was significant at the 0.001 level. The fracture traces are not randomly distributed with respect

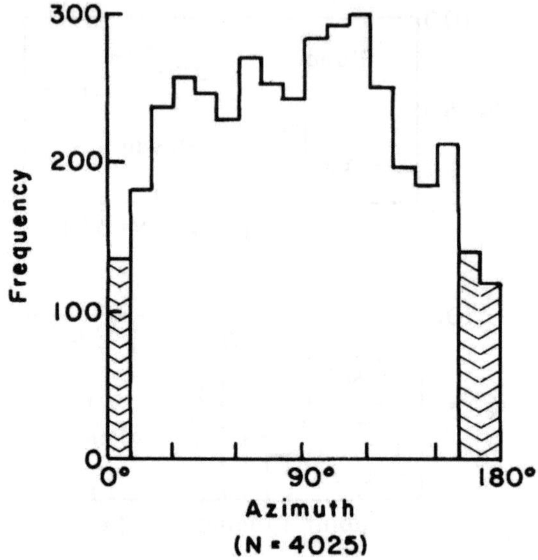

Figure 10.4 Orientation histogram of all fracture traces. Note minor sets near 0° and 180°.

to direction. The same test, run by one-quarter quadrangle areas, was significant to at least the 0.05 level in 10 of 16 areas with more than 100 fracture traces, including six of eight quarters of the Mammoth Cave and Horse Cave quadrangles. In about one-half of the areas the distribution of mapped fracture traces does not differ significantly from random. Preferred orientation is not as pronounced as for the joints.

Binomial tests were run for significant tendency for the one-quarter quadrangle areas to show frequencies above or below their median in each 10 direction sectors. A significant proportion of the areas had trace maxima at 30–40° and at 90–130° azimuth. These directions of fracture trace concentrations are parallel to the major joint sets.

TOPOGRAPHIC RELATIONS

Sinkhole Plain

The Sinkhole Plain exhibits varied topography, which is correlated with the stratigraphic horizon on which it is developed. In the area west of the Park City 7.5-min quadrangle, west to Bowling Green and beyond, the Sinkhole Plain is very wide. The dip averages only 1:135, and the St. Louis, Ste. Genevieve, and Girkin limestones are exposed over a width of 16 km or more. The outcrop belt and the Sinkhole Plain narrows to only 6.4 km near Park City, where the dip averages 1:55.

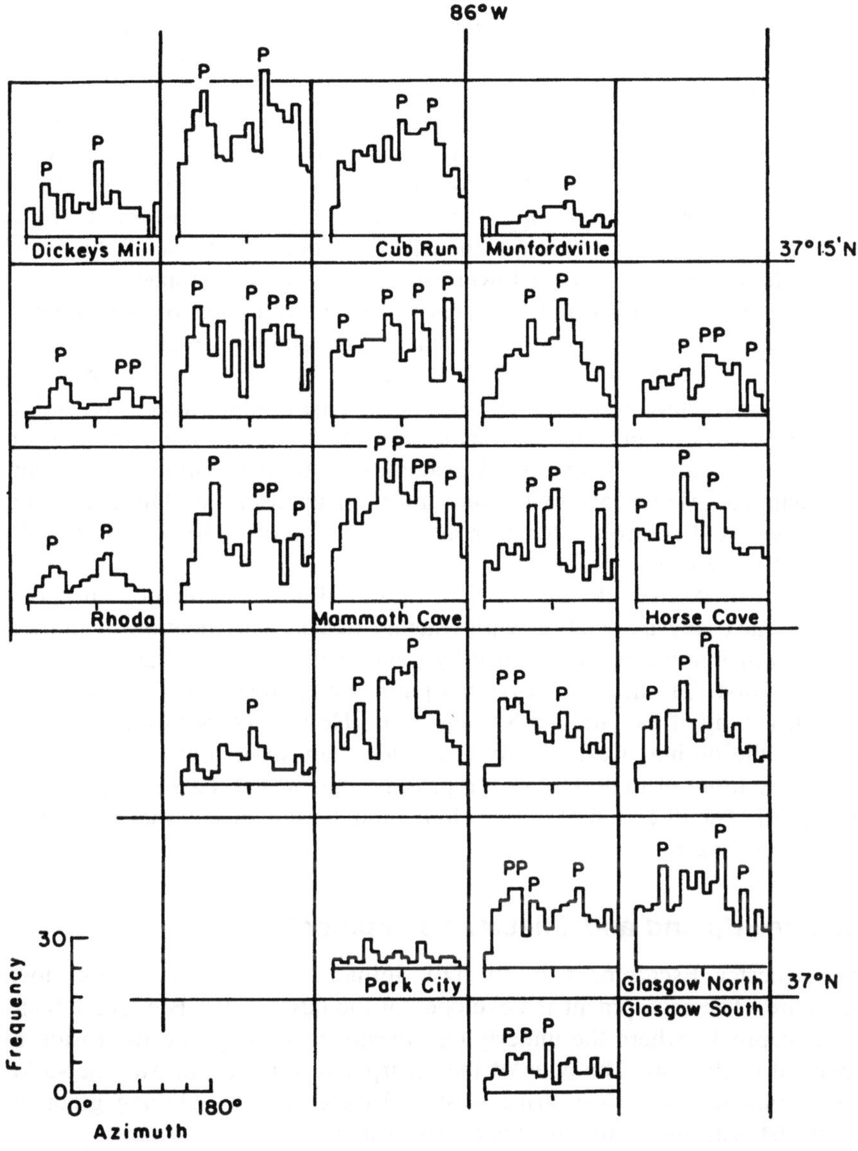

Figure 10.5 Orientation histograms of fracture traces. Note prominent sets at 30–40°, 70–80°, and 90–120°.

The southern border of the Sinkhole Plain is characterized by low relief and surface drainage, which disappears beneath the doline karst to the north. This area is developed on shaly lower St. Louis beds. Howard (1968) refers to the area as a dip slope cuesta. The streams follow the dip north and northwest. The sinks of these streams are the point sources of recharge for the karst aquifer at stratigraphic horizons low in the St. Louis Limestone.

The area between the sinking streams and the Chester Escarpment is the Sinkhole Plain proper. In this area without surface streams, Howard (1968) was able to detect subtle stratigraphically controlled escarpments, most notably developed on the Lost River chert of the upper St. Louis Limestone in areas of low dip. Over the area of the plain the karst aquifer receives internal runoff through thousands of dolines. Most of this water enters the upper St. Louis or lower Ste. Genevieve limestones.

The effects of joints and fractures on the morphology of the Sinkhole Plain is very limited. A very few reaches of the sinking streams are straight and were mapped as fracture traces, and the stream course is probably directed by fracture zones in these reaches. Some fracture traces are straight drainage lines leading into dolines or soil tonal lines running through the long axis of dolines. Some good examples of aligned dolines were found and mapped. About 75% of fracture traces mapped in the plain are related to dolines, and control of location and orientation of these dolines by joint and fracture zones is probable.

However, it is significant that 80% of all dolines examined on air photos did not have any association with fracture traces. The location of most of these dolines might be determined by joints that permit infiltration of water into the karst, but they are individual joints not expressed as fracture traces.

In his study of the Graham Springs basin, Howard (1968) found that only 10% of the dolines were clearly elongate, suggesting control by fractures and/or by trend of underlying cave passage into which the sink might have collapsed. The direction of doline elongation did not closely match the trend of cave passages.

Chester Upland and Chester Escarpment

The Chester Escarpment is strongly influenced by regional and local structure. The escarpment is developed at the edge of the Big Clifty Sandstone caprock, where the underlying limestones are exposed to erosion by removal of the cap. The edge of the escarpment roughly follows the strike, paralleling the east–west strike west of Park City and swinging gradually northeast with the changing strike east of there.

West of Park City the escarpment is developed on a zone of locally steep dip (1:30), and the escarpment is straight and parallel to the structure contours, with little dissection and rare outliers. Where the dip is low, northeast of Park City, the escarpment is dissected into ridges and there are outlying knobs with and without preserved caprock. Dissection is greater in local anticlinal upwarps, and outliers are more common in areas of synclinal downwarp (Howard, 1968).

The flexures in the structure of the Chester Upland do not influence the topography strongly. The karst valleys like Doyel Valley and Houchins Valley are descendents of former surface tributaries to Green River. They were developed down the northwest dip, but otherwise their locations were not dictated by known structural factors. The dip does direct drainage off the caprock in a preferred northwest direction, particularly in the perched Haney Limestone aquifer on Flint Ridge.

The caprock is well jointed, and many joint sample sites in Deike's (1967) study were in the caprock. These joints controlled the orientation of many small cliff faces. Straight stream alignments, mapped as fracture traces, reveal fracture control of some reaches of the surficial drainage, as in the head of Three Sisters Hollow and Floating Mill Hollow on the flanks of Flint Ridge.

The Green River and Barren River cross the region at varied angles to the regional strike and dip, and show no major structural control. Local flexures in the bedrock have not influenced the details of the river courses. For example, the bend at Pike Spring on Green River carries the river farther updip at a place where the structure contour bends the opposite direction around a minor anticline. Updip bends of the Green River are the first locations where the caprock was breached by erosion, thus each was at one time a new possible outlet for groundwater.

Several reaches of the Green River or the base of its valley walls were mapped as straight fracture traces, implying control of parts of the river by fracture zones. Most of these were not related to springs along the river. Exceptions include Garvin Spring, which reaches the river at a straight reach, and one of the springs in the Hidden River complex, which seems to lie where a fracture trace leads away from the river onto the upland.

RELATIONSHIP OF CAVES TO FRACTURES

Primary permeability of the limestone aquifer is very low, and groundwater must utilize secondary openings. Of these, bedding partings are the most important in the Mammoth Cave area. They are continuous over wide areas and occur frequently throughout the stratigraphic section, although some beds are locally 6 m thick. Joints are scattered and of limited extent, and do not by themselves provide continuous flow paths through the rock.

Water enters the aquifer at the top of the limestone at the edge of the caprock in the Chester Upland, by infiltration at doline drains into various stratigraphic horizons in the karst valleys and the Sinkhole Plain, and at the base of the aquifer at sinks of streams along the southern edge of the plain. Groundwater travels as much as 24 km to the springs at base level along the rivers. While some of the flow paths are in the phreatic zone, many are in the vadose zone for much of their length.

Orientation of Cave Passages

For a study of the orientation of cave passages, 1:1200 map sheets of the Flint Ridge part of Mammoth Cave showing about 32 km of cave passages were available (Brucker and Burns, 1964). The only map of Mammoth Cave available was Kaemper's 1908 map at a scale of about 1:4000, showing about 55 km of passages. Nineteen smaller caves distributed over the area were located and surveyed for a total of about 8 km of passage. The Flint Ridge maps were drawn from sketches of variable accuracy, leaving some uncertainty in the identification of straight elements of cave passages. The small scale of the Kaemper map made it impossible to identify short, straight reaches of passage.

The orientation of every segment of cave passage that appeared straight on the maps was measured. Segments longer than 45 m were rare, and those longer than 30 m, 15 m, and 7.5 m were tabulated. The maps also revealed straight trends of passages not necessarily straight in detail. The orientation of trends longer than 30 m and longer than 90 m were tabulated if the passage did not deviate laterally by more than 10% of the trend length. Interpretation of these features was liberal, possibly exaggerating the extent of straight elements in the caves.

Flint Ridge quadrangle map 112, which shows about 5.3 km of passage, was used to compare these five ways of defining a straight part of the cave passage. The five orientation histograms are similar (Fig. 10.6). Kendall's coefficient of concordance was calculated to see to what extent each method would assign the same rank to each direction category. The coefficient, $W = 0.621$, is significant at the 0.001 level. The test suggested that any of the methods could be used, and that omitting data on short segments, as must be done for the Kaemper map, will not alter the results significantly.

In Table 10.4 it can be seen that, at most, some 40–60% of the total cave passage is straight by any of the methods of measurement. Long, straight elements are less common than short ones. Straight segments less than 7.5 m long cannot be identified on the maps except in very narrow passages, but field work suggests they constitute less than 10% of the length of such passages. Straight passage segments account for no more than 40% of the passage on the maps.

A detailed map of 2.13 km of the tourist routes on the historic tour in Mammoth Cave shows less straight passage than the general maps suggest, in this case 29% straight segments longer than 7.5 m. Field examination of this and other parts of the cave show evidence of joint control of about 10% of the passage length even though it does not appear straight. In this part of Mammoth Cave, 41% of the passages are straight, showing evidence of joint control (Fig. 10.7). The amount of oriented passage is similar in other parts of the Mammoth Cave system. A detailed map of 1.45 km of passage

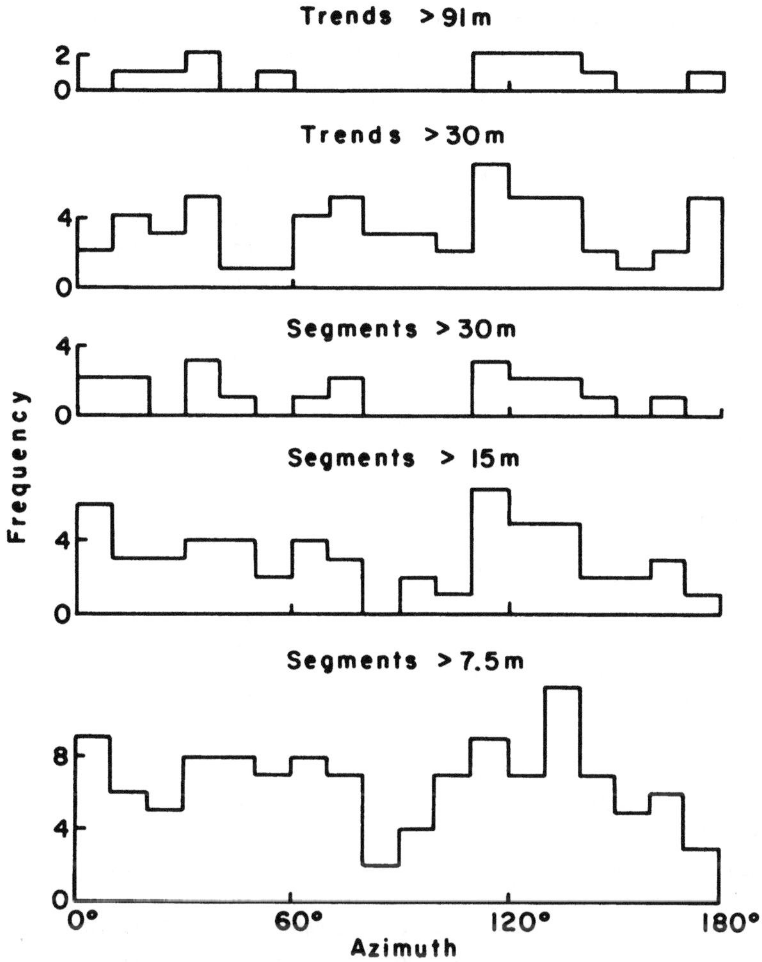

Figure 10.6 Orientation of straight trends and segments for 5334 m of passage on Flint Ridge quadrangle map 112.

on several levels in the Boone Avenue area shows that 24% of the passage is composed of straight segments longer than 15 m.

Figure 10.8 shows the length of straight segments measured in Turner Avenue. Those shorter than 7.5 m were not counted. Straight segments from 9 m to 21 m long were common, longer segments were rare. Long, straight passages such as might be expected following joint zones or fracture traces are not found.

The most striking contrast between the Mammoth Cave region and other

Table 10.4 Percentage of Passage-Length Straight Segments (s) or Trends (t)

Cave[a]	*s >7.6 m*	*s >15.2 m*	*s >30.5 m*	*t >30.5 m*	*t >91.4 m*	*t >304.8 m*
127	39	27	(13)	54	38	
128	34	34	23	49	38	
113	46	30	18	47	35	
112	42	29	15	66	31	
111	32	19	6	55	25	
95		28		71		
96		24		52		
97		32		50		
98		35		60		
81		22		70		
82		28		65		
67		28		55		
68		34		64		
Totals: (%)						
Flint Ridge	39	28	13	60	31	25
Mammoth		(26)				17
Long	32	24		64		
Parker	34	22		47		
Indian	39	26		68		
Hiseville	35	21		68		
Running Branch	30	23		51		
70 Sink	30	14		22		
Cedar Sink	45	35		64		
L & N	53	40		63		
Renick	38	26		50		
Prewitt Knob	44	36		57		
Total Range	30–46	19–40	6–23	47–71	25–38	

		Main Avenue	*Small canyons*	*Combined*
Quadrangle 111	s > 15.2 m	30%	13%	19%
	s > 7.6 m	37%	29%	32%
	t > 30.5 m	56%	54%	55%

[a]Quadrangle numerals taken from Brucker and Burns (1964).

karst regions is in the proportion of joint-controlled cave passages. Table 10.5 presents this contrast in terms of straight passages longer than 7.5 m. Of the 220 cave maps studied, only 29 had a smaller percentage of straight passage than the Flint Ridge system. The data from other regions were treated as samples from a normal population of percentage of straight passage. The probability that Flint Ridge could have come from the same population as the other regions is only 0.0003.

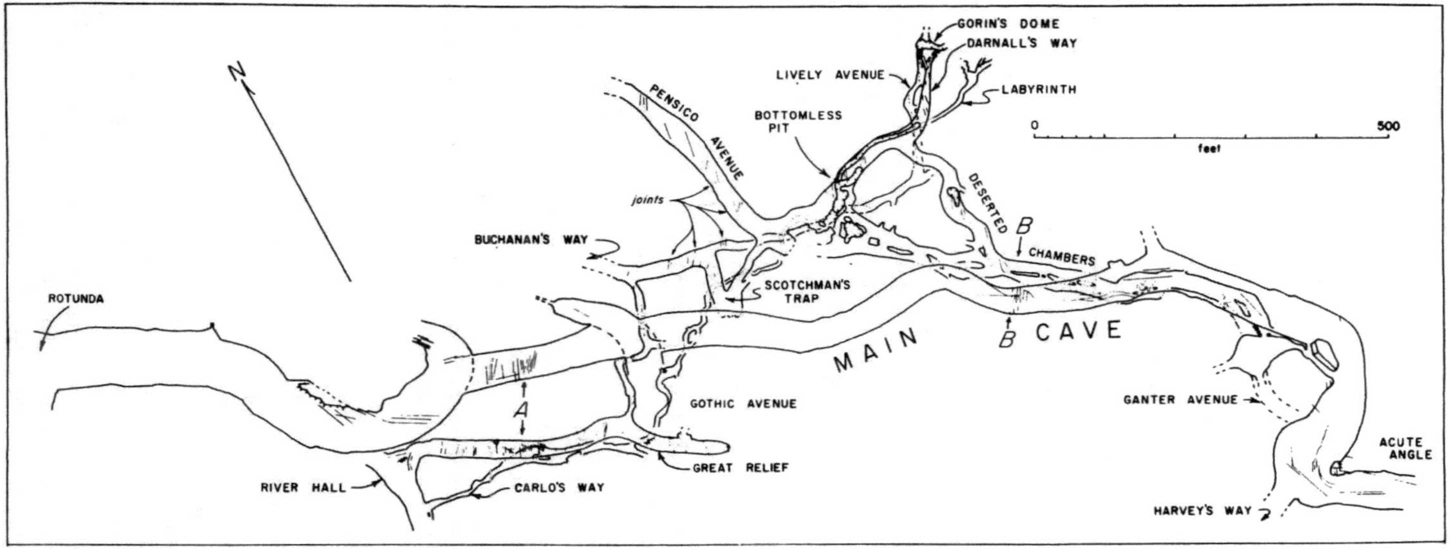

Figure 10.7 Map of the western end of Mammoth Cave.

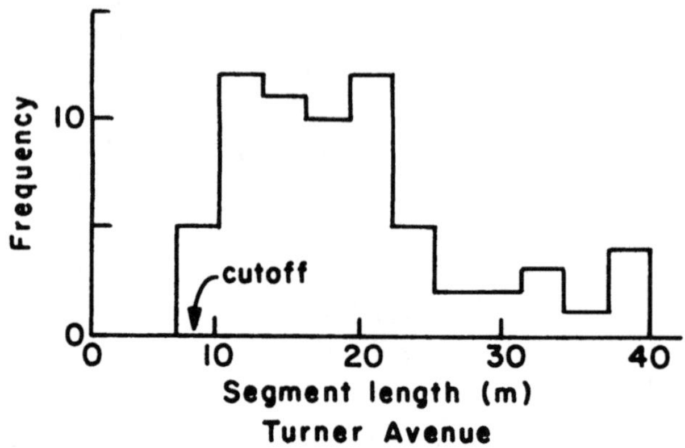

Figure 10.8 Histogram showing length of straight-passage segments.

The caves of other regions are often composed almost entirely of straight passage segments meeting at angles. The largest cave system in Virginia, Breathing–Butler, is a rectilinear maze with 98% of its 16 km (1960 length) of passage disposed in straight segments (Deike, 1960). It was developed slowly under artesian conditions in well-jointed rock. Even long, winding avenues similar to those of Mammoth Cave are largely made up of straight reaches with the angular bends rounded by stream action (see Buchanan Saltpeter Cave in Douglas, 1964, p. 495).

Most similar to the situation in the Mammoth Cave region is that of County Clare, Ireland, where the caves are downdip piracies (really diversions) of surface streams. The dip is 2–3°, and the caves are very recent and generally small in size, though of considerable length. Their passages are developed on bedding (elliptical tubes) with sinuous canyons cut below. The

Table 10.5 Percentage of Straight-Passage Segments in Other Karst Regions Compared to That in the Mammoth Cave Area

Area	*Number of Caves*	*Total Length Considered*	*Percent Straight*
Virginia	31	46.6 km	84
West Virginia	36	28.5 km	80
Spain	42	14.1 km	75
France	91	45.7 km	71
North Africa	6	5.2 km	77
Great Britain	3	9.1 km	83
Ireland	11	18.1 km	51
Flint Ridge	—	32.6 km	39

Statistics of other regions, percent straight passage considered a normal distribution: $\overline{X}$ = 74.5% and s = 10.44%. Comparison to Flint Ridge: $Z = (39-74.5)/10.44 = 3.39$*** and $P = 0.0003$. *Note:* level of significance *** = 0.001.

nonoriented parts of these caves consist of smooth curves and irregular passage directions as at Flint Ridge. Nonetheless, about 10% more straight passage occurs than in Flint Ridge. In Flint Ridge nonstraight passages resemble nothing so much as the bends of a river, with occasional abrupt direction changes. Particularly in narrow canyons, the passages wander about in a way not resolvable into straight segments. Passages with irregular walls or with wide rooms are rare. By contrast, irregular widenings or rooms account for most of the nonoriented cave in many areas. In some cases it seems from maps that these irregular features might also be joint controlled, but in most cases their irregularities are probably due to the way in which solution has attacked the lithology present. On the other hand, the caves in the Mammoth Cave area have remarkably parallel (if not straight) walls, with widenings due to solution erosion by curving streams.

Figure 10.9 illustrates the orientation of the straight cave segments. The most prominent orientation lies at the 100–140° azimuth, but concentrations of segments also occur at 10–50° and possibly at 150–160° and 60–70°. Chi-square tests against a rectangular distribution (Table 10.6) show that the cave orientation is not random for the large systems that compose Mammoth Cave. Tests on smaller caves individually often fail to demonstrate preferred orientation, probably due to small sample size. A one-sample-runs-test to show that the oriented passages cluster together in certain directions producing fewer runs than random order would produce, yielded significant results at the 0.025 level in seven of the eight largest of the small caves.

Orientation histograms of the Mammoth Cave system show similarities to those of joints and fracture traces in the immediate Mammoth Cave area (Fig. 10.10). Statistical tests comparing the caves with joints and fracture traces show that the data come from populations with different distributions. The difference lies not so much in the direction of orientation as in the relative frequency of different directions. Most obviously, the

Table 10.6 Chi-square Tests Against Rectangular Distribution in Straight Cave Segments

Cave		*Number of Observations*	*Degrees of Freedom*	χ^2	*P*
All caves		973	17	109.2***	
Mammoth Cave		492	17	96.0***	
Flint Ridge		388	17	99.7***	
Small caves:	s > 15 m	93	17	12.1	0.75
Small caves:	s > 7.5 m	171	17	16.0	0.55
Indian:	s > 7.5 m	19	2	10.52**	
Parker:	s > 7.5 m	33	5	7.91	0.17
Long:	s > 7.5 m	31	5	5.58	0.36

Note: level of significance: * = 0.05; ** = 0.01; *** = 0.001.

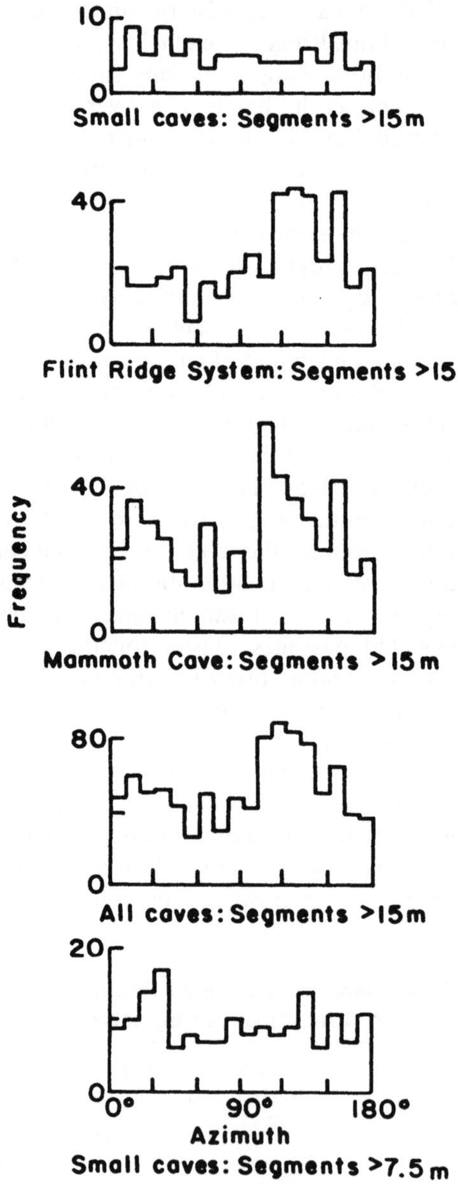

Figure 10.9 Orientation histograms of straight cave segments. Note prominent sets at 100–140°.

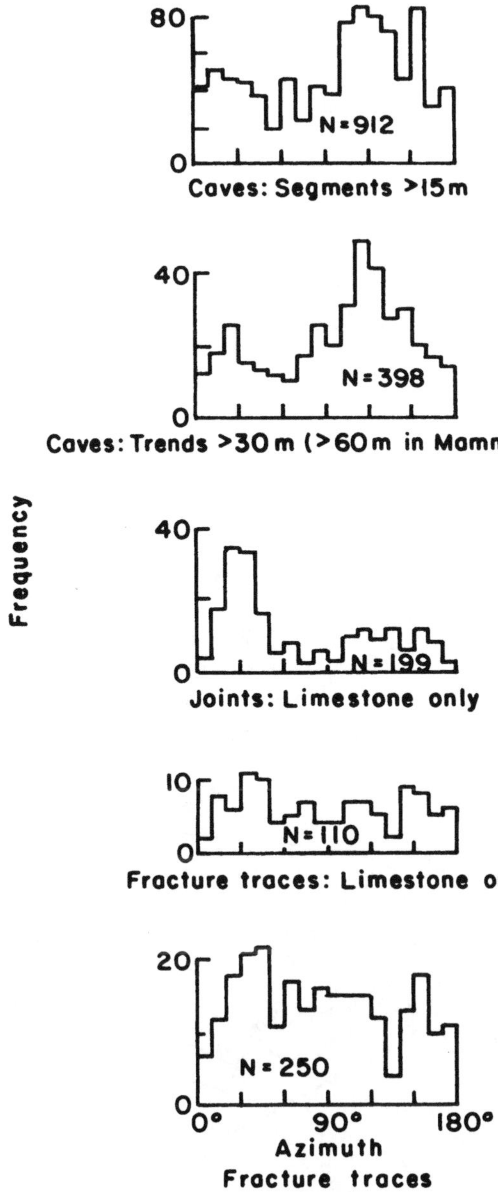

Figure 10.10 Orientation histograms of the largest caverns and adjacent joints and fracture traces.

major northeast joint set has the greatest frequency of joints (and of fracture traces in the local area), but oriented cave passages are most common in a southeast–northwest direction, parallel to a much less numerous joint set.

To eliminate the differential amplitude of the orientation maxima, the

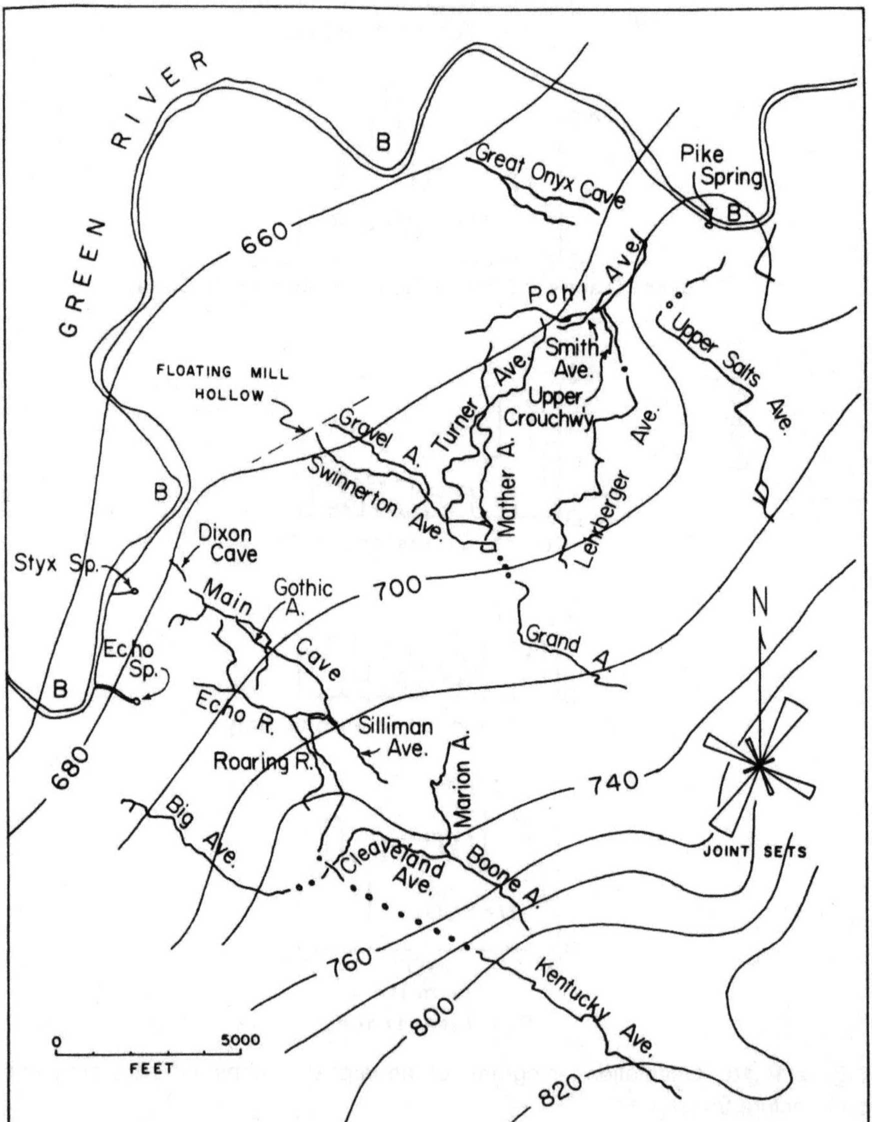

Figure 10.11 Map of major passages of Mammoth Cave and the Flint Ridge system showing structure.

Table 10.7 Probability That Joints and Caves Have the Same Directional Distribution in the Major Joint Set Directions

Northeast 350–60°	*Southeast* 80–145°
$\chi^2 = 116.5$***	$\chi^2 = 6.48 \pm 0.4$

Note: level of significance: *** = 0.001.

joints and cave segments were separated and compared as unimodal distributions corresponding to each major joint set. The directions, 60–80° and 145–170°, were not incorporated in the data, to eliminate the effects of possible minor joint sets.

Chi-square tests of these data in Table 10.7 indicate that the cave segments with northeast orientation do not belong to the same population as the joints in that direction. Although these passages are probably controlled by the northeast joint set, the joints are closely clustered at 20–40° azimuth, but the cave passages developed subequally in directions from 350–50°, with some concentration closer to north than the commonest joints. Other factors than joint frequency in the set are at work.

In the direction of the southeast–northwest joint set the distribution of cave segment orientation cannot be distinguished from that of the joints. The common oriented cave passages in this direction are apparently directed by the joints in this set.

The dominant use of the southeast–northwest joint set by straight cave elements must be accounted for by the nature of the joints, the direction of the dip or strike, the direction of the hydraulic gradient, or by combinations of these factors. The joints in this set have greater horizontal and vertical extent. Because the fractures in the rock must aggregate continuity to permit water movement, the greater extent can be quite important. The irregular openings along the surface roughness of this joint set probably offer more paths for initial water movement compared to the planar northeast joint set.

The trend of most cave passages is down the dip of the regional structure at various angles to the strike, which is also the direction toward Green River (Fig. 10.11). The groundwater flow directions as mapped by Quinlan and Ray (1981) are predominantly downdip and nearly parallel to the northwest–southeast joint set except in the Hidden River groundwater basin. The hydraulic gradient favors the use of this joint set in the area where the Mammoth Cave system is developed.

Caves of the Sinkhole Plain

All the caves beneath the eastern Sinkhole Plain studied by Deike (1967) trended downdip, as do the sinking streams at the southern margin of the plain. Some of the caves are in ponors of these streams, and others are closer to the Chester Escarpment. One exception seems to be a major southwestern tributary of Hidden River Cave, which trends parallel to the strike for 1.1 km. It is along the major route of the Hidden River groundwater basin.

In much of the area south of Cave City and Horse Cave the groundwater flow is downdip north toward the major conduit of the Hidden River basin. The observed use of the northeast joint set (and the minor 150–160° set) under the Sinkhole Plain is due in part to those sets trending close to the northerly dip direction in this part of the plain.

Caves of the Chester Upland

Because there are many kilometers of surveyed cave passage beneath parts of the Chester Upland, more can be seen of the effects of structure on groundwater flow there. For flow through the rock to begin there must be both recharge and some discharge points. Green River crosses the upland in the immediate Mammoth Cave area in a southwest direction parallel to the regional strike. As a result, while the river eroded the gorge it first breached the Big Clifty Sandstone caprock at the updip river bends. These isolated places where the limestone was exposed were each potentially new discharge points. The first such place near the Mammoth Cave system was the bend at Pike Spring, the next was near Echo River Spring where the caprock is 3 m to 6 m lower than at Pike Spring, and the last was at Turnhole Spring where the caprock is 15 m lower than at Echo River. The relationship of the river to the structure, by determining the location of these groundwater discharge points, also influences the trend of caves, which develop to transmit water to these points.

Figure 10.11 shows the diverse relationships of the major passages of the Mammoth Cave system to the general structure, to the Green River, and to each other. The highest passages, interpreted as the earliest to develop, trend toward the updip river bends. These include Collins Avenue (7.6 m beneath the caprock, not shown in Fig. 10.11) and Upper Salts Avenue (15 m lower still), which trend toward the bend at Pike Spring, and Gothic Avenue (21 m below the caprock) and Main Cave–Kentucky Avenue (6.1 m below Gothic Avenue), which trend to the updip bends near Echo Spring. Much of the present groundwater flow, for example Echo–Roaring River, is toward these same river bends. These passages appear to trend down the regional dip.

Many of the intermediate-level passages are tubes that take less direct routes to Green River. Typical are Cleaveland–Big Avenue and Turner Avenue. These passages are not dip-oriented, and Deike (1967) observed that Cleaveland Avenue seemed to follow the strike in part. The detailed relationships of many passages to the structure were defined by Palmer (1981; see Chap. 11). The most common, straight cave trends are parallel to the southeast–northwest joint set, which trends toward the Green River and down the regional dip.

Detailed Relationships of Cave Passages to Fractures

In the caves there is a gradation in passage types between tubes and canyons. Tubes of many sizes, up to 18 m wide, are usually elliptical in cross section, widest at a bedding parting, and ungraded but nearly horizontal throughout their length. These tubes developed at the top of the phreatic zone, the water table following the line of least resistance in response to the hydraulic gradient. Canyon passages have a tube, sometimes no more than a widened bedding parting, at the top, but the passages are principally downcut canyons with subparallel walls often almost as wide as the tube above. The tubes, and thus the passage roofs, trend down the local dip (Palmer, 1977). Canyon floors are graded by the vadose streams that cut them. The water follows lines of least resistance under the direct influence of gravity. Canyons range from shaft drains less than 30 cm wide to regional drains 15 m wide, and they can be 23 m or more high.

One of the largest tubes in Mammoth Cave is Cleaveland Avenue, where beds above and below the initial bedding partings were enlarged to as much as 4.6 m high and 15 m wide. Its large size, achieved while base-level control continued, implies a long period of stable base level. Deike (1967) pointed out that the tube was horizontally and stratigraphically confined, and followed the strike around a fold seen on regional maps. Palmer's (1981) detailed work revealed how close are the control of horizontal water table and local folds. Figure 10.12*B* shows typical exposures of joints in the roof and walls of Cleaveland Avenue. The floor is covered with thin sediment, and no joints are exposed. The exposed joints are developed almost exclusively in the beds just above the bedding parting at the widest part of the passage. In Figure 10.12*B* a swarm of short joints trends about 35° azimuth across the tube. One rough joint at about 120° is approximately parallel to the passage at the west end of the figure, but otherwise the passage is little related to the joints present.

For 425 m west of Figure 10.12*B* there are swarms of joints from 7.5 m to 15 m wide crossing Cleaveland Avenue that are from 20 m to 30 m apart, always with the same relationship to the passage. Farther west the avenue is parallel to the joint swarms for 300 m and has several long, straight segments. The direction in this part of Cleaveland Avenue is probably joint controlled, but joint exposure is poor, and the passage does not follow a visible swarm of joints.

The base-level Cleaveland tube continues west as Big Avenue and probably as Marshall Avenue in Lee Cave, a strike-oriented tube more than 8 km

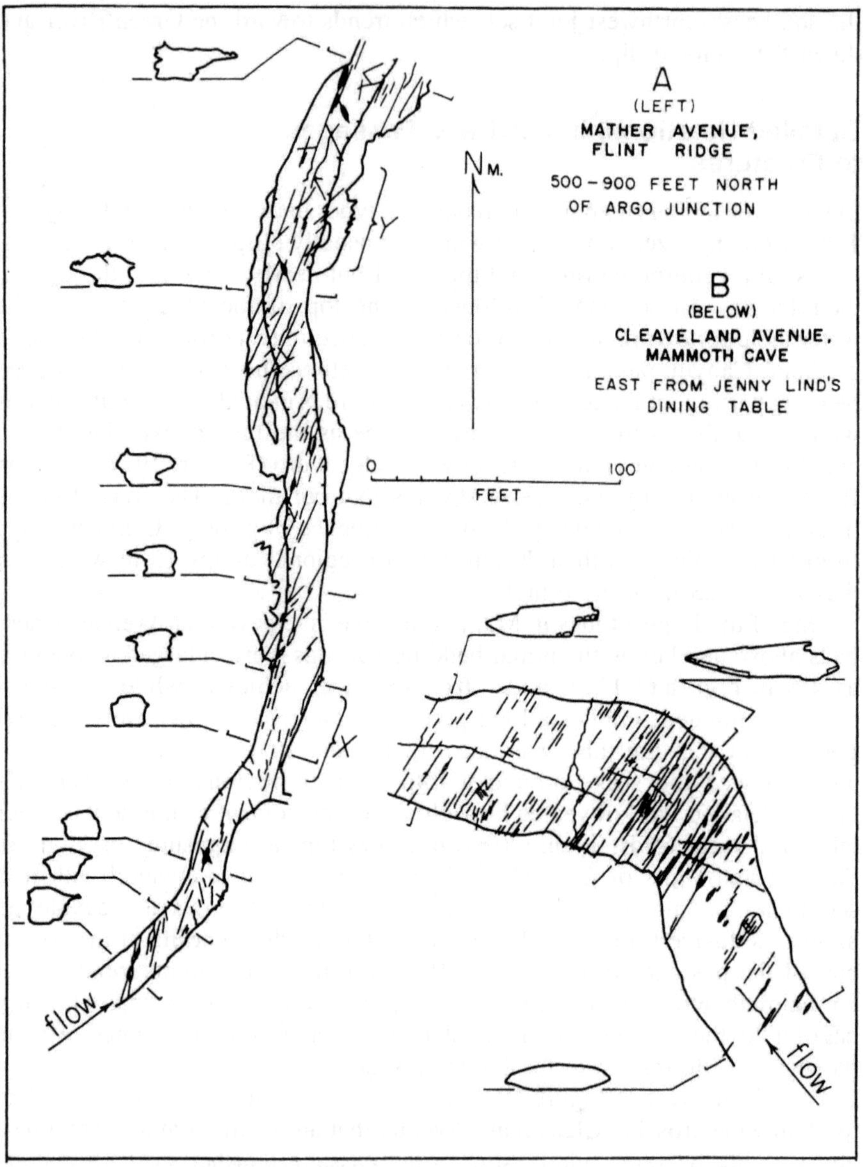

Figure 10.12 Details of joints exposed in selected passages.

long leading toward Turnhole Bend. Along this route the tube passes only 800 m from a bend of Green River. To reach the river, which lies in the downdip direction, the water would have needed to develop a route that rose about 5 m through the beds along joints. But joints are so discontinuous that the route of least resistance lay along the bedding to updip Turnhole Bend, about 4 km away.

Turner, Mather, and Swinnerton avenues are successive downstream branches of Grand Avenue (Fig. 10.11); Turner Avenue is a tube. Mather Avenue is a tube with a wide canyon that angles somewhat downdip. Swinnerton Avenue tube diverges northwest from the floor level of Mather Avenue. Palmer (1981) reported that Swinnerton Avenue loops up and down somewhat and changes horizon along joints at one point. Swinnerton Avenue trends parallel to the northwest joint set for 2.3 km and has 6% to 8% more of its length disposed in straight, probably joint-directed segments than does Turner or Mather avenues, which trend northeast. This is a result of the greater utilization of the northwest joint set by groundwater movement.

Figure 10.12*A* illustrates joints exposed in Mather Avenue upstream from the Swinnerton Avenue junction. In this area Mather Avenue is an irregular tube 4.3–9.8 m wide and 1–2 m high, with a canyon 3.4–4.6 m wide and 1.5–2.4 m deep. Joints exposed are commonly isolated from one another and less than 6 m long, although one joint is exposed for 17 m. Many of the joints do not extend to the passage walls and are confined to the bed exposed in the roof. Ceiling pockets and wall facets express some of the joints. Much of the passage seems to follow the exposed joints or en echelon groups of the joints trending about 20° azimuth, although the joints actually exposed in the passage cannot be proved to have extended downward to the bedding plane at the widest part of the tube, where solution began. The segment marked X is the only one that appears straight on the map. The area shown in Figure 10.12*A* was chosen to illustrate the nature of joint control. Most of Mather Avenue has far fewer joints and less evidence of joint control than this reach.

Main Cave–Kentucky Avenue in Mammoth Cave Ridge is a large downdip canyon directed to the updip reach of Green River upstream from Echo River Spring. Passages from the Rotunda to River Hall on the historic tour are shown in Figures 10.7 and 10.13. The passages below Main Cave are of varied type and in many different stratigraphic horizons. Harvey's Way and Lost Avenue are successive cutoffs from Main Cave, starting as shallow canyons in its floor. Deserted Chambers, Lively Avenue, and Great Relief Hall carried groundwater from nearby Houchins Valley at a later time. Carlo's Way is an abandoned shaft drain.

Exposed joints shown in Figure 10.7 illustrate the lack of joint control seen in much of the cave. Joints exposed in Main Cave are scattered and only occasionally lie parallel to the passage. The curved fractures around Acute Angle could be due to the stress on the wide roof span. Deserted Chambers seems to parallel en echelon joint groups in part. Lively Avenue and Great Relief have exposed joints at 30° azimuth, which cross the passage at angles from 20° to 90°, but the passages seem to lie parallel to groups

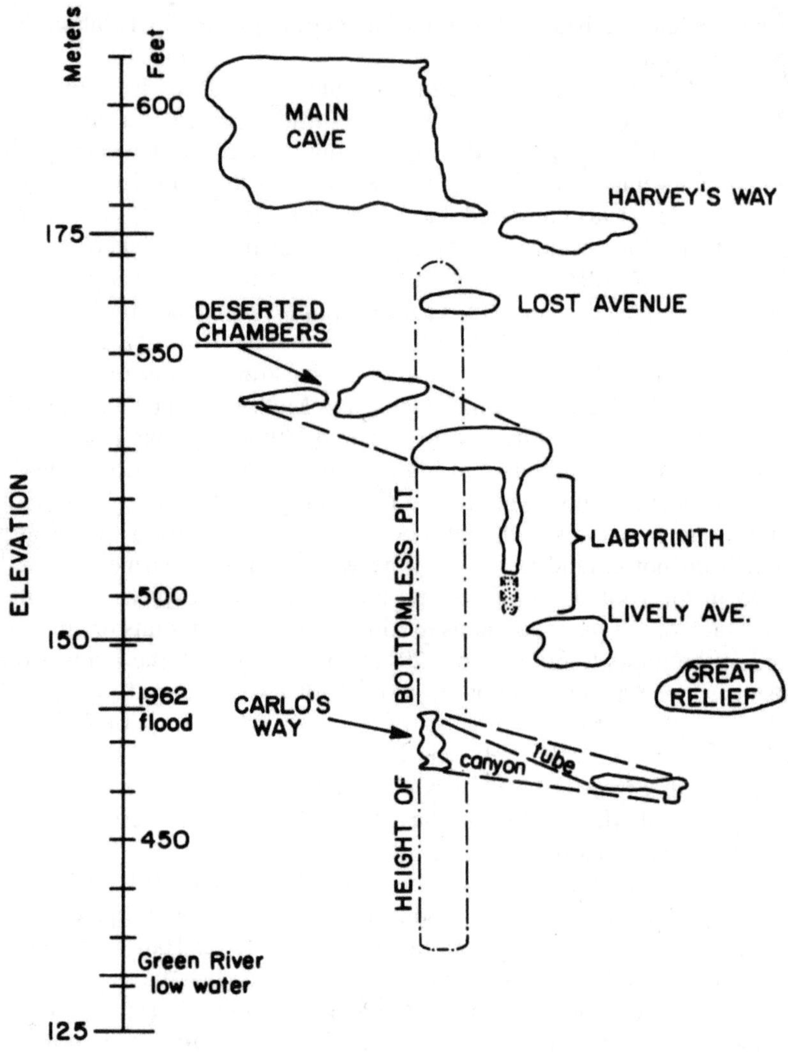

Figure 10.13 Composite cross section of passages shown in Figure 10.7 at the west end of Mammoth Cave.

of these joints, even though the individual joints might be perpendicular to the passage. Carlo's Way has almost no exposed joints, and only 12% of its length is straight. Table 10.8 summarizes the extent of oriented joint-controlled passages shown in Figure 10.7.

Boone Avenue is a dip-oriented tributary to Cleaveland Avenue. Detailed investigation of some 1 km of canyons and tubes on several levels, lying 365–610 m upstream from Cleaveland Avenue along Boone Avenue,

Table 10.8 Percentage of Straight Passage or Passage with Visible Joint Control in a Selected Passage: Main Cave Region of Figure 10.11

	Length (m)	*Straight >15.2 m*	*Straight >7.6 m*	*Total Straight or with Visible Joint Control*
Main Cave	730	31%		47%
Deserted Chambers	610	31%	39%	45%
Lively Avenue to Great Relief	550	20%	27%	40%
Carlo's Way	240	0%	12%	12%
Total	2130	25%	29%	41%

Straight trends <30.5 m = 55%.

revealed tubular elements concordant with bedding and only 25% of the passages joint controlled (Fig. 10.14). Joints that lie parallel to nonstraight as well as straight passages and groups of short en echelon joints are exposed. Straight reaches of Rose's Pass, which are parallel to the major joint sets, have few exposed joints. Joint control here is probable, but the joints must have been confined to a thin horizon that has been completely removed by solution. In small canyons throughout the cave there is limited joint control of the initial tube and very little joint influence on the downcut canyons below.

The other major element of the caves is the vertical shafts, cut downward by solution near the edge of the caprock where drainage off the cap enters the limestone aquifer. Vertical shafts might reach depths of 45 m or more. Most of the shafts cut through varied lithology without effects except in details of solution features of the shaft walls. Shaft drains often develop on several successive bedding partings as the shaft is deepened. The shafts have been deepened one bed at a time by solution of the floor (Palmer, 1981). Even if fractures fixed the initial location of a shaft, the fracture usually would not continue through more than one or two beds; therefore, general lack of fracture control of vertical shafts is the norm. Shaft drains begin along bedding partings as do other canyons. In the drains a few examples of very short new routes, directed by joints or by bedding, were described by Deike (1967).

Thickness of beds and character of bedding partings must affect the occurrence of breakdown in the passages, with the most massive beds supporting large ceilings without collapse. Joints also affect breakdown because many fallen blocks are bounded by joints, such as at Giant's Coffin in Mammoth Cave. Fragments of canyon walls sometimes separate along joints where there is no evidence that the joints influenced the solutional development of the canyon. The effects of structure on collapse features have not been investigated in any detail.

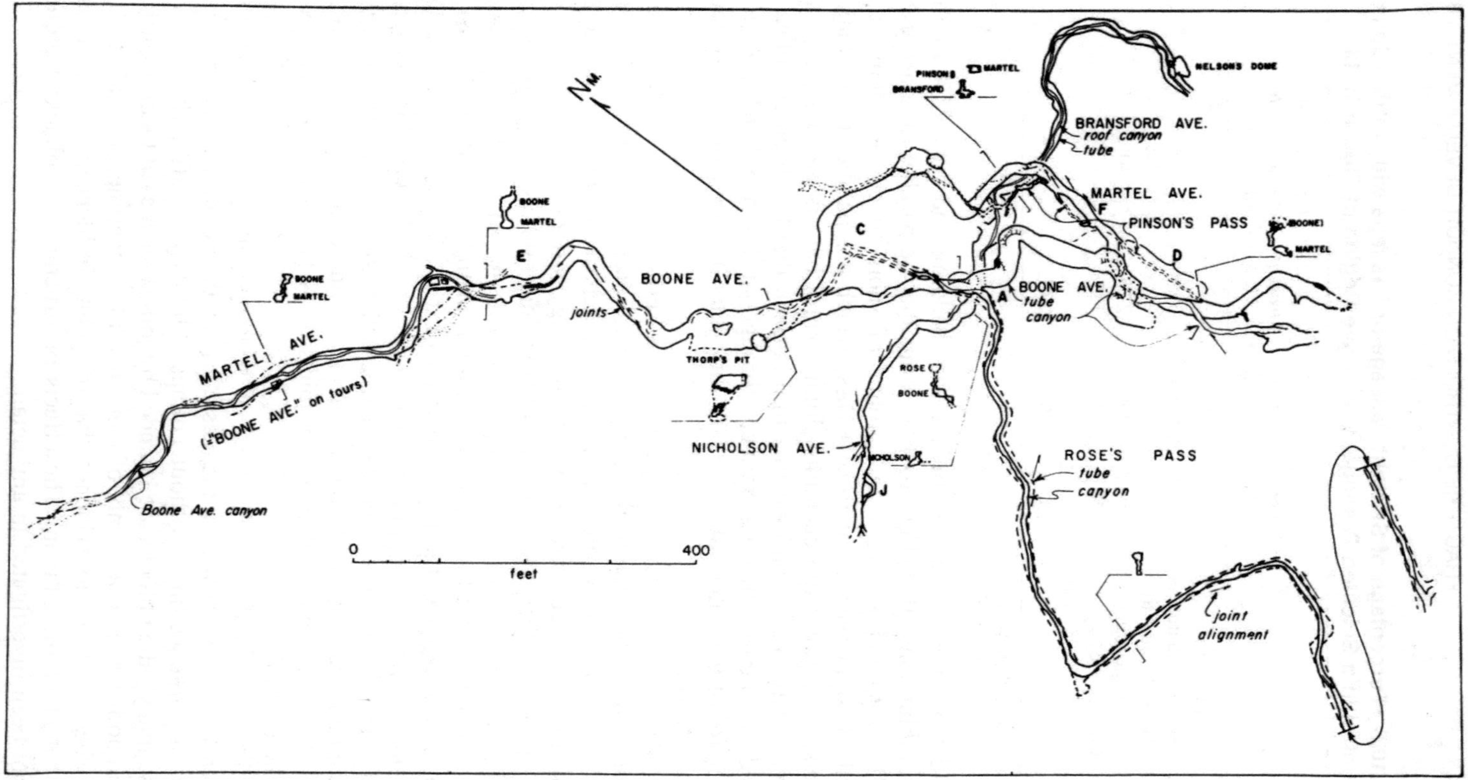

Figure 10.14 Map of Boone Avenue area, Mammoth Cave.

SUMMARY

Groundwater in the limestone is transmitted through solution conduits that are developed along planes of secondary permeability. The groundwater must integrate continuous flow paths using bedding partings and joints and fractures. The resulting cavern passages are a series of low-gradient branching tunnels with steeper cutoffs and modern, vertical shafts connecting them. They are often directed to structurally favored (updip) river bends. The passages follow lines of least resistance through the rock along the hydraulic gradient, with various relationships to dip, folds, fractures, and river location depending on local conditions.

Joints are sparse and short. Major sets include northeast joints, which are planar and often in swarms. The northeast set is rarely followed by cave passages and often ignored. A less numerous set at 100–120° azimuth contains rough-surfaced joints and is favored by the hydraulic gradient and dip in much of the area. This set is most paralleled by straight cavern passages, and statistical tests show no significant difference between the distribution of joints and straight passages in the direction of this joint set.

Fracture traces show less well-defined orientation patterns than joints, but are most developed in the direction of the southeast–northwest joint set, which is the direction of many cave passages. Only a small proportion of dolines and no cave passages are aligned on mapped fracture traces, but the use of en echelon and other joint groups in some passages and the existence of straight trends in crooked passages suggests the influence of joint zones.

The joints that influence the development of the conduits are most commonly located in the bed immediately above the bedding parting, which is the primary path for water movement. Often this influence results in a short, straight length of conduit parallel to the joint direction; however, longer reaches of conduit that trend in the joint directions, but are not straight, are also developed. Passages in the caves are observed to follow joints or groups of joints, sometimes en echelon. The aquifer has somewhat greater permeability in the direction of the major joint sets, which direct groundwater flow locally.

Most individual joints are limited to one or a few beds and do not interconnect. The result is that water seeking a through path must use the bedding planes. Joints might locally exert considerable control on the initial course of a tubular passage. If the course is parallel to a group of scattered joints, the joints taken together could determine the course of the water flow as the passage enlarges to encompass several of them. Often, however, the joints are more or less isolated, and the path of the water follows vagaries of the bedding planes through much of its course. When passages change from one bedding parting to another, often rising through the beds in a downdip and downstream direction, this vertical movement is accomplished by use of joints. As passages reach larger size they probably outgrow the influence of the joints to some extent. The deviations that then develop are due to the hydraulic properties of the stream.

A considerable portion of the length of the caverns is not straight and shows no evidence of fracture control. Fractures that the initial passage crossed without notable effects continued to be ignored, except for solution pocketing extending into the fractures for distances usually less than 30 cm. In some parts of the caves no joints are seen. In all such parts of the system, openness and surface features of the bedding and local structures most influenced the passage route, and hydraulic factors might have had more influence than in better jointed areas.

The canyon passages show less joint influence than the tubes. The route followed by a canyon is somewhat joint directed because the tube whence the incision began was partly so directed. But during incision the hydraulic properties of the stream take over and produce a meandering pattern. The stream might make use of short joints during incision. Only rarely are straight joint-determined segments of canyons seen. They are seldom high, straight reaches, and are usually isolated, straight walls and segments that extend horizontally for no more than a few meters and vertically for no more than the thickness of one or two beds. These are the effects of isolated joints. The impact of jointing was felt anew when an entirely new route was begun to continue the downward evolution of the cavern system in response to denudation of the land and lowering regional base level.

Most vertical movement of water is downdip along bedding, but the vertical shafts at the edge of the caprock carry recharge directly down into the aquifer. The shafts are often located originally along vertical joints, but continue to develop downward through unjointed beds.

Relatively poor joint development is common to the Interior Lowlands, and irregularly curving cavern passages with modest joint control are the rule. In most karst regions the caves are much more rigorously and completely joint controlled even where bedding plays a major role in initiating development.

REFERENCES

Blanc, A., 1958, Répertoire bibliographique critique des Études de relief karstique en Yougoslavie depuis Jovan Cvijic, *C.N.R.S. (Centre de Documentation Cartographique et Géographique) Mém. et Doc.* **6:**135–227.

Bretz, J H., 1942, Vadose and phreatic features of limestone caverns, *Jour. Geology* **50:**675–811.

Brucker, R. W., and D. P. Burns, 1964, *The Flint Ridge Cave System,* Washington, D.C.: Cave Research Foundation, Map Folio.

Deike, G. H., III, 1960, Origin and geologic relations of Breathing Cave, Virginia, *Natl. Speleol. Soc. Bull.* **22:**30–42.

Deike, G. H., III, 1967, The development of caverns in the Mammoth Cave region, The Pennsylvania State University, Ph.D. dissertation, 235p.

Douglas, H. H., 1964, *Caves of Virginia*, Falls Church, Va.: Virginia Cave Survey, 761p.

Ewers, R. O., 1966, Bedding-plane anastomoses and their relation to cavern passages, *Natl. Speleol. Soc. Bull.* **28:**133–140.

Howard, A. D., 1968, Stratigraphic and structural controls on landform development in the central Kentucky karst, *Natl. Speleol. Soc. Bull.* **30:**95–114.

Kaemper, M., 1908, Map of the Mammoth Cave, Kentucky, on file at Mammoth Cave National Park, reprinted in 1981 by Cave Research Foundation, Yellow Springs, Ohio, 1 sheet.

Lattman, L. H., 1958, Technique of mapping geologic fracture traces and lineaments on aerial photographs, *Photogramm. Eng.* **24:**568–576.

Palmer, A. N., 1977, Influence of geologic structure on ground-water flow and cave development in Mammoth Cave National Park, Kentucky, U.S.A., *Internat. Assoc. Hydrogeology Mem.* **12:**405–414.

Palmer, A. N., 1981, *A Geological Guide to Mammoth Cave National Park*, Teaneck, N.J.: Zephyrus Press, 210p.

Quinlan, J. F., and J. A. Ray, 1981, Groundwater basins in the Mammoth Cave region, Kentucky, *Friends of the Karst Occasional Pub. No. 1*, Map.

Silliman, B., Jr., 1851, On the Mammoth Cave of Kentucky, *Am. Jour. Sci. Arts* **11:**332–339.

11

STRATIGRAPHIC AND STRUCTURAL CONTROL OF CAVE DEVELOPMENT AND GROUNDWATER FLOW IN THE MAMMOTH CAVE REGION

Arthur N. Palmer

Within the broad spectrum of karst processes, cave development in the nearly horizontal limestones of the east-central United States is often considered to be among the most easily understood. Yet, close examination shows the geologic influence upon such caves to be every bit as complex as that in the most intensely deformed regions. Furthermore, these relationships are masked by a subtlety that defies all but the most careful mapping. This chapter outlines the conclusions from a 20-year study in the Mammoth Cave area aimed at clarifying the influence of local stratigraphy and geologic structure on passage patterns, as well as the relationship of caves to the geomorphic history of the region.

FIELD DATA

Stratigraphic units in the region have been described previously from surface exposures by Weller (1927), Haynes (1964), and Pohl (1970). Geologic structure is shown on the U.S. Geological Survey (USGS) geologic quadrangles of the area. However, none of this mapping extended underground. Except for a few scattered observations, not even the major formational contacts had been identified in caves.

The exploration and mapping of caves in the region has expanded greatly in recent decades, making it possible to relate underground geologic observations to those on the surface. The project on which this chapter is based began in 1970. A hand-level survey was made in the northeast part of the Mammoth Cave system (Crystal Cave), including all passages known at that time. Plan-view maps provided by the Cave Research Foundation (CRF) were used for horizontal control. Crystal Cave was ideal for this study because it spans a greater vertical range than any comparable part of the system and because it is close to the Green River Valley and therefore sensitive to changes in fluvial base level. The complexity and high degree of interconnectivity of its passages provide a fine three-dimensional view of local stratigraphy and structure.

Approximately 20 km of passages were surveyed, including a total of 3221 horizon measurements at 991 level stations. Elevations were obtained for geologic contacts, major bedding-plane partings, passage ceilings and floors, sediment levels, and past and present water levels. A topographic survey was made over Crystal Cave with a tripod-mounted Brunton compass and tape to locate recharge points and springs. Geologic maps and continuous passage profiles have been drawn from these data, allowing graphical and statistical analyses of the various interrelationships. Since this initial phase, an equal amount of leveling has been done for this project in major passages throughout Mammoth Cave, using a hand level, tripod-mounted surveyor's level, and a flexible water-filled U-tube manometer.

The mean vertical error on closed loops was 0.018% with the hand level and approximately half as much with the other leveling methods. The CRF base maps were made with hand-held Brunton compasses, with a mean horizontal error of about 1%. These degrees of accuracy are adequate for all interpretive purposes.

RELATIONSHIP OF CAVES TO STRATIGRAPHY

The Mammoth Cave system extends through a stratigraphic section of about 130 m, from the middle of the St. Louis Limestone nearly to the top of the Girkin Formation. The detailed stratigraphic column shown in Chapter 2 (Figure 2.2) omits the lower 35 m, which is exposed only in parts of the cave system of limited accessibility. The missing section consists mainly of dolomitic limestones with numerous beds, lenses, and irregular nodules of chert in discontinuous zones. Most passages are in the Ste. Genevieve Limestone and Paoli Member of the Girkin Formation. Only Collins Avenue in Crystal Cave and a few minor upper levels of Mammoth Cave and Roppel Cave lie stratigraphically above the Paoli Member, although breakdown in the main passage of Salts Cave and its distributaries extends upward about 10 m from this unit. The northwesterly regional dip brings most of the St. Louis Limestone below base level in the Green River Valley, but many low-level passages in the southern (updip) parts of Mammoth Cave and Roppel Cave are located in the middle beds of the St. Louis Limestone. The Lost River chert bed, a distinctive 3-m marker near the top of the St. Louis Limestone (according to Pohl, 1970), which is usually included in the Ste. Genevieve Limestone can be traced throughout many of the lowest passages in central parts of Mammoth Cave and northern Roppel Cave. It is absent in Echo River (Mammoth Cave), except for a few isolated nodules.

Lithologic units are more easily distinguished within the caves than on the surface because the extensive, interconnected passages provide a broad continuity of exposure. Also, variations in composition and texture are accentuated by the almost total dominance of chemical over mechanical weathering under moist subaerial conditions.

As a result, it was possible to subdivide the stratigraphic section in Crystal Cave into about 65 distinct lithologic units. Surprisingly, despite variations in thickness and texture, each could be traced throughout the entire Mammoth Cave system and outlying caves over an area equivalent to two 7.5-min topographic quadrangles.

The resulting section is shown in Figure 11.1, in which only the major units are identified. Stratigraphic nomenclature follows the recommendations of Pohl (1970), with some minor adjustments. The contact between the Karnak and Joppa members is rather arbitrary, as is the case almost everywhere else in the Illinois basin. The Aux Vases Member does not extend all the way to the top of the Ste. Genevieve Limestone, in accordance with Swann (1964), leaving an unnamed interval referred to tentatively as "Levias" (Palmer, 1981). However, this unit represents only a small fraction of the Levias Member as mapped in Indiana and should be renamed after more precise interpretation of depositional environments. The position of the Sample Member agrees with that in the Cub Run quadrangle immediately north of Mammoth Cave as mapped by Sandberg and Bowles (1965).

Controlling horizons for initial passage development in Crystal Cave are also shown in Figure 11.1. Of the 20 km of passages in Crystal Cave, 55% originated as vadose canyons or perched tubes, and the remainder were originally of phreatic origin. The initial level of development is considered to be the ceiling in vadose passages and the widest part in phreatic passages, unless there is convincing evidence otherwise. Vadose entrenchment below initial phreatic tubes is not considered in Figure 11.1. The distribution of passages shown here is not representative of the entire Mammoth Cave system or other caves in the region and is intended only to show the lithologic settings favored by each type of passage. Several observations can be made from the stratigraphic data:

1. Slightly more than half the passage length in Crystal Cave, both canyons and tubes, originated along the contact between contrasting lithologies.
2. Initial canyon development was concentrated along the upper contact of impure grainstone, dolomite, or shaly limestone, but also within sparite and micrite units.
3. Tubes tend to be concentrated along the upper contacts of impure grainstones, shaly limestones, and micrites, and within relatively pure limestones.

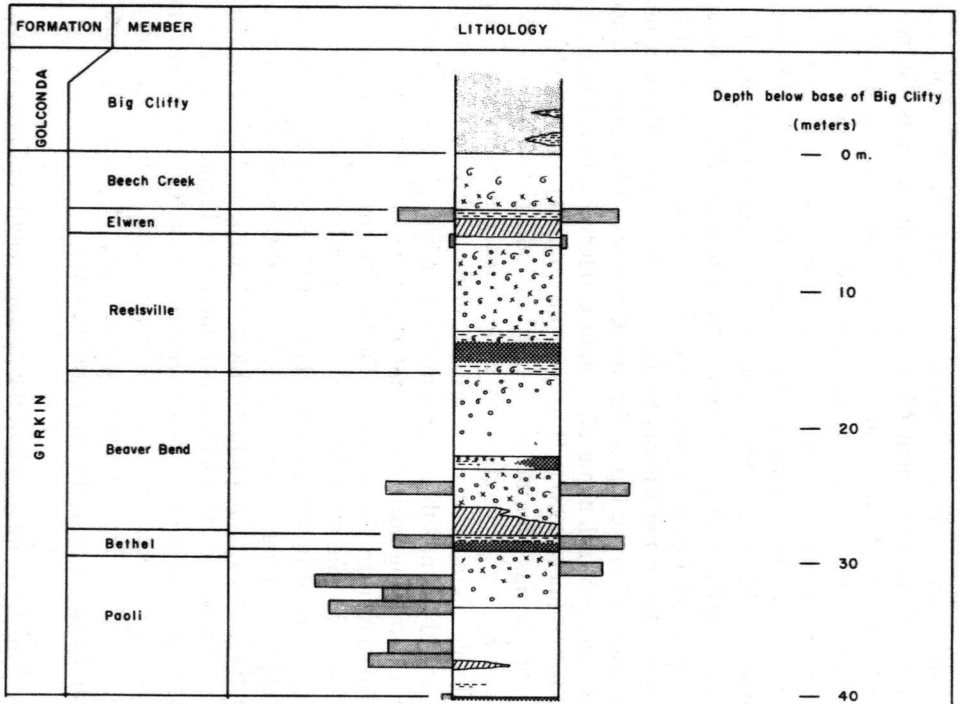
FORMATION
MEMBER
LITHOLOGY
GOLCONDA
GIRKIN
Big Clifty
Beech Creek
Elwren
Reelsville
Beaver Bend
Bethel
Paoli
Depth below base of Big Clifty
(meters)
0 m.
10
20
30
40

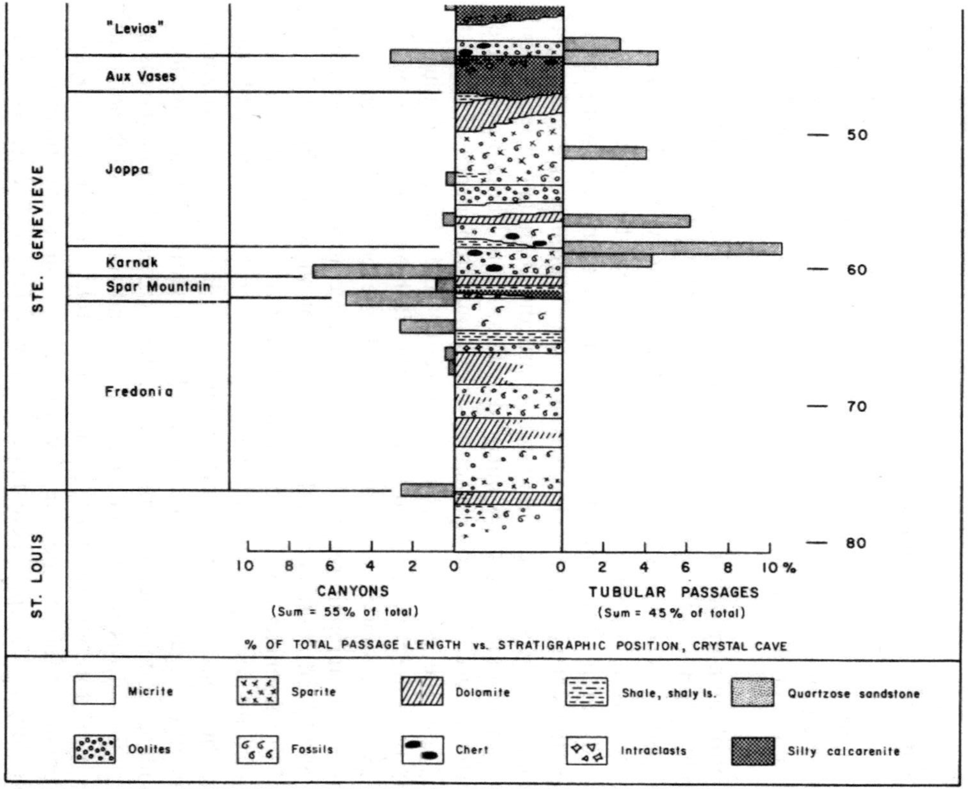

Figure 11.1 Stratigraphic column for the Mammoth Cave system, showing the controlling horizons for passages in Crystal Cave (northeast Flint Ridge). Controlling horizons are assumed to be the ceilings of canyon passages and the widest parts of tubes.

Development of a passage along a lithologic contact should not be considered a form of stratigraphic perching, unless the underlying unit is distinctly more resistant and has resisted vadose entrenchment. Most contacts, regardless of lithology, are favorable horizons along which parting can occur. In Crystal Cave less than 40% of the canyon length and 20% of the tube length are situated along the upper contacts of decidedly low-permeability units such as shale.

An important relationship implicit in Figure 11.1 is the high degree of concordance between cave passages and strata, which causes strong peaks and wide gaps in the histograms of passage frequency. The prominent bedding and paucity of faults and major joints are responsible. Moreover, the great variation in lithology, including minor shaly and silty interbeds, promotes the development of hydraulically transmissive bedding-plane partings. The impact on prediction of groundwater-flow paths is strong (Palmer, 1986).

The controlling horizon for either a canyon or a tube is commonly a major bedding-plane parting. In approximately 60% of Crystal Cave the most prominent parting is between beds of contrasting lithology. In a few places, however, the partings deviate from such a contact several tens of centimeters into the body of a lithic unit (Fig. 11.2). Traced along the walls of a passage, partings can be observed to wax and wane in prominence and might disappear entirely. A single parting rarely exerts the dominant influence on passage origin for more than a few hundred meters. Diminishing prominence of one parting is usually accompanied by increasing prominence of at least one other, and the control of passage elevation and shape is gradually transferred from the former to the latter. In this fashion nearly all lengthy passages exhibit a subtle discordance in which the ceilings rise and fall over a limited stratigraphic range. Stratal discordance appears to be greatest in zones of bedded chert. This relationship is not shown in Figure 11.1, because of the absence of bedded chert in Crystal Cave. In Mammoth Cave and Roppel Cave some passages are perched on chert beds in the St. Louis Limestone, but commonly only for short distances. Passages in chert zones more commonly began their development along a parting within limestone and have merely intersected the chert during their growth. The main reason for this relationship is that the chert tends to bond tightly to the adjacent limestone beds, forming irregular contacts that rarely separate as partings.

Most passages follow the same stratigraphic unit along their entire traversable length, except where upper or lower levels diverge as tributaries or diversion routes, or where a canyon deepens downstream in steps. Although migration of passages along lithologic units is not common, several major passages in the Mammoth Cave system do so, with stratigraphic discordance as much as 25 m. Sharp jogs from one bed to another are usually hosted by prominent joints (Fig. 11.3). Figure 11.4 contrasts the strong

Figure 11.2 Waterfall Trail, in Crystal Cave, is a vadose tube perched on the shaly Aux Vases Member for much of its length. Shown here is the downstream phreatic section, which deviates from the contact along a prominent bedding-plane parting in the overlying "Levias" Member. The top of the Aux Vases Member is shown at the level of the subject's right foot. Headward vadose entrenchment has formed the canyon below the initial tube.

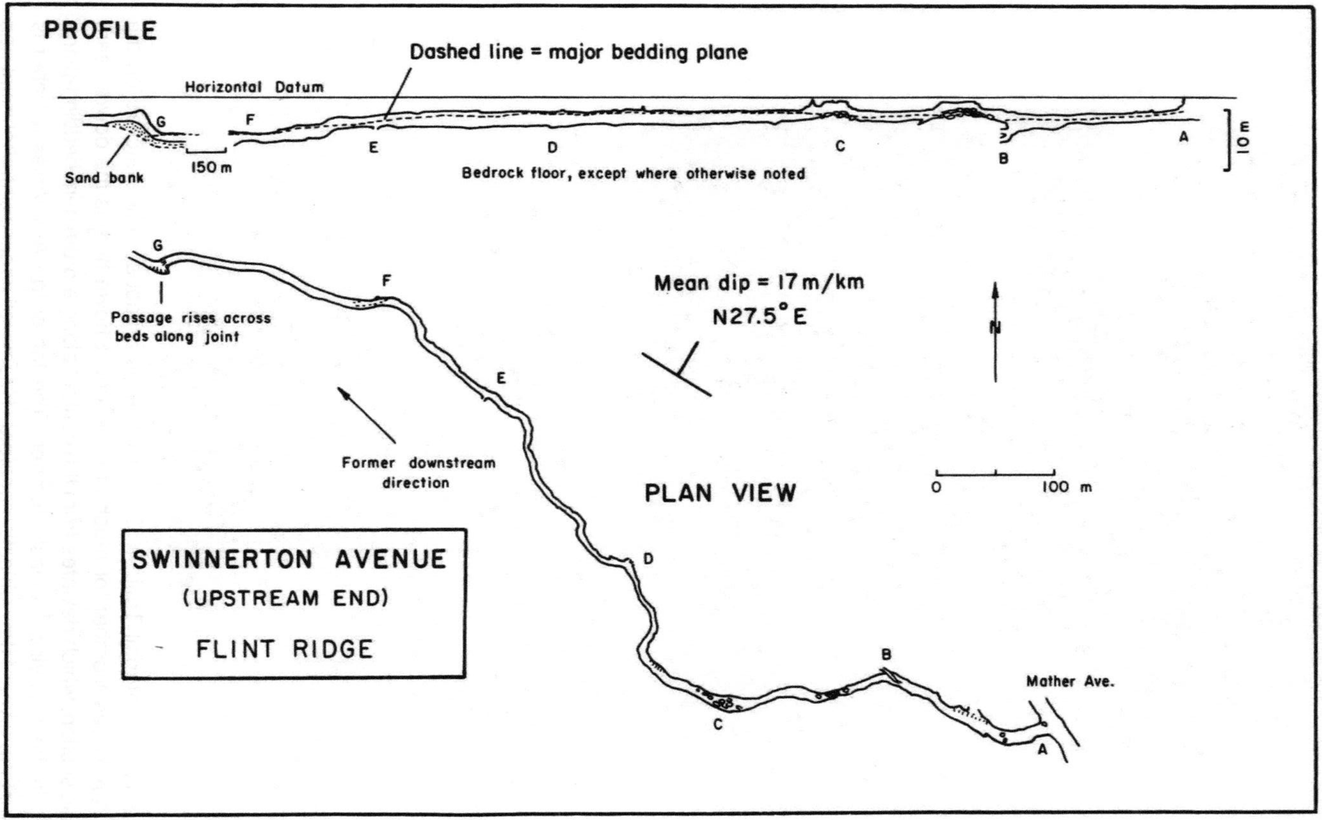

Figure 11.3 Map and profile of Swinnerton Avenue in Flint Ridge, showing the relationship between strata, bedding-plane partings, and passage pattern.

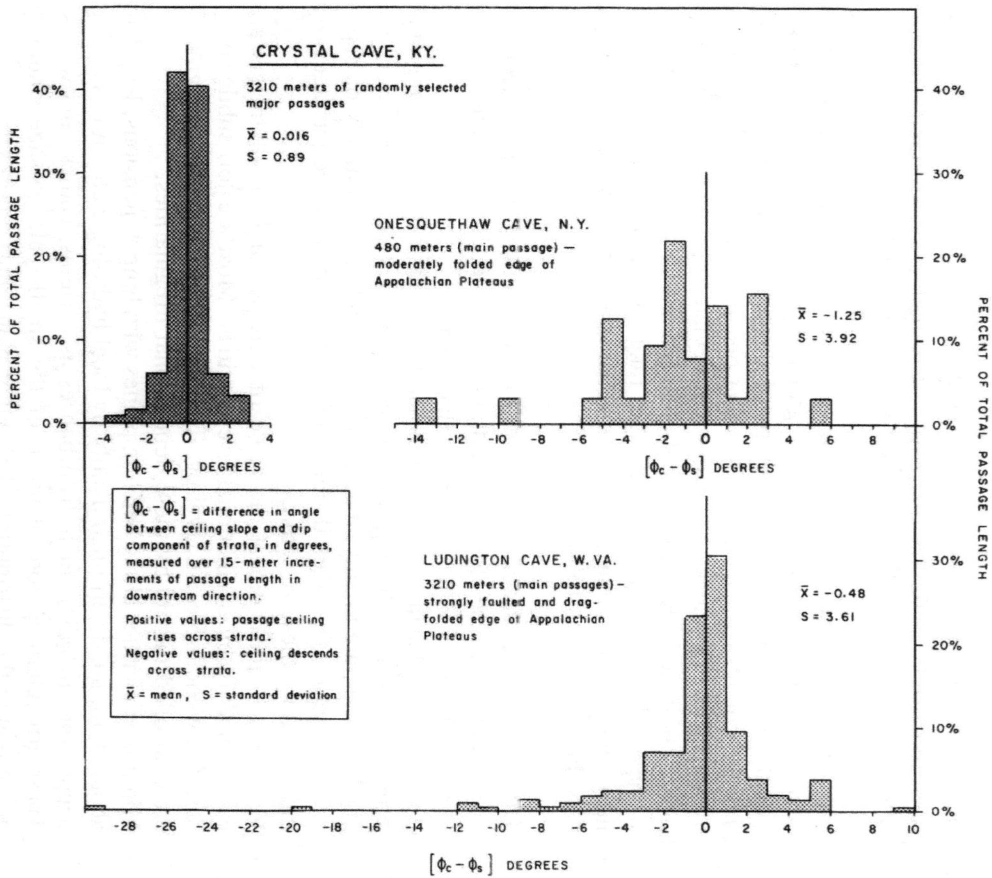

Figure 11.4 Discordance between solutional passage ceilings and strata in Crystal Cave, contrasted with the greater amount in caves of the Appalachian Mountains. Deviation from zero indicates discordance to the local strata. Negative values represent ceiling gradients descending across the strata in the downstream direction and positive values represent ceiling gradients rising across the strata.

stratal concordance of passages in Crystal Cave with the much greater discordance of caves in folded and faulted strata in the Appalachian Mountains.

RELATIONSHIP BETWEEN CAVE PATTERNS AND STRATAL DIP

The general northwesterly trend of the Mammoth Cave system has long been cited as an example of the effect of regional dip on groundwater flow and cave development. However, one of the main purposes of the leveling survey was to determine the influence of local variations in dip and to show why many of the passages in the same area trend at large angles to each other.

Within the limits of Crystal Cave, an area of approximately 2 km^2, the mean dip is approximately 11 m/km in the direction N 24° W, as shown by planar regression analyses of the leveling data on widely exposed contacts. In detail, however, each rock unit possesses a slightly different dip direction and magnitude that can deviate significantly from the regional dip (Fig. 11.5). In several places there are distinct reversals of dip toward the south. The structures that control groundwater flow are local, and it is meaningless to attempt to relate cave trends to the regional dip. Even the carefully drawn structural contours on the USGS maps of the region are not sufficient to portray the control over cave development because they represent contacts above and below the cavernous zones that deviate significantly in structure from the limestone beds, just as the dips of individual limestone beds differ from one another. Structural irregularities range in scale over several orders of magnitude. These can be grouped into four somewhat arbitrary categories diminishing in scale as follows:

1. The regional dip toward the center of the Illinois basin is of tectonic origin and determines the distribution of major landscape features, such as the Pennyroyal Plateau and Chester Upland.
2. Broad domes, basins, and flexures large enough to appear on regional structural maps, with long dimensions up to several kilometers or tens of kilometers, are mainly of tectonic origin and less commonly of depositional origin. The overall trends of entire caves or cave passages are influenced by these structures.
3. Variations in local dip at the scale of tens or hundreds of meters are caused mainly by depositional irregularities. Most are too subtle to be detected without precise leveling. Trends of individual passage segments are controlled mainly by these depositional irregularities.
4. Minor irregularities along bedding planes with long dimensions, less than about 10 m, are determined mainly by local hydrologic conditions during deposition, including such disturbances as currents, eddies, and waves. These irregularities control much of the small-scale passage sinuosity and possibly the distribution of bedding-plane anastomoses.

The smaller bedding structures are superimposed on the larger ones, and this hierarchy also applies to the resulting passage features. Leveling surveys show that passage sinuosity, most conspicuously that of vadose canyons, is strongly controlled by variations in local dip direction. The local trend of a passage is influenced only by structures larger than the passage width because these structures are capable of directing the flow of the solvent water. Assuming that passages originate as small individual tubes or anastomotic arrays and that the passage grows wider with increasing discharge, control gradually shifts from the smallest structures to larger ones as the passages develop. At any stage, however, the influence of structural features of different scales can be seen in the superposition of small-scale sinuosity on broader trends.

Determining this control is not as simple as it might appear because the measurements are limited to the very passages whose trends are under investigation. It is difficult to tell whether the observed dip within the passage represents the true dip. The approach used in this study was to calculate the regression plane from data points on the dominant parting or contact within each passage, thus determining the mean dip direction and magnitude. The apparent dip between successive leveling stations within the passage was then compared with the dip component predicted from the regression plane. Where the passage follows an actual dip that is steeper than the predicted dip, it is probable that the passage is selectively following a local steepening in that direction. An example is shown in Figure 11.5, where Waterfall Trail in Crystal Cave makes a sudden bend that appears to be along the strike at an altitude of 167.3 m (549 ft), but which is actually down the dip of a local trough. On the other hand, where the actual dip is the smaller of the two, the results are ambiguous. Such a passage could still be following the steepest available dip, or it could be independent of the dip.

In the widest passages, particularly the main passages of Mammoth Cave, the tripod-mounted surveyor's level or U-tube manometer is able to distinguish the exact dip from measurements on opposite walls. If measurements of controlling beds are spaced closely enough, structural contours can be drawn showing the exact relationship between dip and passage trend (Fig. 11.6).

The influence of stratal orientation on cave patterns is greater than that of fractures, which appear to have only local influence (see Deike, 1967). Although geologic mapping and statistical analysis of the resulting data are not yet complete, it appears that the initial openings for cave development are either the partings between beds or fractures confined within favorable beds. Powell (1976) has shown that in many Indiana caves that appear superficially to be controlled by bedding-plane partings, the true control is actually strongly fractured beds, particularly thin dolomite units. More speculatively, George (1984) correlated lineaments on aerial photographs with abrupt passage bends and breakdown zones in the Mammoth Cave area.

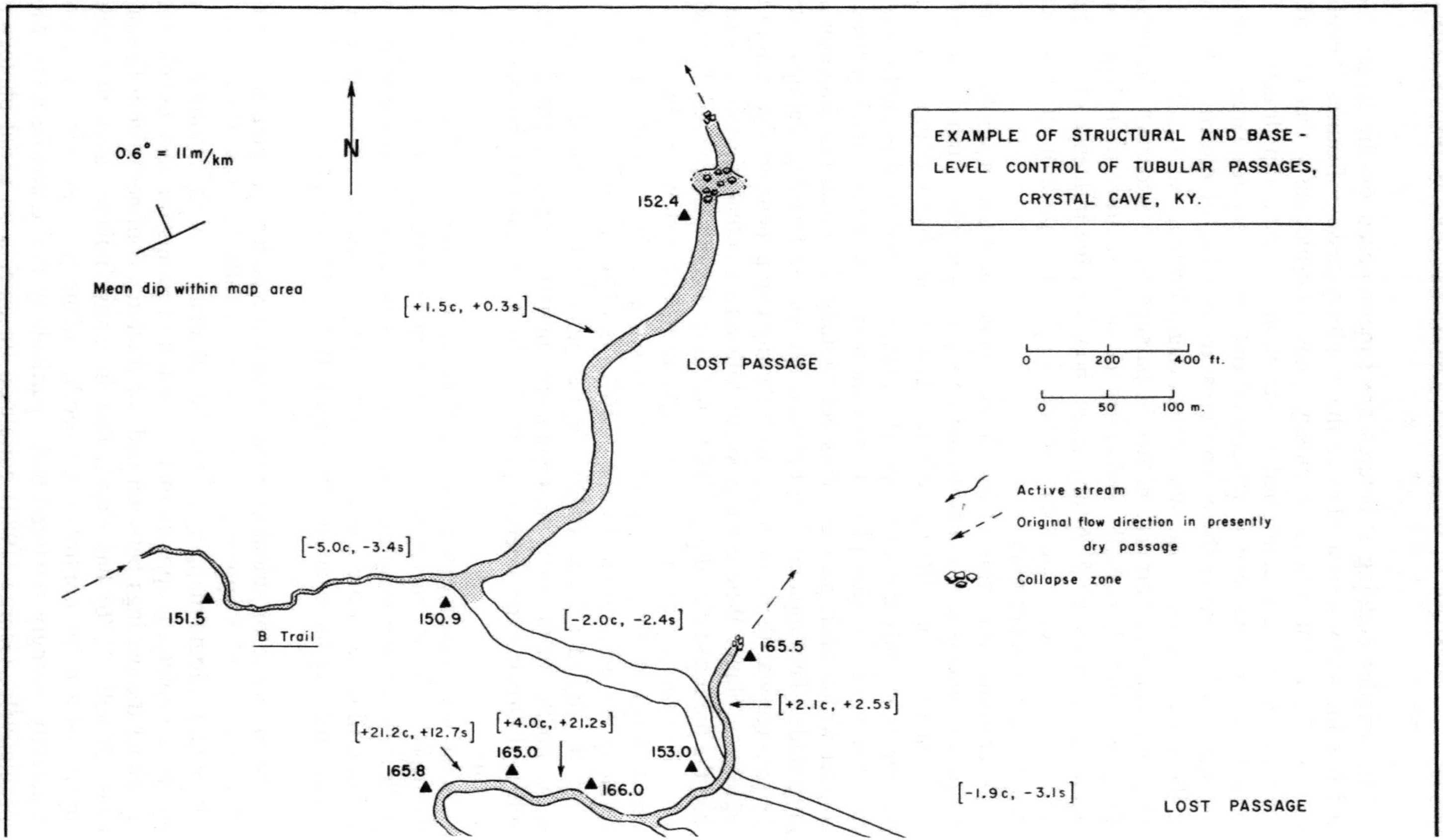
EXAMPLE OF STRUCTURAL AND BASE-LEVEL CONTROL OF TUBULAR PASSAGES, CRYSTAL CAVE, KY.
0.6° = 11 m/km
Mean dip within map area
N
0 200 400 ft.
0 50 100 m.
Active stream
Original flow direction in presently dry passage
Collapse zone
LOST PASSAGE
LOST PASSAGE
B Trail
152.4
151.5
150.9
165.5
165.0
165.8
166.0
153.0
[+1.5c, +0.3s]
[-5.0c, -3.4s]
[-2.0c, -2.4s]
[+2.1c, +2.5s]
[+21.2c, +12.7s]
[+4.0c, +21.2s]
[-1.9c, -3.1s]

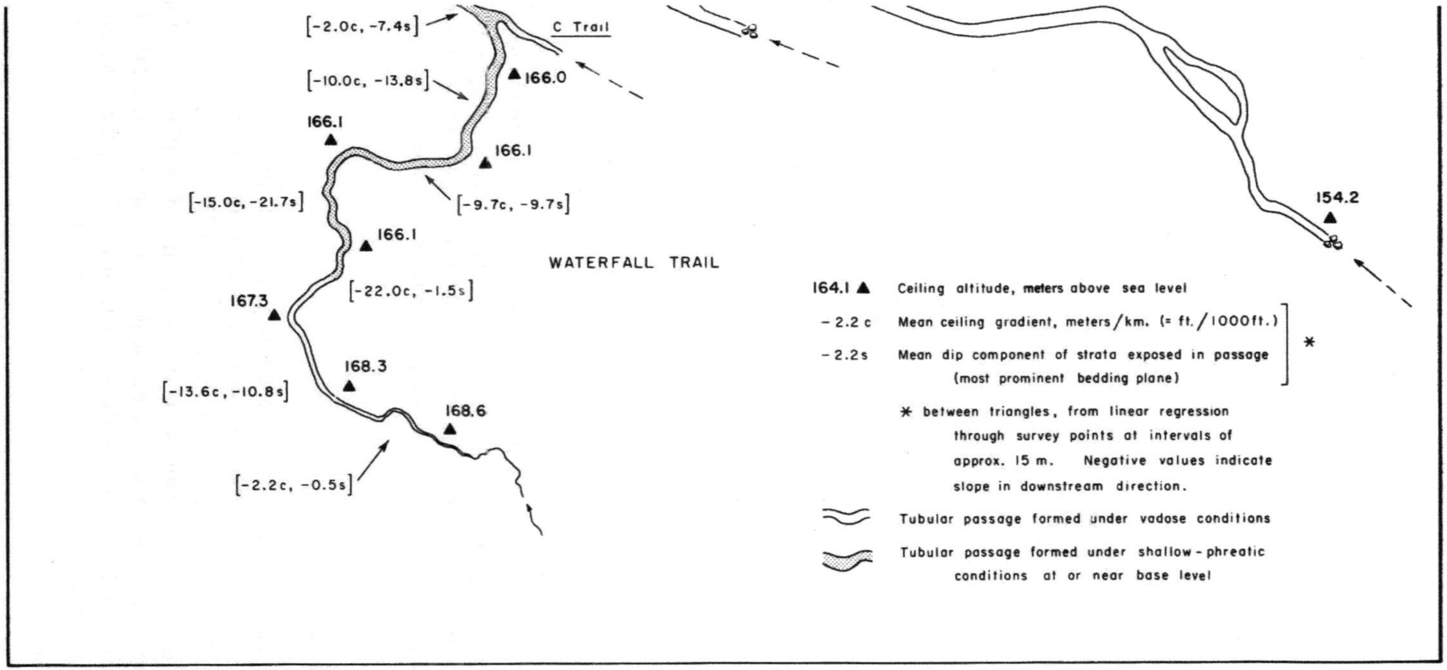

Figure 11.5 Map of Waterfall Trail and the Lost Passage in Crystal Cave, showing the relationship of vadose and phreatic sections to local bedding altitude.

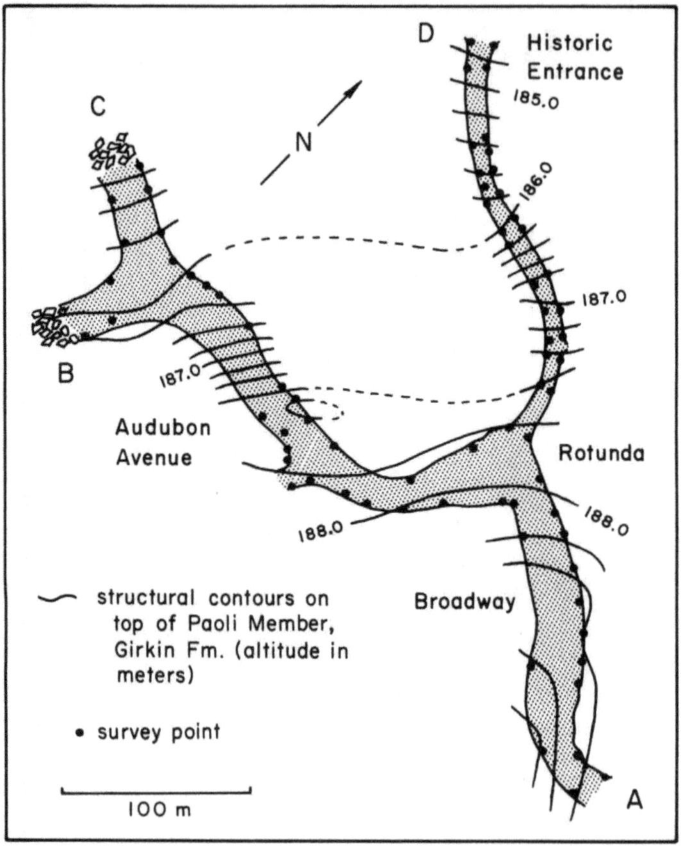

Figure 11.6 Map of the Rotunda area of Mammoth Cave, showing the relationship of passage pattern to stratal dip. Broadway and Audubon Avenue (**A–B**) originated as a strike-oriented passage at or near the water table. The passage at **C** and Houchins' Narrows (**D**) are later vadose diversion routes extending directly down the dip.

Some local passage segments are entirely joint controlled. However, throughout most of the cave the fractures are smaller in scale than the passages that intersect them. Nearly all fractures in the Mammoth Cave area have scales equivalent to Group 4 in the hierarchy of bedding structures described above, with only a few at the scale of Group 3. Their effect on passage trends is similar to that of the equivalent bedding-plane irregularities, but only where the fractures are more prominent than adjacent bedding-plane partings.

Trends of Vadose Passages

Virtually all canyon passages trend down what appears to be the steepest component of the dip. Their ceilings and the partings that control them both decline persistently in the downstream direction, implying that the canyon trends have been dictated almost entirely by gravitational flow along the available openings. In most vadose canyons, upward solution from the bedding-plane parting along which the initial flow took place is negligible (Fig. 11.7). Some tubes show the same downdip pattern and lack of upward solution, implying a vadose origin. Most of these are perched on distinctly insoluble or resistant beds. Those with no obvious stratigraphic perching might have been floored with clastic sediment, which can temporarily retard entrenchment. Small relict downdip tubes are common; most of these seem to have been formed by vadose water and then quickly abandoned before entrenchment could take place. Many of these merge upstream or down-

Figure 11.7 Most canyons in the Mammoth Cave system began their development along a single well-defined bedding-plane parting but have enlarged only by vadose entrenchment, with negligible solution of the ceiling. Such passages are almost invariably oriented downdip. The Girkin–Ste. Genevieve contact is the controlling horizon in this passage.

stream into the tops or bottoms of canyons, where it is obvious that the tubes represent temporary deviations, rather than phreatic conditions, for the entrenching vadose water.

Many canyons possess seemingly erratic patterns, with a high degree of sinuosity and individual segments oriented at large angles to the regional dip direction. The leveling surveys show that the sharpest changes in direction between successive passage segments are accompanied by local steepening of dip in that direction. Nearly all canyon segments in which flow has occurred at large, even obtuse, angles to the regional dip direction are actually oriented down the dip of local structures.

Perhaps unexpectedly, most of the places where the beds exposed in a canyon gain elevation in the downstream direction are oriented approximately in the same direction as the regional dip. In such areas of dip reversal the canyon ceilings generally show a short section of adverse gradient. Vadose water eventually encountered local structural basins and ponded under perched phreatic conditions, spilling over the lowest threshold, which was most commonly in the direction of the regional dip. Entrenchment below this level eventually caused open-channel flow to prevail. In the Mammoth Cave region such areas of ponding above base level have been so slight and short-lived that there is little impact on passage morphology. Near base level, however, cave development under these conditions can result in an alternation between tube and canyon morphology over distances of several hundred meters.

Within the boundaries of the leveling survey, the intermediate-scale structures, so influential upon passage trends, are mainly elongate structural troughs and ridges oriented parallel to the regional strike. These structures are chiefly depositional irregularities such as channels, bars, and banks. It is a compelling thought that these features, formed in response to hydrologic conditions more than 300 million years ago, exert such a strong influence on the hydrology of today.

Trends of Phreatic Passages

Tubular passages that show solution of ceilings upward from their controlling beds or bedding-plane partings, are presumably of phreatic origin. Nearly all have slightly undulating profiles with some adverse segments (oriented upward in the original downstream direction). Their overall gradients are much less than those of canyons, as low as a few tens of centimeters per kilometer. Individual upward-sloping segments are rarely abrupt enough to be detected visually, although some passages contain sharp upward jogs as much as ten meters. The maximum measured total relief of downward loops is about 23 m. Such conditions imply "shallow-phreatic" cave development, which should be expected in an area of such low-relief structure and sparse faulting.

The majority of tubular passages are oriented nearly parallel to the local strike of the host beds. Favorable horizons that conduct vadose water to the water table continue to influence the flow in the phreatic zone. Because of the downward decrease in width of most partings and fractures, the most favorable flow routes are at or near the phreatic zone, except where lithologic variation imposes deep flow (as in artesian aquifers), or where structural deformation produces deep zones that happen to be more favorable to flow than the shallow routes. Phreatic water selectively following favorable beds or partings will tend to avoid penetrating to great depth if possible. The resulting path of most efficient flow is along the strike to the nearest point of intersection with an outlet valley. Such paths are rarely the most direct, but groundwater-flow efficiency is dictated much more strongly by parting width than by hydraulic gradient, so the water is able to sacrifice considerable directness in favor of the widest openings. With these relationships in mind, it is possible to understand the meandering of phreatic passages, even those that appear to wander aimlessly.

This relationship holds true even in passages that are slightly discordant to the beds. Swinnerton Avenue, in Flint Ridge, for example, begins as a dip-oriented diversion route in the floor of Mather Avenue, loops downward in several places to maximum depths of 4 m, and rises intermittently to within 20 cm of its initial ceiling elevation (Fig. 11.3). The regression plane through the most prominent geologic contact exposed in the passage has a mean dip of 17 m/km in the direction N 27.5° E. Despite numerous small-scale variations in structure within Swinnerton Avenue, the mean passage trend is 82° to the mean dip (i.e., nearly parallel to the overall strike). The structure surveyed in this passage differs significantly from the N 10–30° W dip direction shown for the Big Clifty Formation on the USGS base map (Haynes, 1964). From geologic maps alone, it is easy to misinterpret the structural control of caves in such low-dip areas.

PASSAGE INTERRELATIONSHIPS

The ability to distinguish so clearly the vadose or phreatic origin of cave passages in the Mammoth Cave region is a great advantage in sorting out the complex groundwater-flow history of the region. Some typical passage relationships are described here, where spatial or temporal changes of flow character have created distinctive passage interrelationships. These provide some of the most persuasive clues to the geomorphic history of Mammoth Cave, as described in Chapter 12.

It is impossible to present all the structural data for Mammoth Cave, or even a relatively small part of it such as Crystal Cave, without extensive maps. Therefore the following discussion is limited to a few selected areas in which the passage interrelationships are particularly clear.

The contrast between vadose and phreatic influences is most clearly

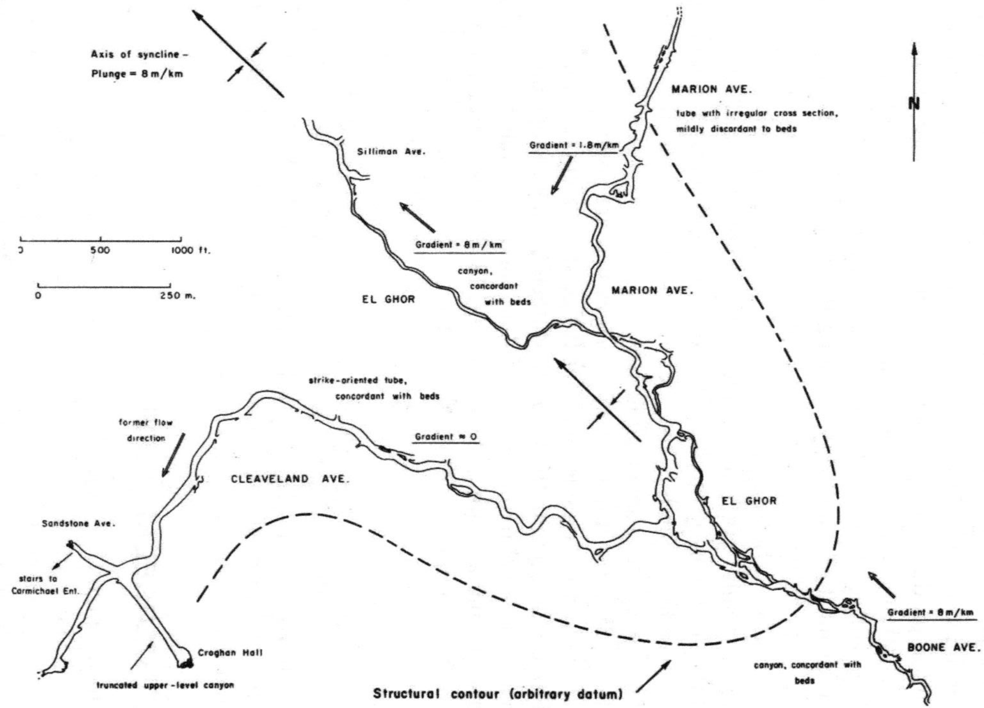

Figure 11.8 Map of Cleaveland Avenue and related passages, showing dip-oriented canyons following the plunge of a local syncline, with tubes oriented approximately along the strike.

Figure 11.9 Cleaveland Avenue, in Mammoth Cave, a nearly strike-oriented tube formed at the piezometric surface.

shown in Cleaveland Avenue and its related passages, all of which are presently inactive (Figs. 11.8 and 11.9). Cleaveland Avenue has one of the most perfect tubular cross sections in the entire region. For a length of 1.5 km it follows the exact strike of the local beds, following every nuance in the structure. The elevation of the solutional ceiling varies less than 1 m over this distance. In the former upstream direction the passage bifurcates into (1) an irregular tube (Marion Avenue), which, though slightly discordant to the structure, maintains a nearly horizontal profile subparallel to the strike, and (2) a multilevel canyon, Boone Avenue (Fig. 11.10), which extends updip toward the original water source. All three passages are contemporaneous. The junction between the Boone Avenue canyon and the Cleaveland–Marion Avenue tube represents the point where downdip vadose water once joined strike-oriented phreatic flow at the water table. Boone Avenue follows the axis of a broad syncline, a further indication of its vadose origin.

The lower levels of Boone Avenue bypass Cleaveland Avenue and continue down the axis of the syncline as a canyon known as the Pass of El Ghor. Evidently, when the base level and water table dropped below Cleaveland–Marion Avenue level, the vadose inflow abandoned the phreatic tube in favor of a downdip direction. Thus, both the spatial and temporal relationship between vadose and phreatic passages are shown. The leveling survey shows a slight reversal of dip immediately northwest of Cleaveland Avenue,

Figure 11.10 Boone Avenue, a multilevel dip-oriented canyon that was formerly a vadose tributary of Cleaveland Avenue.

but this is only a minor flexure that was apparently overridden by the large amount of groundwater flow.

The Pass of El Ghor leads directly to Silliman Avenue, a tube at the subsequent water-table level. In contrast to the other passages described here, it does not change to a strike direction at the vadose–phreatic transition, but instead meanders with an overall direction close to that of the dip. This direction simply afforded the most efficient outlet to the Green River Valley. There is no true dip control of any of the passage segments. Eventually it descends to join Echo River, which lies at the present base level, and farther downstream the tube rises back upward in several sharp steps to a breakdown-and-fill choke almost at the original Silliman Avenue elevation. The total vertical relief of this downward loop is 23 m.

As described previously, some dip-oriented passages are tubes, rather than canyons, particularly where there is perching on resistant or insoluble beds or where the dip of the controlling structure is slight. The transition from vadose to phreatic conditions is still detectable as a reduction in gradient, as a change from a pervasive downdip trend to one that is independent of the dip (with a strong preference toward strike orientation), and from a continuous downward gradient to an undulatory one. One or more of these clues might be missing, but more commonly they are all present.

For example, Waterfall Trail in Crystal Cave is a wide tube perched on a resistant shaly and dolomitic bed at the top of the Aux Vases unit (Figs. 11.2 and 11.5). Its upstream section trends down the steepest available dip, which is quite varied in both direction and magnitude. Several hundred meters from its downstream end the passage deviates upward from the resistant bed along a prominent bedding-plane parting in the overlying unit and changes to an approximate strike direction. The latter section has a gently undulatory profile with a mean gradient of less than 0.5 m/km. The upstream parts still contain an active perched stream, but entrenchment has abandoned the former phreatic section. The entrenching vadose water has created a series of downdip canyons that diverge from the original phreatic passage trend.

The same relationships can be seen in the tubular Lost Passage, at a lower level in Crystal Cave (Fig. 11.5), now entirely relict except for a few independent streams fed by intersecting passages. The original vadose water in the upstream part of the tube was perched on a low-dip contact of only 3 m/km, where downcutting could have been inhibited by an armor of clastic sediment on the floor. Eventual entrenchment of this upstream section produced a keyhole-shaped cross section, and underlying vadose canyons parallel this same trend. However, vadose streams today cut discordantly across the downstream (phreatic) part of Lost Passage, following independent downdip courses.

In the Rotunda area of Mammoth Cave (Fig. 11.6) the main passage follows the strike around a sharp northward-plunging anticlinal nose. This passage was apparently formed by the initial phreatic flow at or near the water table. A younger diversion passage (leading to the Historic Entrance) was created by water draining from the original passage precisely down the local dip. Where the dip direction changes, so does the passage trend.

SUMMARY

It should be clear from the preceding information that it is not possible in such a low-dip region to use regional dip (or even the mapping of surface geology) to explain the orientation of cave passages. It is necessary to measure the dip of the controlling beds or partings within the actual area, with data points spaced no farther apart than the passage meanders. The local dip varies greatly from bed to bed because of variations in stratal thickness. In places where the mean dip is steeper, these irregularities would have little effect on groundwater-flow patterns. Furthermore, a steeper dip is usually accompanied by increased faulting and folding, which tend to disrupt the typical diminution of fracture width with depth. Paradoxically, the gentler the dip, the more important it is in controlling cave patterns.

Interpretation of cave origin—that is, whether a given passage was originally vadose or phreatic—involves more than the simple distinction between canyon passages and tubes. Many tubes, such as upstream Waterfall Trail, are simply perched vadose passages that have not yet been entrenched by their flow. Many canyons, such as Collins Avenue, originated as phreatic tubes that have been entrenched over their full width. Although precise mapping is required to do so, the distinction between vadose and phreatic origin can best be determined by the relationship to the stratal dip. Ideally, downdip canyons change downstream into tubes having no consistent downdip trend, which is commonly oriented nearly along the strike.

This transition zone, or "piezometric limit" (Palmer, 1972), occurs where gravitational flow changes to flow in which the gravitational influence is offset to varied degrees by the downward increase in pressure within the phreatic zone. In this context piezometric limit is more appropriate than water table or potentiometric surface because piezometric limit connotes only the local conditions that existed while the passage was undergoing the majority of its enlargement. The elevation of the piezometric limit in each major passage provides a major clue to the geomorphic history of the Mammoth Cave system (see Chap. 12).

The great percentage of dip-oriented passages in the Mammoth Cave system shows that at least half of its passages, including some of the largest, are of vadose origin. In phreatic passages the original ceilings rarely penetrated more than a few meters below the water table. It is possible that some developed entirely at the water table and were only partly water filled throughout most of their evolution, with upward enlargement only during floods. Such shallow development is not necessarily the rule in other parts of the Illinois basin, especially in faulted areas or where the limestone is interrupted by extensive gypsum beds.

Perching of groundwater flow above base level is unusually strong in the Mammoth Cave area. Although much of the descent of vadose passages takes place within the first few hundred meters, where shafts cluster along the axes of dry valleys and along ridge flanks, perching of vadose passages has been observed to extend for distances as great as several kilometers. Thus, the prevalent interpretation of vadose water descending straight down to the water table from its point of infiltration is far from accurate. Any models of karst hydrology, whether digital hardware, or conceptual, must take this fact into account. Most importantly, it must be considered when tracking underground contamination because it provides a means for water to cross not only topographic divides, but also groundwater divides identified from well data (Palmer, 1986).

Strike-oriented tubes illustrate the principle that the width of the initial openings far outweighs the hydraulic gradient in controlling the path of greatest efficiency in the phreatic zone. These openings, which developed at the contemporary base level, are the primary paths of groundwater flow, and they do not represent diversion from primary dip-oriented tubes as the result of new outlets developing in the strike direction. The tendency for strike-oriented phreatic flow is strongest in well-bedded rocks such as those of the Mammoth Cave region, but is diminished in structurally deformed strata (Palmer, 1986).

REFERENCES

Deike, G. H., III, 1967, The development of caverns in the Mammoth Cave region, The Pennsylvania State University, Ph.D. dissertation, 235p.

George, A. I., 1984, Cave passage modification changes in relation to lineaments, Mammoth Cave National Park, Kentucky, *Cave Research Foundation Annual Rept.*, pp. 19–24.

Haynes, D. D., 1964, Geology of the Mammoth Cave quadrangle, Kentucky, *U.S. Geol. Survey Geol. Quad. Map GQ-351.*

Palmer, A. N., 1972, Dynamics of a sinking stream system: Onesquethaw Cave, New York, *Natl. Speleol. Soc. Bull.* **34:**89–110.

Palmer, A. N., 1981, *A Geological Guide to Mammoth Cave National Park*, Teaneck, N.J.: Zephyrus Press, 210p.

Palmer, A. N., 1986, Prediction of contaminant paths in karst aquifers, in *Proceedings of the Environmental Problems in Karst Terranes and Their Solutions Conference*, Dublin, Ohio: National Water Well Association, pp. 32–53.

Pohl, E. R., 1970, Upper Mississippian deposits of south-central Kentucky, *Kentucky Acad. Sci. Trans.* **31:**1–15.

Powell, R. L., 1976, Some geomorphic and hydrologic implications of jointing in carbonate strata of Mississippian age in south-central Indiana, Purdue University, Ph.D. dissertation, 168p.

Sandberg, C. A., and C. G. Bowles, 1965, Geology of the Cub Run quadrangle, Kentucky, *U.S. Geol. Survey Geol. Quad. Map GQ-386.*

Swann, D. H., 1964, Late Mississippian rhythmic sediments of the Mississippi Valley, *Am. Assoc. Petroleum Geologists Bull.* **48:**637–658.

Weller, J. M., 1927, The geology of Edmonson County, *Kentucky Geol. Survey Ser. 6* **28:**246p.

12

GEOMORPHIC HISTORY OF THE MAMMOTH CAVE SYSTEM

Arthur N. Palmer

In comparison with surface features, a cave is able to preserve far more information about the local geomorphic history and hydrologic evolution. Caves are extremely sensitive to environmental conditions while they are forming, and, in relict cave passages, physical weathering and similar disruptive agents are almost totally absent, allowing environmental clues to be preserved long after those on the surface have vanished. This is particularly true in a complex multilevel cave such as the Mammoth Cave system.

This chapter illustrates how the pattern of passages in Mammoth Cave can be used to interpret the late Tertiary and Quaternary history of the region. Further information on this topic is given by Deike (1967), Miotke and Palmer (1972), Miotke (1975, 1976), and Palmer (1981, 1985).

INTERPRETATION OF CAVE LEVELS

As a cave system evolves, the largest passages tend to cluster at certain levels. At least five distinct levels can be distinguished in Mammoth Cave, all but one of which is represented by several major passages. Within any given level the various passages lie within different strata, which shows that stratigraphic control, if any, is not dominant. The effluent Green River Valley lies approximately downdip from the recharge areas, eliminating the possibility of widespread perching behind structural thresholds. The only remaining possibility is that the clustering of major passages reflects periods of rather static fluvial base level.

The influence of base level on the development of a cave is indicated by the elevation of the piezometric limit in each of the major passages. As described in Chapter 11, this is the point where the passage morphology changes from vadose to phreatic, reflecting the conditions that prevailed during solutional enlargement. Not every passage contains a recognizable piezometric limit, and in many the point is inaccessible because of breakdown or erosional truncation.

With the solutional passage ceilings as reference, piezometric limits at each passage level in Mammoth Cave cluster within a very narrow vertical range. Scatter is only 1–2 m, which is surprising when one considers the great seasonal fluctuation in water level, as well as survey inaccuracy. No more than four piezometric limits have been mapped at any given level, but they are distinct enough to allow some bold conclusions. Their tight clustering implies a very low-relief piezometric surface within all passages reaching the phreatic zone at any given time, which, in the absence of geologic control, would require a fluvial base level at the same elevation.

The piezometric surface within an actively forming cave passage evolves toward a position very close to the fluvial base level, unless there is geological interference. During the earliest stages of passage development the piezometric surface undoubtedly lies higher; however, as the passage enlarges, the hydraulic efficiency increases so much that groundwater can be transmitted with only a barely detectable head loss, except during floods. As a result, the piezometric limit is a fair indication of the position of base level when the passage was forming. These relationships are not so clear in caves fed by large sinking streams prone to violent flooding, or where the geologic structure is complex. Stratigraphic perching of an entire flow route usually precludes base-level control entirely. Mammoth Cave, with its large number of passages, well-defined geology, and simple passage morphologies, is the ideal place to study passage levels. The best examples are located within 3 km of the Green River. Farther away, in the updip direction, base-level control is not so clear, especially where there is perching on chert beds in the St. Louis Limestone.

DESCRIPTION OF CAVE LEVELS

The major levels in the Mammoth Cave system are described in Figure 12.1, from highest (oldest) to lowest and labeled alphabetically for convenience. These levels were first identified in 1972 during the geologic leveling survey of Crystal Cave, the northeast part of the system (see Chap. 11). Since then the same levels have been mapped throughout the rest of the system and in a few nearby caves.

Unless stated otherwise, all passages described here are relict. The use of "upstream" and "downstream" applies to the conditions that existed when the passage was forming. Present-day altitudes cited here are probably not equivalent to what they were when the passages were actively forming, but from the standpoint of passage origin, only relative altitudes are significant.

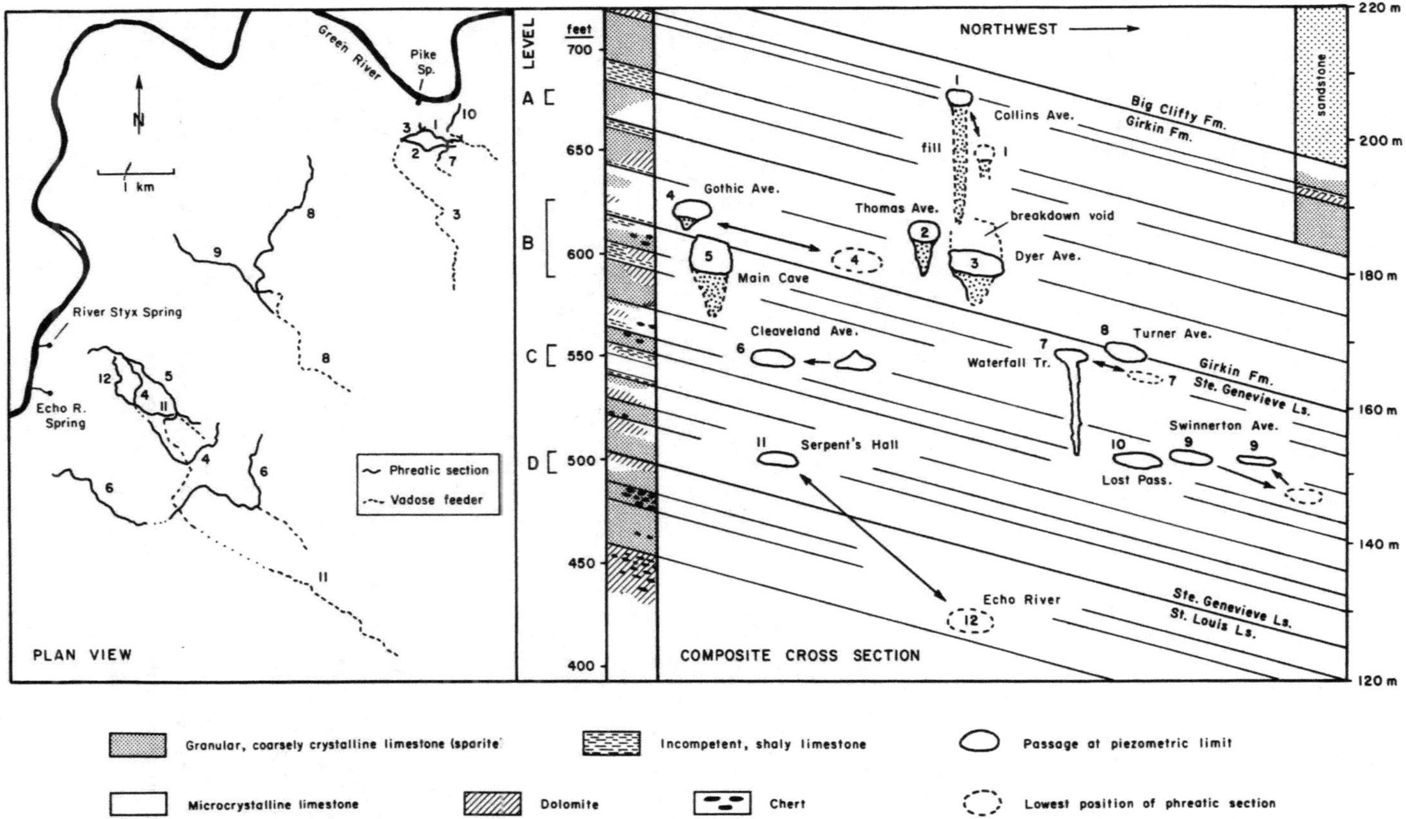

Figure 12.1 Levels of major passages in Mammoth Cave system.

Figure 12.2 Collins Avenue, in Crystal Cave, is the only known representative of level A in the Mammoth Cave system. The passage is filled nearly to the ceiling with clastic sediment elsewhere, but at this point the sediment has been carried away by the water that formed Dyer Avenue (level B, seen at left), which intersects Collins Avenue at the floor level. In the foreground is the tomb of Floyd Collins, the person who discovered Crystal Cave.

Level A, at an altitude of 210 m (690 ft), is represented by only one passage, Collins Avenue in Crystal Cave. It is a wide ungraded tube, entrenched downward by vadose water to a depth of 20 m and later filled almost to the ceiling with stratified sand and gravel. The sediment has been exhumed where Collins Avenue was intersected by a later passage, and at this point nearly the entire height of the canyon is visible (Fig. 12.2). The solutional ceiling lies within 5 m of the Big Clifty Sandstone caprock over most of its distance. Its profile rises and falls along its length with a vertical range of about 7 m, and the downstream end is slightly higher in elevation than the upstream termination in sediment fill. Consequently, there is no accessible piezometric limit, and it is possible to conclude only that this segment of passage began to form while below the level of the nearby Green River.

Level B consists of wide canyons and tubes at 180–190 m (590–620 ft) (Fig. 12.3). Main Cave in Mammoth Cave is an example. Every passage at this level shows evidence of partial or complete filling with detrital sediment at one time (Fig. 12.4). These passages formed after the Big Clifty Sandstone had been breached by the Green River over its entire course through the Mammoth Cave area. Their average stratigraphic position is in the Paoli Member in the lower quarter of the Girkin Formation. Apparently, the upper 75% of the Girkin was exposed but remained noncavernous until the level B passages formed, with Collins Avenue (level A) one of the few exceptions.

Levels C and D, at 168 m (550 ft) and 153 m (500 ft), respectively, contain the clearest and most numerous examples of the transitions from vadose canyons to phreatic tubes (Fig 12.5). All of the phreatic sections are wide tubes, most of which contain relatively little sediment. Cleaveland Avenue in Mammoth Cave is the best example of such a passage (Fig 11.9).

Below level D are several minor, poorly developed levels. The phreatic sections are mainly tubes with little or no entrenchment below the original openings (Fig. 12.6). Many of these passages lie below present base level as the result of late Quaternary alluviation, which reached thicknesses of approximately 10–15 m in the Green River Valley.

PARAGENESIS: PROS AND CONS

Downward phreatic loops tend to accumulate clastic sediment transported by cave streams. Where the floors and lower walls of such a passage are shielded by sediment, dissolution is limited to the upward direction (Renault, 1970). As the passage enlarges upward, more sediment is deposited on the floor, so the average flow velocity remains rather constant at the threshold of sediment transport. In extreme cases, a high canyon might be produced entirely in the phreatic zone, with upward solution terminating only at the water table (Ford and Ewers, 1978). Paragenetic canyons can easily be

Figure 12.3 Dyer Avenue in Crystal Cave is typical of level B, a wide canyon partly filled with clastic sediment.

Figure 12.4 Clastic sediment completely filling a passage at level B in Mammoth Cave.

mistaken for vadose canyons that have been partly filled with sediment (Fig. 12.7). In fact, the last stages of aggradation of a vadose canyon are hydrologically almost identical.

Sediment in most caves is rarely the product of either aggradation or paragenesis. It accumulates mainly as sand and gravel in active stream passages and as clay and silt in areas of rather static flooding. Collapse or similar blockage can cause local filling of a passage with sediment. However, only a gradual rise in base level or paragenesis can account for the tens of meters of stratified sand and gravel observed in levels A and B of Mammoth Cave.

Paragenesis is a compelling prospect for the origin of these levels because a paragenetic passage requires only a single phase of development. Moreover, deposits caused by a rise in base level can be spotty unless the rise in base level is slow and uniform and the passages are large enough to allow the flow velocity to fall below the critical tractive force. Nevertheless,

Figure 12.5 Great Relief Hall, in Mammoth Cave, is typical of the tubular passages at levels C and D.

the evidence in Mammoth Cave is strongly against paragenesis. It is important to come to a decisive conclusion, because the geomorphic and hydrologic implications of these two possible origins are radically different. The following evidence in the Mammoth Cave system argues in favor of vadose entrenchment and aggradation due to a fluctuating base level:

1. Many passages are filled to the ceiling with sediment (see Fig. 12.3).
2. Cut-and-fill structures and sinuous channels in the sediment suggest open-channel flow.
3. The ceilings of all passages in level B and most of Collins Avenue (level A) are perfectly concordant with the strata, implying that the initial solution took place at those horizons.
4. Breakdown blocks identified stratigraphically occur in the sediment as much as 7 m below their points of origin, indicating rather large passage heights at the time of breakdown.
5. Some sediment-filled passages in level B are wide, low, sediment-filled tubes rather than canyons.
6. Known phreatic loops in Mammoth Cave contain only scattered, thin, irregular deposits with little upward solution above them.

Figure 12.6 Eyeless Fish Trail is an active tubular passage at base level in Crystal Cave. At the time this photograph was taken backflooding from the Green River was causing a reversal of the normal flow direction and the water velocity was several tens of centimeters per second in what is usually the upstream direction.

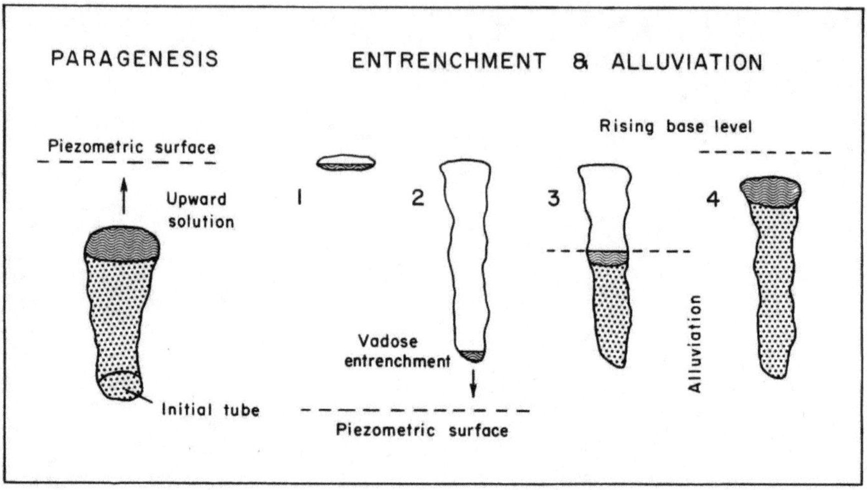

Figure 12.7 Comparison between paragenesis and fluctuating base level in the origin of sediment-filled canyons.

Intuitively, it seems doubtful that *all* passages on levels A and B would contain thick fill without some regional significance. A strong criterion for paragenesis is the upward propagation of meanders in the downstream direction (Ewers, 1985), but this has not been documented in Mammoth Cave. Finally, there is no apparent reason for the initial flow to have descended deep beneath the Green River, given the prominent beds dipping in the same direction as the flow.

CORRELATION OF CAVE LEVELS WITH SURFACE FEATURES

The first step in interpreting the significance of the cave levels was to correlate them with the surrounding landscape (Miotke and Palmer, 1972). The most striking observation was that levels A and B lie at or above the level of the Sinkhole Plain and of the karst valleys in the Chester Upland at grade with it. These levels include the largest passages in the cave system, which are typically wide canyons and tubes containing great thicknesses of clastic sediment. The underlying passages are much more complex, with numerous intersecting canyons, shafts, and tubes reflecting the increase in the number of recharge points as dissection of the Chester Upland progressed.

The Pennyroyal Plateau, although controlled in a few local areas by resistant beds, is predominantly an erosional surface that developed during the late Tertiary period as the result of slow fluvial entrenchment interrupted by periods of aggradation (Powell, 1964; Miotke and Papenberg, 1972;

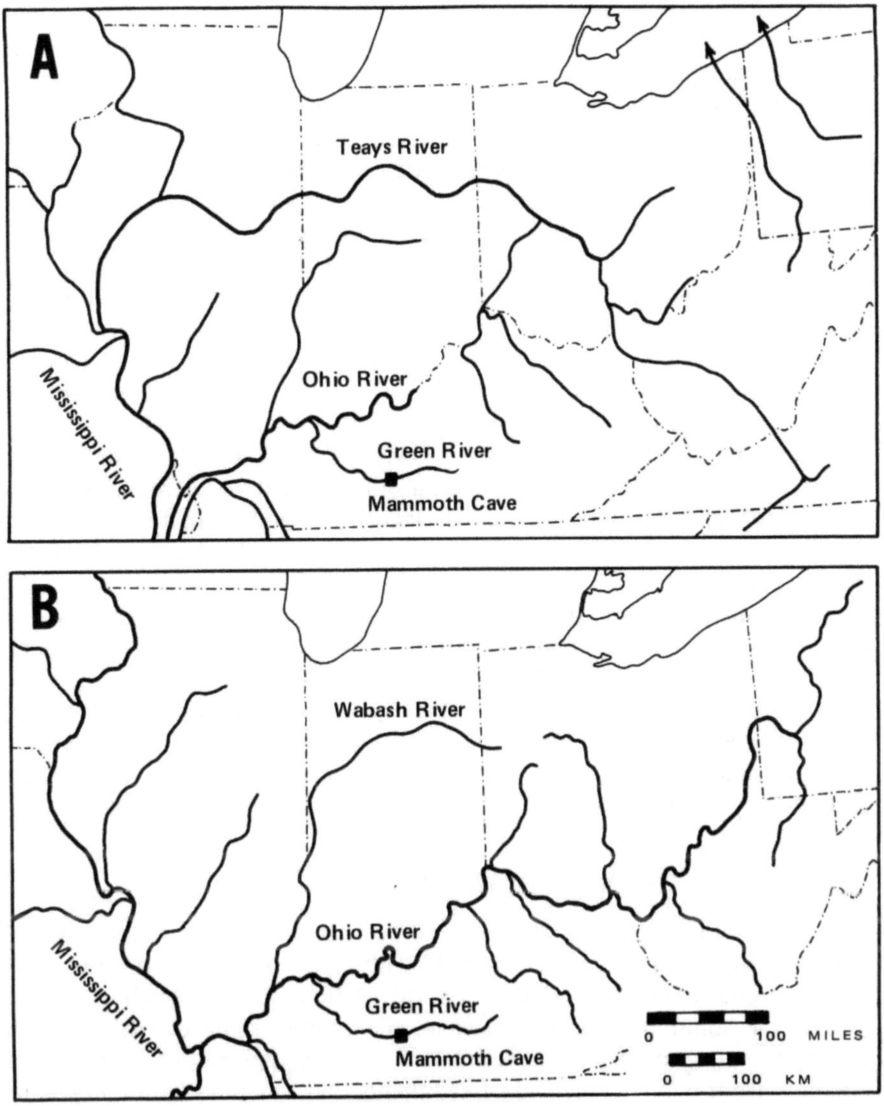

Figure 12.8 Pre-glacial (*A*) and post-glacial (*B*) patterns of the Teays River and the Ohio River. *(Modified from Thornbury, 1965)*

Palmer and Palmer, 1975). At the time most of the surface drainage from the Appalachian Mountains drained through the Teays River, located north of the present Ohio River (Fig. 12.8). Early Quaternary diversion of the Teays drainage into the Ohio River by continental glaciers greatly increased the discharge in the Ohio River, causing rapid entrenchment. The Green River, a tributary of the Ohio River, did likewise, creating the steep-walled valley that cuts through the Pennyroyal Plateau. In the Chester Upland the change in entrenchment rate is apparent from the abrupt narrowing of the valley below 180 m.

The difference in character between the upper levels (A and B) and those below was caused mainly by this change in fluvial entrenchment rate. During the Tertiary period, degradation of exposed limestones was able to keep pace with the slow, gradual erosion in river valleys, creating the low-relief Pennyroyal surface. Surface drainage prevailed on this surface. Only in the Chester Upland, with its resistant caprock, were there thicknesses of limestone above base level sufficient to allow extensive underground drainage. Surface water from the Pennyroyal and from minor valleys in the Chester Upland sank into limestone ridges, creating the sparse but large upper-level passages of the Mammoth Cave system. The slow fluvial entrenchment occurred in the subsurface as well, creating wide canyons. Periodic intervals of aggradation are shown by the fact that every passage at levels A and B received extensive detrital fill, much of which remains today. These deposits correlate with fluvial, lacustrine, and colluvial deposits at the Pennyroyal level, some of which are thick enough to appear on local geologic maps.

Rapid Quaternary entrenchment disrupted this pattern. The large passages of level B were abruptly abandoned in favor of lower flow routes, developing many narrow canyons. The distinctive tubes at levels C and D apparently represent periods of relatively static base level that interrupted periods of base-level instability. The Green River Valley contains well-developed terrace remnants at level D (Miotke and Palmer, 1972). A similar terrace undoubtedly existed at level C, but only scattered evidence remains.

ABSOLUTE AGE DETERMINATIONS

It is difficult to assign absolute ages to erosional geomorphic events, except indirectly by dating the related deposits. Thus, the history of surface erosion is still vague in most parts of the world. Although it is equally difficult to date the solutional origin of a cave, the chemical and detrital deposits that it contains can be preserved intact for long periods of time. Dates obtained from paleomagnetism of sediments and uranium–thorium content of travertine, though few, strongly support the geomorphic interpretations in the preceding paragraphs. On the other hand, no fossil evidence has yet been found that is old enough to be used in dating the cave.

Paleomagnetism of Cave Sediments

Schmidt (1982) collected samples of detrital cave sediments from the various levels in the Mammoth Cave system and analyzed them for polarity of remnant magnetism. All deposits below level C (168 m) showed normal polarity. At and above that level most of the deposits showed reversed polarity, interspersed with normal samples. The lowest reversed deposits almost certainly represent the Matuyama polarity epoch, which ended 730,000 years ago. Although a passage obviously predates the sediment within it, both are probably the product of the same flow. The intervening time between the origin of level C and the sediment deposition was probably not long. This places level C in the early Quaternary, as predicted by the geomorphic evidence. The normally polarized sediments below represent the Bruhnes normal polarity epoch, indicating late Quaternary cave development.

Passages at the higher levels A and B can be interpreted in several ways because of the complexity of pre-Bruhnes magnetic events. Schmidt (1982) favored ages of 0.97–1.67 million years for level B and as much as 2.13 million years for level A, assuming a uniform rate of river entrenchment throughout the history of the cave. However, the enormous size of the passages and their relationship to broad erosional plains on the surface appear to represent much more developmental time than that of the lower levels. The sedimentary record could be discontinuous enough that the upper levels, or at least level A, correlate with polarity epochs older than those suggested by a linear extrapolation of age vs. elevation. Either way, level A is almost certainly of late Tertiary age and level B is no younger than early Quaternary.

Radiometric Dating of Speleothems

Uranium–thorium dating of travertine is usually limited to about 350,000 years ago (Harmon et al., 1975), although recent refinements in this and other techniques show promise of extending the range of precise dating of cave deposits nearly tenfold. Hess and Harmon (1981) dated travertine in the lower levels of the Mammoth Cave system and found the limit of the existing uranium–thorium technique was exceeded in samples from level D. Samples from an altitude of 143 m, below level D, showed a maximum age of 141,000 years. Travertine not only postdates the cave in which it occurs, but it is deposited by vadose seepage that has no relationship to the cave-forming waters, so the time interval between them can be considerable. The level D samples were obtained from an area of stalactites with matching stalagmites, which indicate deposition after the cave had been abandoned by its stream.

This evidence, combined with the normal paleomagnetism of detrital sediments, shows the range of possible dates for level D to be 350,000–700,000 years ago, or mid-Quaternary. Hess and Harmon (1981) projected their age data into other levels and suggested an early Quaternary age for the highest. This estimate, extrapolated on the basis of elevation, probably underestimates the age of the uppermost passages.

PALEOCLIMATIC STUDIES

Harmon et al. (1978) examined isotopes of oxygen, hydrogen, and carbon in travertine from the Mammoth Cave system. Fractionation between ^{18}O and ^{16}O is a function of temperature, which allows an approximate estimate of paleotemperatures. For this interpretation to be valid, isotopic equilibrium is required during deposition. Analysis of the other isotopes showed that equilibrium was likely. Because the samples were also dated radiometrically, as described above, brief glimpses of the Quaternary climatic history could be obtained. Temperatures were approximately the same as those of today during the interval 105,000–230,000 years ago, interrupted by periods of 6–12° C colder at 130,000–160,000 years ago and 195,000–215,000 years ago. The more recent cold period correlates with the Illinoisan glaciation, but the older cold period falls within what is usually considered

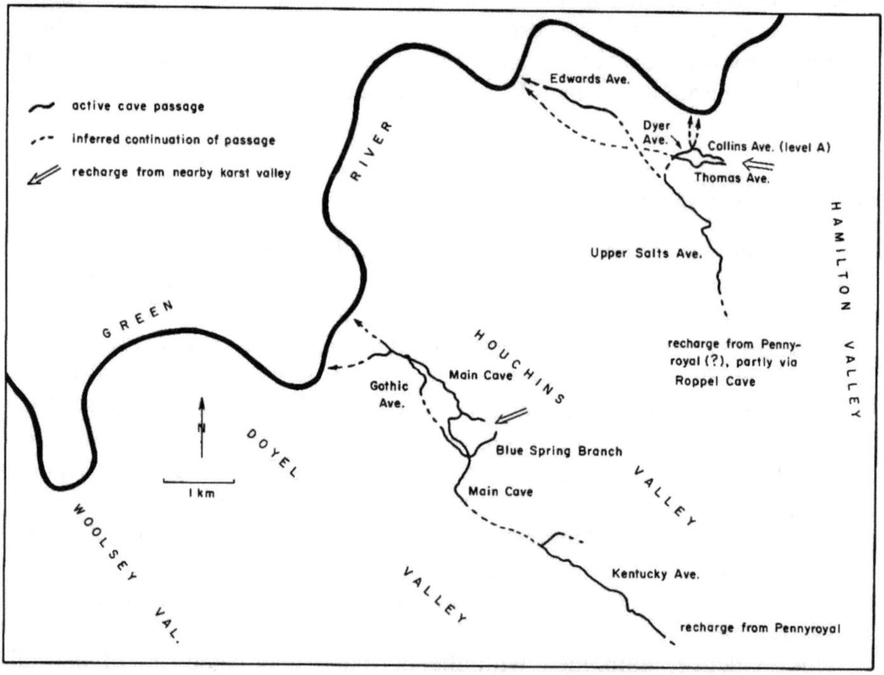

Figure 12.9 Pattern of passages at levels A and B in the northwest part of the Mammoth Cave system. The Tertiary pattern of the Green River is inferred from the valley configuration. All passages except Collins Avenue are at level B.

the Yarmouthian interglacial. Such evidence corroborates the recent view that the Pleistocene glacial history is more complex than originally thought.

EVOLUTION OF THE MAMMOTH CAVE SYSTEM

The following history is a reconstruction based on the passage configurations, relationships to surface geomorphology, and absolute age dating. Many details are still missing, but the overall sequence of events is fairly clear. Only the major passages are described here because the others either share the same history or are small vadose passages with little geomorphic significance. This evolution is shown diagrammatically for the northwest part of the Mammoth Cave system in Figures 12.9 and 12.10. No attempt is made here to give a comprehensive history of the entire cave system; instead, what follows outlines only the major geomorphic events and provides examples of how groundwater-flow paths evolve through time.

Interpretation of the geomorphic history of the Mammoth Cave system requires reconstructing flow paths from passages that have been segmented by breakdown and fill since their origin. Their downstream continuation

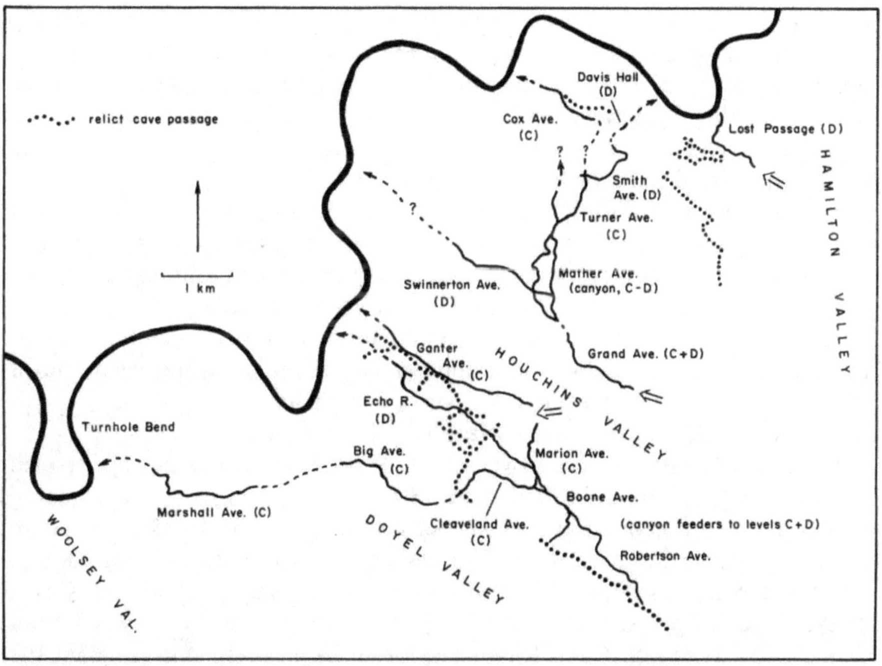

Figure 12.10 Pattern of major passages at levels C and D in the northwest part of the Mammoth Cave system.

to the Green River is easily inferred in most cases, but their upstream continuation and recharge areas are more difficult to deduce. In general it is possible to interpret the size of the recharge area from the size of the passage and the indicators of flow velocity within the passages. Although recharge to a given passage can be from several disparate sources, the major source is usually clear from the location of areas of exposed limestone at or above the altitude of the passage. In accordance with the principles given in Chapter 11, projection of known passages in the updip direction helps to determine the original source areas for water.

Late Tertiary History

The first significant groundwater flow through the limestones of Mammoth Cave took place only when the insoluble Big Clifty Sandstone caprock was breached by the Green River late in the Tertiary period. The first major cave passage to form was Collins Avenue (level A), which was the subsurface cutoff for water draining into the river from a nearby tributary, Hamilton Valley (Figs. 12.2 and 12.9). The solutional ceiling of Collins Avenue lies at an altitude of 200–210 m (655–690 ft), only 5–15 m below the top of the Girkin Formation. Its ungraded profile shows that the passage originated below the contemporary level of the Green River. Slow, uniform entrenchment created a wide canyon in the floor of the initial tube, but this was later filled nearly to the ceiling with stratified sand and gravel as the result of regional aggradation. This alternation between entrenchment and aggradation is a typical result of climatic fluctuations, with entrenchment during the humid parts of the cycle.

Level B formed during the last stages of Teays drainage, prior to the present configuration of the Ohio River, in the late Tertiary or early Quaternary (Fig. 12.9). It corresponds in level, and almost certainly in age, to the nearby low-gradient surface of the Pennyroyal Plateau. The wide canyons with deep sediment fill suggest a continuation of the slow entrenchment, alternating with aggradation, that apparently prevailed during the origin of level A. The low-relief Pennyroyal surface with its thick, residual depositional overburden accords well with this interpretation.

Level B actually involved two stages of development at closely spaced altitudes, approximately 190 m and 180 m. These stages are considered part of the same geomorphic event because in many passages they merge to form a single large canyon. The early passages at level B are Gothic Avenue and its upstream and downstream correlatives in Mammoth Cave. These passages are fed by recharge from Houchins Valley between Flint Ridge and Mammoth Cave Ridge and drain to a spring not far from the present Historic Entrance and Thomas Avenue in Crystal Cave, which was a successor to Collins Avenue but drained to a spring several kilometers west of Crystal Cave. Later passages at level B include (1) Kentucky Avenue and Main Cave in Mammoth Cave, which were apparently fed by water from the Pennyroyal Plateau and merged with the relict passages near the Historic Entrance; and (2) Upper Salts Avenue, in Flint Ridge, which drained an unknown area to the southeast (probably through part of Roppel Cave), to two successive springs north of Flint Ridge. Upper Salts Avenue originally continued through Crystal Cave as Dyer Avenue, which connects Thomas and Collins avenues (Fig. 12.3). Later it diverted to a more westerly course, forming Edwards Avenue in Great Onyx Cave and discharged at roughly the same location as Thomas Avenue.

By the time the level B passages formed, at least 75% of the Girkin Formation had been exposed in the Green River Valley throughout the Mammoth Cave area. Almost all of the passages at this level are located in the Paoli Member, which is no more favorable to solution than the overlying parts of the Girkin. The great time gap that must have existed between the origins of Collins Avenue and the level B passages is most likely the result of the much greater flow distances in the latter. The short, direct path of Collins Avenue required relatively little time to develop into a cave passage. In contrast, the higher flow routes that anteceded the level B passages never enlarged into detectable solution conduits.

The phreatic ceiling of Collins Avenue lies about 30 m above the local level of the Sinkhole Plain, and therefore the Green River must have been even higher. This remarkable fact indicates that between the origin of Collins Avenue and the final stages of level B, not only did the Green River erode downward at least 30 m, but the entire surrounding Pennyroyal surface was lowered by the same amount. Although there is still some disagreement among researchers as to whether Collins Avenue is paragenetic, the interpretation of the Green River level is not affected. Since the establishment of the present Ohio River drainage perhaps 1 million years ago, the vigorously eroding Green River has achieved a net downward entrenchment of about 55 m, but with very little lowering of the Pennyroyal surface. In consideration of the much slower entrenchment rates during the Tertiary, the age of Collins Avenue could be on the order of 10 million years.

Quaternary History

Although development of level B probably extended into the early Quaternary, the geomorphic effect of Pleistocene glaciation was not felt in the Mammoth Cave region until the Teays drainage was diverted into the Ohio River, initiating the period of rapid fluvial entrenchment (Fig. 12.8). Cave passages below level B date from this period.

The drop in the water table below level B must have been rapid, in comparison with previous rates, because the cave streams at this level abandoned the passages entirely instead of entrenching their existing courses. The newly formed passages were typically narrow canyons that reached the water table at an altitude of 168 m (550 ft) (level C, Fig. 12.10), where they became wide tubes. The uniformity of this altitude throughout the cave system suggests a long-term stable base level.

Groundwater diverted from Kentucky Avenue in Mammoth Cave formed a series of downdip canyons that culminated in Boone Avenue, a major tributary of Cleaveland Avenue at level C (see Chap. 11). Another branch of Cleaveland Avenue was Marion Avenue, a phreatic tube at the same level, which received its water from Houchins Valley. Cleaveland Avenue continued downstream through what are now truncated passage segments, Big Avenue in Mammoth Cave and Marshall Avenue in Lee Cave, exiting at Turnhole Bend on the Green River. Water from Main Cave diverted to form Ganter Avenue, a tube that discharged at a spring near the present Historic Entrance.

In Flint Ridge, much of the water that originally formed Salts Cave appears to have diverted to form Grand Avenue and Turner Avenue, segments of a lengthy tube that traversed the entire length of Flint Ridge, discharging in the vicinity of Great Onyx Cave possibly through Cox Avenue. Waterfall Trail in Crystal Cave (see Chap. 11) also originated at that time; its upstream sections still contain an active stream perched on shaly limestone at the top of the Aux Vases Member of the Ste. Genevieve Limestone.

Renewed entrenchment caused most of the level C passages to be abandoned in a similar manner, forming new tubes at 152 m (500 ft) (level D, Fig. 12.10). Water from Cleaveland Avenue and related passages formed the vadose canyon known as the Pass of El Ghor, which culminated in the tubular Silliman Avenue, Serpent's Hall, and Echo River, a downward loop of which is still active at present base level. Water from the Ganter Avenue catchment formed Great Relief Hall (Fig. 12.5), a tributary of Echo River apparently formed by diversion of water from Ganter Avenue. The spring for these passages was near the present Historic Entrance. In Flint Ridge, Mather Avenue and Swinnerton Avenue were formed in sequence by the water diverted from Turner Avenue. Mather Avenue is a wide canyon more or less parallel to Turner Avenue, but Swinnerton Avenue, which exits from the floor of Mather Avenue, is a phreatic tube that conducted water northwestward to an entirely new spring location on the Green River about one-half the distance between Great Onyx Cave and the Historic Entrance of Mammoth Cave. At the same level, the Lost Passage in Crystal Cave was formed at least partly by water from the same karst valley that once fed Collins and Thomas avenues.

The lack of consistent levels below level D suggests that late Quaternary fluctuations in base level were too frequent or erratic to allow long-term stability of the water table. Passages formed at this time are responsible for the present linkages between Mammoth Cave, the Flint Ridge Cave system, Proctor Cave, Morrison Cave, and Roppel Cave.

The most recent event of regional significance has been alluviation, probably post-Wisconsinian, to depths of 10–15 m in the Green River. The lowest passages have been flooded by this rise in base level, and some older passages that had previously been abandoned were reactivated. It is possible that some sub-base-level passages have been blocked by detrital sediment. During a rise in base level, sediment filling is probably not the usual cause of diversion to other passages, but rather the result of slackening of flow due to overflow into higher and more direct passages. In most water-filled conduits groundwater is able to maintain its flow unimpeded by accumulated sediment, because, as shown by the many still-active alluviated springs throughout the central United States, the critical tractive force is easily exceeded for erosion of all but the largest grain sizes as the passage cross sections diminish.

Geomorphic Implications

Although entrenchment of the wide Tertiary canyons appears to have kept pace with valley deepening, levels C and D represent the only times when downward cave development was able to overtake the more rapid Quaternary valley deepening. There are no discernible intermediate levels, even minor ones, between levels B and C, or between levels C and D. Apparently valley erosion during the Quaternary took place in discrete steps, with periods of relative stability interrupted by sudden entrenchment. Valley deepening probably took place by headward erosion, rather than by uniform incision, so that at any fixed point the lowering of base level would occur in sudden pulses as the nickpoint migrated upstream past the point. The few scattered levels below D might indicate diminishing rates of entrenchment or more frequent fluctuations in river level.

When one considers the rapid fluctuations in climate, runoff, sediment accumulation, and river levels during the Quaternary period, it is surprising that such well-developed levels as C and D could have formed. It is possible that rises in base level allowed them to enlarge under phreatic conditions after their original development, as has been the case with the most recently formed passages, but the precise nature of the piezometric limits at levels C and D shows that any such extension of their solutional histories had little geomorphic impact.

Throughout the history of the Mammoth Cave system the water from a given catchment area has periodically shifted spring locations over great distances. For example, during its flow history the water from the Kentucky

Avenue area has shifted from the Historic Entrance Area to Turnhole Bend, 6 km downstream, and then back to approximately the original location. These shifts reflect the influence of geologic structure, as illustrated in Chapter 11. Persisting throughout these fluctuations are certain favorable sites for springs, where water has exited at the same general spots along the Green River Valley. Some of these sites are located on southerly river meanders, which provide the most direct outlet for water in the plateau.

The present springs for the cave system are shown in Figure 12.1. Pike Spring, which now drains most of Flint Ridge, is in approximately the same location as the original spring for Collins Avenue. Once a spring and its conduit system develop, stress within the surrounding bedrock tends to widen additional fissures, which in turn become enlarged by solution. A cave trend can perpetuate itself in this way.

More commonly, however, the spring locations are determined by geologic structure. For example, the Historic Entrance lies at or near the axis of a broad syncline. Most spring locations have been determined by the strike orientation of the shallow-phreatic passages that feed them. The former springs for Cleaveland Avenue, Swinnerton Avenue, and the Lost Passage are examples.

CONCLUSIONS

If examined independently, neither the evolutionary history of a cave system nor that of the surrounding karst landscape can be interpreted so thoroughly as when a combined study is made of both. Relict cave patterns contain a much more precise record of hydrologic and geomorphic changes than any erosional surface features, but unless they are related to the surface landscape, many of the subterranean clues cannot be properly evaluated. As shown in this chapter, events hundreds of kilometers distant can sometimes have a profound effect on cave origin. Now that such a close correlation has been made with surface features, solutional caves can be interpreted more boldly in the future.

From a hydrologic standpoint, it is useful to examine the evolution of a complex system like Mammoth Cave to help understand the behavior of karst aquifers, even those that are inaccessible. Although this chapter gives only a brief overview, it is apparent that the origin, pattern, and evolution of groundwater flow paths can be clarified greatly by the study of caves.

REFERENCES

Deike, G. H., III, 1967, The development of caverns of the Mammoth Cave region, The Pennsylvania State University, Ph.D. dissertation, 235p.

Ewers, R. O., 1985, Patterns of cavern development along the Cumberland Escarpment in southeastern Kentucky, *Caves and Karst of Kentucky*, P. H. Dougherty, ed., Kentucky Geol. Survey Spec. Pub. 12, Ser. IX, pp. 63–77.

Ford, D. C., and R. O. Ewers, 1978, The development of limestone cave systems in the dimensions of length and depth, *Canadian Jour. Earth Sci.* **15:**1783–1798.

Harmon, R. S., P. Thompson, H. P. Schwarcz, and D. C. Ford, 1975, Uranium-series dating of speleothems, *Natl. Speleol. Soc. Bull.* **37:**21–33.

Harmon, R. S., H. P. Schwarcz, and D. C. Ford, 1978, Stable isotope geochemistry of speleothems and cave waters from the Flint Ridge–Mammoth Cave system, Kentucky: implications for terrestrial climate change during the period 230,000 to 100,000 years B.P., *Jour. Geology* **86:**373–384.

Hess, J. W., and R. S. Harmon, 1981, Geochronology of speleothems from the Flint Ridge Mammoth Cave System, Kentucky, U.S.A., *Proc. 8th Internat. Congress Speleol. (Bowling Green, Ky.)* **2:**433–436.

Miotke, F. -D., 1975, *Der Karst im zentralen Kentucky bei Mammoth Cave*, Jahrbuch der Geographischen Gesellschaft zu Hannover für 1973, 360p.

Miotke, F. -D., 1976, *Die Höhlen im Mammoth Cave-Gebiet/Kentucky*, Würtzburg: Böhler Verlag, 133p.

Miotke, F. -D., and A. N. Palmer, 1972, *Genetic Relationship Between Caves and Landforms in the Mammoth Cave National Park Area*, Würtzburg: Böhler Verlag, 69p.

Miotke, F. -D., and H. Papenberg, 1972, Geomorphology and hydrology of the Sinkhole Plain and Glasgow Upland, central Kentucky karst: Preliminary report, *Caves and Karst* **14:**25–32.

Palmer, A. N., 1981, *A Geological Guide to Mammoth Cave National Park*, Teaneck, N.J.: Zephyrus Press, 210p.

Palmer, A. N., 1985, The Mammoth Cave region and Pennyroyal Plateau, in *Caves and Karst of Kentucky*, P. H. Dougherty, ed., Kentucky Geol. Survey Spec. Pub. 12, Ser. IX, pp. 97–118.

Palmer, M. V., and A. N. Palmer, 1975, Landform development in the Mitchell Plain of southern Indiana: Origin of a partially karsted plain. *Zeitschr. Geomorphologie* **19:**1–39.

Powell, R. L., 1964, Origin of the Mitchell Plain in south-central Indiana, *Indiana Acad. Sci. Proc.* **73:**177–182.

Renault, P., 1970, *La Formation des Cavernes*, Paris: Presses Universitaires de France, 126p.

Schmidt, V. A., 1982, Magnetostratigraphy of sediments in Mammoth Cave, Kentucky, *Science* **217:**827–829.

Thornbury, W. D., 1965, *Regional Geomorphology of the United States*, New York: John Wiley & Sons, 609p.

AUTHOR CITATION INDEX

SUBJECT INDEX

N
M A M M O T H C A V E
P A R K
Joppa Ridge
Jim Lee Ri

MAMMOTH CAVE SYSTEM

MIX
Papier aus verantwortungsvollen Quellen
Paper from responsible sources
FSC® C105338

If you have any concerns about our products,
you can contact us on
ProductSafety@springernature.com

In case Publisher is established outside the EU,
the EU authorized representative is:
Springer Nature Customer Service Center GmbH
Europaplatz 3, 69115 Heidelberg, Germany

Printed by Libri Plureos GmbH
in Hamburg, Germany